The Cytoskeleton

An Introductory Survey

M. Schliwa

Springer-Verlag Wien New York

Prof. Dr. Manfred Schliwa
Department of Zoology
University of California
Berkeley, California, U.S.A.

With 88 Figures

ISSN 0172-4665

ISBN 3-211-81884-7 Springer-Verlag Wien — New York
ISBN 0-387-81884-7 Springer-Verlag New York — Wien

Für Usch

Preface

The term *cytoskeleton* has become firmly established in today's scientific vocabulary. Indeed, it is difficult to believe that only ten years ago, it was virtually non-existent. Since then, the modern field of research on the structural organization of the cytoplasm has turned into one of the most productive and rapidly expanding research areas in Cell Biology today. Considerable progress has been made towards the identification of the various structural components of the cytoskeleton and their interactions with one another and with membranes. The first attempts to understand, in molecular terms, complex cellular processes such as shape changes, locomotion, division, and organelle movements have been made. And it is now apparent that the cytoskeleton has impact on other biological processes such as the control of gene expression, protein synthesis, cell cycle regulation, and development.

This monograph outlines the basic properties of the major components of the polymeric filament networks and their interactions and associations. Wherever possible, emphasis is placed on more recent references. Any attempt to cover a research field this complex in an introductory monograph is, by necessity, fragmentary, and oversights or omissions are inevitable. I wish to apologize in advance to all those colleagues who feel that their work is not adequately represented.

This book would not have possible without the help of Ursula Euteneuer and Susan Spath, whose numerous thoughtful suggestions and comments on the final draft essentially made the text readable. To them I am deeply indebted. I also thank Paul Licht, Mona Radice, and Paul Tibbetts for their help in the struggle with the idiosyncrasies of the bibliography program, and the Department of Zoology at Berkeley for providing access to the departmental computer to prepare the list of references. I am grateful to the members of my laboratory for their moral support and indulgence during the months of my preoccupation with this undertaking, and hope that their patience is partly rewarded by the final result. Finally, I wish to thank the many colleagues who kindly supplied illustrative material. They are individually acknowledged in the figure legends.

Berkeley, October 1985   MANFRED SCHLIWA

Contents

Prologue

> At its own level of organization the cell is a unity; and it remains a unity, though with our analytical mental equipment we conceive it more easily as a plurality of discriminated organelles. Those attributes of the whole unit which we are always seeking to project into its components are only true *of the whole,* no matter how far we succeed in resolving the unit into smaller and smaller components.
>
> LAWRENCE E. R. PICKEN

In the past 150 years, conceptions of the structure of the cytoplasm have undergone many revisions and refinements. A recent revision has been the introduction of the concept of the *cytoskeleton.* Although the modern era of this research field is only about a decade old, the roots of the cytoskeleton concept can be traced back to the early days of the cell theory. In fact, the very term "cytoskeleton" was introduced by JOHN T. NEEDHAM more than 200 years ago, but it was mainly used to describe exoskeletal structures of protozoans. Its use in conjunction with studies on *endo*skeletal, fibrillar components is much more recent (*e.g.,* FRANZ 1939). In broad biological terms, the cytoskeleton can be thought of as the integrated system of molecules that provides cells with shape, internal spatial organization, motility, and possibly a means of communication with other cells and the environment. It is also the molecular basis of some of the physical properties of cells such as rigidity and elasticity. Dissection of this complex system has been essential for approaching it experimentally. One way of separating the cytoskeletal apparatus from other cell components is to solubilize less stable cell components in non-ionic detergents. Hence, the cytoskeleton can also be operationally defined as the material that remains after cells are treated with a detergent that removes membranes, organelles, small molecules, and soluble proteins. This cytoskeletal residue retains many of the structural and some of the functional properties of intact cells and is more accessible to a variety of experimental analyses. To-date, "the cytoskeleton" is essentially equated with "the cell residue after detergent extraction".

This book provides an introductory survey of our current knowledge of the cytoskeletal proteins and their interactions. It comprises two major segments, "Components" and "Interactions". The major components of the cytoskeleton are considered first—contractile proteins, microtubules, intermediate filaments, and some "exotic" proteins (sections I–V). Following some general comments about their three-dimensional organization (section VI), the interactions of these components with each other and cell membranes will be discussed in some detail (sections VII and IX). How these interactions contribute to a coherent framework and an integrated cytoplasmic matrix will also be addressed (section VIII).

Modern cytoskeletal research is in a phase of explosive growth and rapid change. Its goal is to obtain a comprehensive and detailed understanding of cytoplasmic structure at the molecular and supramolecular levels. This monograph traces some of the findings that established the foundation for the more recent developments in this field of cell biological research.

Part A. Components

I. Cytoplasmic Contractile Proteins

I.1. Historical Aspects

Cytoplasmic contractile proteins have been studied extensively since the late 1960ies, when they were recognized as major components of the cytoplasm of essentially all eukaryotic cells. However, cytoplasmic contractility was considered an important property of living cells already 150 years ago, when the concept of the cell as the unit of life was first emerging. In 1835, DUJARDIN proposed that the cytoplasm is structured and has both elastic *and* contractile properties. He described the sticky material that emerges from squashed cells as a homogeneous, hyaline, living substance ("gelée vivante") and named it the "sarcode". DURJADIN's idea of the homogenous sarcode was replaced at the turn of the century by a new concept which viewed the cytoplasm as a heterogenous and highly structured substance. Observations during the first half of this century added the notion that the cytoplasm has a characteristic "viscosity" (HEILBRUNN 1926). However, once the complexities of the relationship between the mechanical and contractile properties of the cytoplasm and its supramolecular architecture were realized, the thought that the cytoplasm could be characterized by a single physical parameter, its viscosity, was abandoned. In 1939, LEWIS developed the remarkably accurate view that the cytoplasm is pervaded by "extremely large and asymmetrical molecules which presumably form long fibers on gelation" and also can exert "contractile tension". This view was first substantiated by LOEWY's discovery (1952) that extracts of the slime mold, *Physarum polycephalum*, shared several properties with mixtures of the then recently discovered proteins, actin and myosin (actin was identified by STRAUB, 1942). Like muscle actomyosin, the slime mold extracts changed their consistency upon the addition and hydrolysis of ATP. Subsequently, HOFFMANN-BERLING and WEBER (1953) and HOFFMANN-BERLING (1954) showed that like myofibrils, glycerinated nonmuscle cells would undergo feeble but measurable contraction upon the addition of ATP. The first isolation of crude actomyosin from a vertebrate cell source (blood platelets) by BETTEX-GALLAND and LUESCHER (1959) was followed by the purification of both actin (HATANO and OOSAWA 1966) and myosin (HATANO and TAZAWA 1968,

ADELMAN and TAYLOR 1969) from slime mold plasmodia. These papers mark the origin of the field of cytoplasmic contractile protein biochemistry.

The biochemical approach was soon complemented by a revolutionary morphological approach for identifying actin filaments in thin sections of nonmuscle cells. ISHIKAWA *et al.* (1969) exploited the finding by HUXLEY (1963) that heavy meromyosin (HMM), a tryptic fragment of the myosin molecule, will bind to actin filaments in a characteristic arrowhead configuration. Their report initiated a flood of papers on "actin-like" filaments in many cell types and helped establish that actin is a ubiquitous protein present in all eukaryotic cells. In the mid70ies the development of three new techniques helped advance the field another giant step. Firstly, the application of immunocytochemical techniques to cytoplasmic contractile proteins (LAZARIDES and WEBER 1974) brought the identification of these proteins to the light microscopic level and allowed researchers to obtain an overview of their distribution in large numbers of cells. Secondly, the analysis of gelation and contraction events in cell-free cytoplasmic extracts (KANE 1975) led to the discovery and identification of the bewildering spectrum of actin-associated proteins in nonmuscle cells. Thirdly, the development of an electron microscopic techique for the visualization of filament organization in cells in an almost life-like fashion (BROWN *et al.* 1976) aided in our appreciation of the complexity and intricacy of filament organization in cells and helped establish a term which gave an entire branch of biological research (as well as this monograph) a name.

The sliding filament model of muscle contraction is accepted as the mechanism of force production in eukaryotic cells, but several structural and functional features clearly distinguish muscle from nonmuscle contractile proteins. Firstly, the degree of order of actin and myosin arrangement is vastly different in striated and nonmuscle cells. The efficiency of contraction is related to the degree of order, and both have reached evolutionary culmination in striated muscle. Secondly, under conditions where all of the muscle actin is polymerized both *in vivo* and *in vitro*, only about half of the actin of many nonmuscle cells is found in the polymer form (*e.g.,* BRAY and THOMAS 1975, FOX *et al.* 1984); because all actins are very similar, this observation suggests that muscle and nonmuscle actins have different regulatory mechanisms, rather than altered polymerization properties. Thirdly, the interaction between actin and myosin is regulated in different ways; it is actin-linked (via tropomyosin-troponin) in striated muscle, and myosin-linked (via calmodulin-myosin light chains) in smooth and nonmuscle cells. Fourthly, the extent of polymerization of actin (and possibly also myosin) is subject to rapid changes in nonmuscle cells, whereas muscle thin and thick filaments are essentially permanent. In addition, assembly and disassembly in nonmuscle cells may be fine-tuned to the extent that both may take place simultaneously in different regions of the cell. Fifthly, actin may be responsible for the generation and maintenance of skeletal or structural features of nonmuscle cells in the absence of myosin. These five features indicate that contractile protein organization and function are much more versatile in nonmuscle cells, which makes their study in these cells both difficult and exciting.

There are numerous overview articles, conference proceedings, monographs, multivolume essay series, and edited compilations of original papers that deal with cytoplasmic contractile proteins. Because they can not possibly all be listed here, a brief guide to only some of the more recent works will be provided. Molecular and structural aspects of contraction in striated and smooth muscle are dealt with in multiauthor volumes edited by POLLACK and SUGI (1984) and STEPHENS (1984). Contractility in muscle and nonmuscle cells is discussed in SHETERLINE's (1983) monograph. For cytoplasmic contractile proteins, the first comprehensive overview by POLLARD and WEIHING (1974) is still a valuable source. More recent summaries include the papers by POLLARD (1981), KORN (1982), and STOSSEL (1984). Compilations of review and/or original research articles can be found in the essay series Cell and Muscle Motility published PLENUM PRESS, in Vol. 46 of the Cold Spring Harbor Symposia on Quantitative Biology, and in Molecular Biology of Cell Locomotion (HUXLEY *et al.* 1982). Other pertinent reviews on more specialized topics will be mentioned in due course.

I.2. Actin

I.2.1. Structure

In most eukaryotic cells, actin is the most abundant protein, typically representing 5–10% and occasionally up to 25% of the total protein. Actin exists both as a monomer (G-actin, for gobular) and a polymer (F-actin, for filamentous). The actin monomer is a globular protein composed of a single polypeptide chain 374 (many nonmuscle actins) or 375 (skeletal muscle actin) amino acid residues in length. It is an acidic protein with an isoelectric point of approximately 5.4. Heterogeneity of actins from different cell sources and also within a single cell type was detected initially by isoelectric focusing, which resolved 3 isoforms separated by a single charge shift. The 3 actins were termed alpha, beta, and gamma actin, in descending order of acidity. As a result of extensive amino acid sequencing studies (ELZINGA *et al.* 1973, LU and ELZINGA 1976, VANDEKERCKHOVE and WEBER 1978a,b,c) and nucleotide sequencing of actin genes (FIRTEL *et al.* 1979, FYRBERG *et al.* 1981, NG and ABELSON 1980), several more isoforms of actin were detected. For example, at least six different major actin species are found in vertebrate tissues: one skeletal muscle, one heart muscle, two smooth muscle, and two cytoplasmic. An additional unique cytoplasmic actin (and corresponding gene) not previously found in warm-blooded vertebrates was recently detected in the chicken (CHANG et al. 1984). Amphibia may have as many as eight kinds of cytoplasmic actin expressed to different extents in different tissues (VANDEKERCKHOVE et al. 1981). Apart from a unique sequence in the first 18 amino terminal residues, all actins differ from each other by only a few—generally inconsequential—amino acid substitutions. For example, skeletal actin and cytoplasmic gamma actin show no more than 6% sequence variation, corresponding to 25 amino acid substitutions. Even actin from the slime mold *Physarum polycephalum* still shows 92% sequence homology with bovine skeletal muscle actin. In general, actins from related tissues of different species resemble each other more closely than actins from different tissues of the same species (note the analogy to alpha and beta tubulin heterogeneity; see section II.3.). Thus the amino acid sequence of actin and, therefore, the structure of the protein, has been astoundingly conserved, which can be taken as an indication that many domains of the actin molecule are involved in specific interactions which have not changed much in the course of evolution. In light of this remarkable conservation, the question arises: why have different actin isotypes evolved at all, and what are the possible isotype specificities of actin function? This question can not yet be answered satisfactorily. For all that is known, the different actin isoforms are functionally similar and interchangeable. For example, a cardiac muscle actin gene, when introduced by transfection into a tissue culture cell, will be expressed, and the cardiac actin will partition between the Triton X-100 insoluble and soluble phases to the same extent as the endogenous cytoplasmic (beta) actin (GUNNING et al. 1984). Conversely, many lower invertebrates have only actin genes whose coding sequences more closely resemble cytoplasmic actins than vertebrate muscle actins, but they may use their cytoplasmic-type actin in their sarcomeric muscles (FYRBERG et al. 1981). Thus a muscle actin isotype is not essential for sarcomere formation and function.

The number of actin genes and their distribution in the genome is highly variable from species to species. In some, there ar more copies of actin genes than there are different actins. Yeast has only one actin gene, but *Drosophila* has six (FYRBERG et al. 1980), *Dictyostelium* and the sea urchin approximately 20 (FIRTEL et al. 1978, DURICA et al. 1980, JOHNSON et al. 1983), and mice and humans up to 30 (SORIANO et al. 1982, MINTY et al. 1983). In humans, for example, the two cytoplasmic actin genes are present in multiple copies (PONTE et al. 1983), whereas both skeletal and cardiac actin are encoded by single copy genes. Many of the more than 20 beta-actin sequences in the human genome are pseudogenes. In many species the members of the actin multigene family have been shown by *in situ* hybridization to be highly dispersed (*e.g.,* FYRBERG et al. 1981, SORIANO et al. 1982), although in *Dictyostelium* and echinoderms, some actin genes may be closely linked (McKeown and Firtel 1981, Lee *et al.* 1984). Analyses of the structure of actin-related sequences in species ranging from protozoa to mammals show a substantial degree of evolutionary conservation in terms of the number and position of introns, and the transcriptional orientation of linked genes (*e.g.*, Zakut *et al.* 1982, Johnson *et al.* 1983). These studies indicate not only a high degree of interspecies homology of actin gene sequences, but also substantial conservation of gene organization.

On the basis of extensive sequence analyses, Vandekerckhove and Weber (1984) have developed the following hypothetical scheme of the evolution of actin isotypes. While many invertebrates express only one—the cytoplasmic—set of actins (which may be used even in striated muscle), a muscle-specific actin gene seems to have appeared with the evolution of vertebrates. This isotype still is the only form of muscle actin found in lower vertebrates today, and it is used in skeletal, cardiac, and smooth muscle. Fishes already express two types of muscle actins, one specific for striated and one for smooth muscle. Later in evolution, presumably with the appearance of reptiles, both these muscle actins may have duplicated again to give rise to skeletal and cardiac-specific forms of striated muscle actin, and vascular and stomach-specific isotypes of smooth muscle actin. In agreement with this scheme, the four muscle-specific isoactins seem to be present in only one copy in warm-blooded vertebrates.

The structure of the 41,800 d-actin molecule (molecular weight based on amino acid composition) has been studied recently by electron microscopy and X-ray diffraction of actin : DNAse I complexes (Suck *et al.* 1981), single actin filaments (Egelman *et al.* 1983), or crystalline sheets induced by the Lanthanide, gadolinium (Aebi *et al.* 1980, 1981, Smith *et al.* 1983). The globular actin molecule is not a sphere but rather a distinctly asymmetrical wedge-shaped molecule (dimensions $5.6 \times 3.3 \times 4.0$ nm) with a slight groove that divides the molecule into two lobes. This asymmetry of actin may be important for certain features of actin filaments and their interaction with other molecules.

Given the remarkable conservation of the primary structure of actin, it is not surprising that the basic structure of actin filaments from all plant and animal sources is identical. By electron microscopy of either thin sectioned or negatively-stained material, they appear as long, curved strands approximately 6–8 nm wide (Fig. 1). Their basic structure in negatively stained images is that of a 2-start, double-stranded, right-handed helix with an axial repeat of 36 nm, corresponding to a total of 13 subunits between crossover points along the helix. A somewhat simplistic model of the actin filament depicts it as two strings of ping-pong balls wound around each other in a perfectly helical fashion. Recent evidence, however, suggests a different structure in which there is some disorder in the arrangement of actin

subunits within the filament (EGELMAN *et al.* 1982, 1983). The symmetry of
an ideal filament in which actin monomers are represented by wedges rather
than ping-pong balls is described by an angle between adjacent subunits of
167° and a certain rise per subunit, the value of which depends on how one
positions the monomer in the filament. Apparently, these two parameters

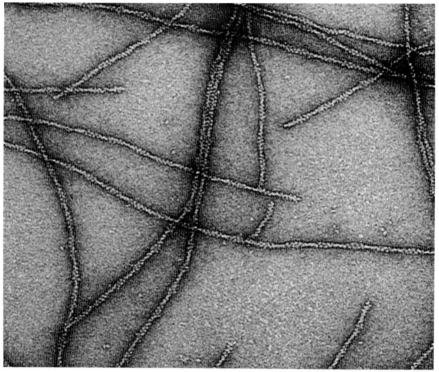

Fig. 1. Negatively stained actin filaments. × 100,000. (Courtesy of J. SPUDICH)

can vary to some degree under different conditions, even within a single
filament (EGELMAN *et al.* 1982). A change in the angle between neighboring
subunits form 167° to 166°, for example, results in a change in the distance
between crossover points along the actin filament from 38.2 nm to 35.5 nm
(assuming a rise per subunit of 2.7 nm). Thus actin filaments possess a
certain degree of structural flexibility which seems to be important for some
of their mechanical properties and their organization into higher order
structures, particularly bundles (DE ROSIER and TILNEY 1984).

    Two different models have been proposed for the arrangement of
subunits within an actin filament. In one, the long axis of the wedge-shaped
actin monomer is oriented parallel to the filament axis, resulting in filaments
about 7–8 nm in diameter (FOWLER and AEBI 1983), while in the other, the

long axis is nearly perpendicular (EGELMAN *et al*. 1983), giving the filament a diameter of 9–10 nm. Using X-ray diffraction of actin filaments in live muscle rather than after preparation for electron microscopy, EGELMAN and PADRON (1984) presented evidence for a native filament diameter of 9–10 nm. However, the question which of the two models is correct is not yet solved (SMITH *et al*. 1984), and it may be settled only if the positions of certain key amino acid residues within the three-dimensional structure of

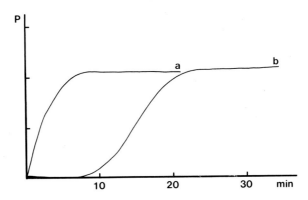

Fig. 2. Schematic diagram of the time course of actin polymerization *in vivo*. Polymerization (*P*, arbitrary units) can be measured by, *e.g.*, light scattering, absorbance at 232 nm, or specific viscosity. The kinetics of assembly of actin alone (*b*) show a slow nucleation step, followed, and overlapped by, a more rapid elongation step. The plateau phase represents steady state conditions. Addition of nuclei for polymerization, such as sonicated F-actin fragments, eliminates the lag phase (*a*)

the actin monomer are determined (for first steps towards this goal, see SUTOH 1984, ELZINGA and PHELAN 1984).

Actin has one binding site for ATP and one high affinity and about 5 low affinity sites for divalent cations (calcium or magnesium). Nonhydrolyzable analogs of ATP, as well as ADP, will also bind to actin. Upon polymerization, the ATP usually becomes hydrolyzed, so that F-actin contains one mole of ADP per mole of actin subunit (KORN 1982). The possible importance of ATP hydrolysis in assembly is discussed in section IV.

I.2.2. Actin Assembly in vitro

Polymerization of actin is usually induced by the addition of 50–100 mM potassium chloride or 1–2 mM magnesium to solutions of G-actin in a low ionic strength buffer. The polymerization of G-actin to form filaments involves at least two, and probably three, different major steps. The two major steps, nucleation and elongation, are probably preceded by an obligatory monomer activation step where subunits undergo a slight conformational change upon addition of salt to resemble F-actin monomers in a G-actin state (RICH and ESTES 1976, ROUAYRANC and TRAVERS 1981, PARDEE and SPUDICH 1982. COOPER *et al.* 1983). Nucleation, which is slow

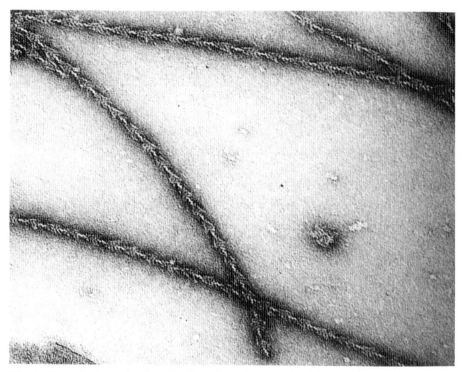

Fig. 3. Negatively stained actin filaments decorated with myosin subfragment 1.
× 100, 000. (Courtesy of J. SPUDICH)

and highly dependent on actin concentration, requires that at least three actin monomers get together to form a stable nucleus that resembles a very small fragment of an actin filament from which elongation can proceed (COOPER *et al.* 1983, FRIEDEN 1983, TOBACMAN and KORN 1983, LAL *et al.* 1984). The existence of a nucleation step is demonstrated by experiments in which small fragments of actin filaments or "seeds" will eliminate the nucleation lag period and will induce rapid elongation when added to a solution of G-actin (Fig. 2). The rate of nucleus formation, which involves several steps whose actual rates have not been measured, is the major variable determining the rate of actin assembly *in vitro* and is thought to be an important regulatory step for *in vivo* assembly as well.

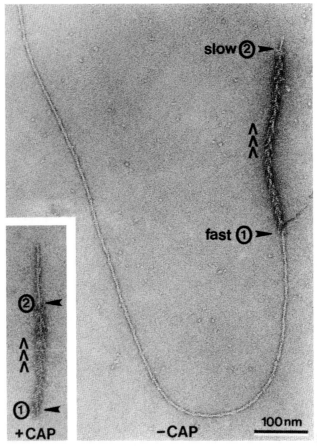

Fig. 4. Actin polymerization from a nucleus decorated with myosin subfragment 1. In the absence of a capping factor such as cytochalasin B or an end-binding protein (— *CAP*), subunits are added at both ends, and the rate of subunit addition is higher at the fast-growing (*1*) than at the slow-growing end (*2*). *Insert*: In the presence of a capping factor (in this case, *Acanthamoeba* capping protein), the nucleus elongates from the slow-growing end only (*2*). ×60,000. (Courtesy of T. POLLARD)

Compared to nucleation, elongation is fast. It occurs by the addition of actin monomers to the ends of the nucleus. In the presence of excess ATP, one molecule of ATP will be hydrolyzed for every molecule of actin added to the polymer. Because of this hydrolysis step, the critical concentrations for subunit addition are different for the two ends (see section IV.). Also, subunits add more rapidly at one end than at the other. The two ends can be identified by decoration with HMM which forms asymmetrical arrowhead complexes with actin filaments (Fig. 3). When fragments of HMM-decorated actin filaments are added to a solution of G-actin, the rate of subunit addition will be faster at the end marked by the arrowhead "barbs" (WOODRUM et al. 1975). At the "pointed" end subunits add more slowly (Fig. 4). Elongation continues until the concentration of monomer is reduced to the so-called critical concentration, and is in equilibrium with the polymer. This general scheme of actin assembly is similar to that of tubulin, and is consistent with the condensation-polymerization model proposed by OOSAWA and KASAI (1962) for the aggregation of linear and helical macromolecules. It might be slightly complicated in the case of actin because of an additional reaction involving the end-to-end addition of filaments. This process is termed "annealing" and is proposed to occur rapidly after solutions of F-actin are sonicated (NAKAOKA and KASAI 1969). However, the contributions of spontaneous filament breaking and annealing to the process of assembly are unknown. (For more details concerning filament assembly, see discussion in section IV.)

The rate constants for the addition and loss of subunits at the two ends of an actin filament have been determined directly by measuring the rates of elongation at the ends of intestinal microvillar cores (POLLARD and MOOSEKER 1981). The rate constant for subunit addition to the barbed end of an actin filament is large and appears to be diffusion-limited. The association rate constant at the pointed filament end is 5–7 times slower, depending on the conditions used to induce polymerization. Using the rate constants determined in these experiments, the efficiency of the polymerization reaction can be estimated. In 0.5 mg/ml of actin, about 70 subunits will be added and two lost per second at the fast-growing end, while 20 are added and one is lost at the slow-growing end, corresponding to a rate of elongation of 0.24 μm/second. This rate appears of the right order of magnitude to be compatible with cellular processes involving rapid reorganization of actin.

I.2.3. Inhibitors

Certain natural compounds interfere with a broad range of cellular activities by inhibiting either actin polymerization or depolymerization. The most widely known of these compounds are the cytochalasins and phalloidin.

I.2.3.1. Cytochalasins

The cytochalasins, of which about five are well characterized, are a group of low molecular weight, heterocyclic compounds that are naturally occuring secondary fungal metabolites. Since the first description of their effects on cells by CARTER (1967), they have been used widely—and often uncritically—as tools for the identification of "actin-dependent processes" in cells, which has given rise to a voluminous literature (summarized in

TANNENBAUM 1978). Their mode of action at the molecular level, however, was not uncovered until 10 years after their discovery. The cytochalasins bind to high-affinity sites (K_d about 10 nM) on F-actin, but not G-actin, at a ratio close to one molecule per actin filament, suggesting that the binding site is at the filament end (BROWN and SPUDICH 1979, FLANAGAN and LIN 1980, McLEAN-FLETCHER and POLLARD 1980b). There is strong evidence that the binding site is on the fast-growing or barbed end of actin filaments, as demonstrated by the inhibition of filament elongation at that end when HMM-decorated filament fragments are used as seeds. This usually results in an inhibition of the rate of actin filament assembly (BRENNER and KORN 1980) at low (submicromolar) concentrations of the compounds. At higher concentrations, the drugs may actually increase the initial rate of elongation by increasing the concentration of cytochalasin-induced actin nuclei available for (pointed-end) polymerization (DANCKER and LOW 1979, BRENNER and KORN 1981). Another action of the cytochalasins is to reduce the average filament length of F-actin (HARTWIG and STOSSEL 1979), which has been interpreted as "breaking" of filaments induced by the drug (MARUYAMA et al. 1980, SELDEN et al. 1980, SCHLIWA 1982). Whether filament breakage is actively induced by intercalation of a cytochalasin molecule into an actin filament along its length, or simply by blockage of filament reannealing after spontaneous breaking, can not be said with certainty. In any case, the effect of cytochalasins on the structure and consistency of actin-based gels *in vitro* and cytoskeletal assemblies *in vivo* is dramatic, usually a complete breakdown or "solation" of the gel, and a disruption of actin-containing structures (Fig. 5). The "labilization" of the actin-based cytoskeleton has been exploited to prepare enucleated cells. When cultured cells treated with cytochalasin are centrifuged, the nucleus is extruded, presumably due to the disruption of the actin-containing cortex (*e.g.,* GOLDMAN et al. 1973, SHAY et al. 1974). Disruption of the actin network is not accompanied by major changes in the G-to-F ratio since neither short-term nor long-term exposure of cells to cytochalasin produces a net depolymerization of actin filaments (MORRIS and TANNENBAUM 1980). (For protein modulators having effects similar to those of the cytochalasins, see section I.3.3.)

I.2.3.2. Phalloidin

Phalloidin and related compounds (phallotoxins) are cyclic peptides (phalloidin is an octapeptide) isolated from toadstools (*Amanita phalloides*) which have effects opposite to those of the cytochalasins. They stabilize actin filaments against a variety of depolymerizing or even denaturing stimuli, lower the critical concentration for assembly, and accelerate the rate of polymerization (LENGSFELD et al. 1974, WIELAND 1977, ESTES et al. 1981).

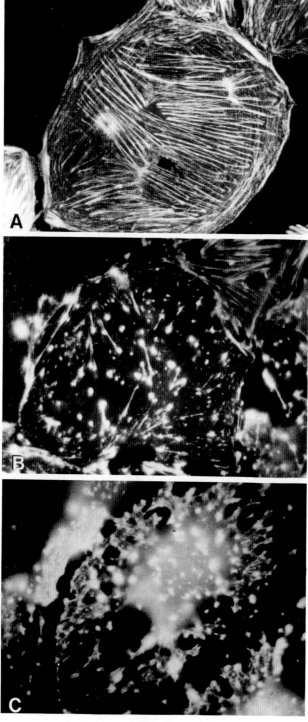

Fig. 5

The precise mechanism of the stabilizing action of phalloidin on actin filaments is not understood, but it seems to enhance assembly by reducing the actin dissociation rate constants at both ends (COLUCCIO and TILNEY 1984).

The general usefulness of phalloidin as an experimental tool has been hampered by its impermeability to most cells, except perhaps hepatocytes. Therefore, investigators had to resort to more tedious approaches, such as microinjection (WEHLAND et al. 1977, 1980). Once injected, phalloidin will stabilize cellular actin filaments against extraction procedures (such as high salt) normally dissolving F-actin, and will interfere with various cellular processes thought to depend on a dynamic actin filament system. Its most useful application, however, seems to be in the form of a covalently labelled derivative for the extremely specific visualization of F-actin by fluorescence microscopy (WULF et al. 1979).

I.3. Proteins Associated with Cytoplasmic Actin

Many of the features that distinguishes the contractile machinery of nonmuscle from that of muscle cells (e.g., three-dimensional arrangement, regulation, functions other than those leading to contraction) are conveyed by a spectrum of proteins that specifically interact with cytoplasmic actin. In contrast to, especially, striated muscle cells, the actin-based filament system of nonmuscle cells exhibits a high degree of temporal and spatial plasticity. Superficially, it appears to be disordered. However, this does not mean that cytoplasmic contractile networks are poorly organized; it is more likely that the degree of order in cytoplasmic networks is just as high as in striated muscle, only that their organization is flexible and transient, and not paracrystalline and permanent.

In principle, the formation and maintenance of actin-based structures can be regulated in two ways: 1. through filament assembly and length specification, and 2. through inter-filament interactions. Both regulatory pathways are established in nonmuscle cells. A still growing number of specific protein factors that function in these regulatory pathways have been identified and characterized. They are collectively referred to as "actin-binding proteins" or "actin-associated proteins". It has become customary in recent reviews of actin-binding proteins to list their most significant

---

Fig. 5. The effect of 1 µg/ml cytochalasin D on cultured BSC-1 cells as seen by fluorescence microscopy with rhodamine-phalloidin. *A* Untreated cell. *B* After 10 minutes in cytochalasin D, many of the stress fibers have already disappeared. *C* After 30 minutes in cytochalasin D, stress fibers are replaced by punctate aggregates, and the cells arborize. × 450

properties in the form of tables which provide an overview of their diversity and complexity (SCHLIWA 1981, CRAIG and POLLARD 1982, WEEDS 1982, KORN 1982). This useful tradition will be continued here, with the understanding, however, that such tabular summaries frequently require unjustified simplification, and that in some cases the grouping of a protein into one of the major categories is arbitrary. Some proteins also have dual properties and therefore appear in more than one category. The three major categories used here for the classification of actin-binding proteins are 1. depolymerizing and polymerization-inhibiting proteins, 2. crosslinkers, and 3. end-binding proteins. Virtually all of the known actin-binding proteins can be assigned to one of these major classes which, in turn, can be subdivided into a small number of subclasses. Indeed, it may turn out that despite their prolific diversity, all actin-binding proteins will fall into one of only about a dozen different functional categories. In the following paragraphs, some of the properties of proteins within these groups will be discussed. For more specific information, the original references should be consulted.

I.3.1. Depolymerizing and Polymerization-Inhibiting Proteins

This category comprises proteins which increase the concentration of monomeric actin, or make actin unavailable for polymerization (Table 1). Such an effect can be achieved in two ways: 1. by sequestering G-actin, thereby shifting the steady-state balance between G- and F-actin, and 2. by increasing the rate of actin disassembly by affecting the overall dissociation rate constant. In the first case, G-actin will bind to a polymerization-inhibiting protein more tightly than to a (growing) actin filament. In fact, in many instances, stable 1 : 1 or 2 : 1 complexes (LEE *et al.* 1984) are formed between G-actin and one of these proteins. In the second case, one should observe an acceleration of the rate of depolymerization of F-actin. While this may be true for some of the proteins listed in Table 1, there is as yet no direct evidence that these proteins bind to F-actin and increase the rate of dissociation under physiological conditions.

Probably the best-studied and also the most widely distributed protein in this category is profilin. Mammalian profilin is a small, basic (pI 9.3) protein whose 142 amino acids have been sequenced (NYSTROM *et al.* 1979). It may not be an evolutionarily highly conserved protein because profilin from *Acanthamoeba* differs significantly from mammalian profilin in amino acid composition, molecular weight, and isoelectric point (REICHSTEIN and KORN 1979). Actin and profilin interact to form a stable 1 : 1 complex, and many of the effects of profilin on actin assembly can be explained by a model in which the complex is inactive in actin filament nucleation or elongation (TOBACMAN *et al.* 1983, TSENG *et al.* 1984, OZAKI and HATANO 1984). KORN (1982) calculates that if the critical concentration for actin assembly in the

Table 1. *Proteins that interact with G-actin and depolymerizing proteins*

Protein	Source	Molecular weight (d)	Comments	References
Profilin	mammalian cells	15,000	forms stable 1:1 complex with G-actin; not calcium-sensitive reduces nucleation and elongation rates	CARLSSON et al. 1977, MARKEY et al. 1978 REICHSTEIN and KORN 1979, OZAKI et al. 1983
	Acanthamoeba, Physarum	12,000		
DNAse I	many cell types	29,000	forms stable 1:1 complex with G-actin; significance of binding to actin unknown	HITCHCOCK et al. 1976
Actin-depolymerizing factor	mammalian brain	19,000	effects similar to those of profilin; accelerates actin depolymerization	BAMBURG et al. 1980
18 K protein	hepatoma cells	18,000	similar to profilin; binds to and shortens actin filaments	OHTA et al. 1984
Vitamin D-binding protein	blood serum	52,000 and 56,000	seems to accelerate actin depolymerization	VAN BAELEN et al. 1980
Serum actin inhibitory protein (SAIP)	rabbit serum	56,000 and 60,000	similar to vitamin D-binding protein	VANDEKERCKHOVE and SANDOVAL 1982
Actin modulator	thyroid	40,000	binds to G-actin (information limited)	KOBAYASHI et al. 1983
Depactin	echinoderms	17,000–20,000	forms stable 1:complex with actin may also sever actin filaments	HOSOYA et al. 1982, MABUCHI 1983

Table 2. *Cross-linking proteins*

Protein	Source	Molecular weight (d)	Comments	References
Actin-binding protein (ABP)	macrophages	2 × 270,000	cross-links F-actin into three-dimensional network; accelerates nucleation; long, flexible molecule; binding regulated by phosphorylation (?)	HARTWIG and STOSSEL 1975, 1981, ZHUANG et al. 1984
Filamin	smooth muscle, many other cells	2 × 250,000	similar properties, but distinct from ABP	WANG et al. 1975, WANG 1977, WALLACH et al. 1978
HMWP	Physarum	2 × 230,000	shares properties with ABP, filamin, spectrin	SUTOH et al. 1984
Spectrins	erythrocytes (spectrin) other cells (fodrin) terminal web (TW 260/240)	240 + 220,000 240 + 235,000 240 + 260,000	forms tetramers (two of each subunit); cross-links F-actin; binds calmodulin, ankyrin	BRANTON et al. 1981 BURRIDGE et al. 1982 GLENNEY and GLENNEY 1983
Fascin	echinoderms	58,000	bundles actin filaments	BRYAN and KANE 1978
Gelactins	Acanthamoeba	23, 28, 32, and 38,000	cause gelation of F-actin	MARUTA and KORN 1977
Fimbrin	intestinal epith.; also other cells	68,000	bundles actin filaments	BRETSCHER 1981, GLENNEY et al. 1981
Villin	intestinal epith.	95,000	bundles actin at submicromolar calcium; is a capping protein at micromolar calcium	BRETSCHER and WEBER 1980, GLENNEY and WEBER 1981
Actinogelin	Ehrlich ascites tumor cells	115,000	calcium-sensitive; forms weak gels with F-actin	MIMURA and ASANO 1979

Protein	Source	Molecular weight	Properties	References
Dictyostelium ABPs	Dictyostelium	120,000	calcium-sensitive; cross-links F-actin into network	CONDEELIS 1981
		90,000	calcium and pH-sensitive cross-linker	CONDEELIS and VAHEY 1982
		30,000	calcium-dependent; bundles actin filaments	FECHHEIMER and TAYLOR 1984
Alpha-actinin	smooth muscle, brain platelets; present in many cell types	2 × 100,000	smooth muscle: calcium-intensive; other cells: calcium-dependent	ROSENBERG et al. 1981, BURRIDGE and FERAMISCO 1982, DUHAIMAN and BAMBURG 1984
Acanthamoeba gelation protein	Acanthamoeba	2 × 90,000	calcium-sensitive cross-linker	POLLARD 1981
Caldesmon	smooth muscle	2 × 150,000	binds actin and calmodulin; binding to actin is calcium-independent, leads to bundles; binding to calmodulin is calcium-dependent and abolishes binding to actin; can inhibit ATPase activity	SOBUE et al. 1981, BRETSCHER 1984, NGAI and WALSH 1984
Vinculin	smooth muscle, fibroblasts; present in many cell types	130,000	bundles actin; also a capping protein	GEIGER 1979, JOCKUSCH and ISENBERG 1981, WILKINS and LIN 1982
HMW MAPs	brain	330,000, 280,000	cross-link actin filaments into network	GRIFFITH and POLLARD 1978, 1982

cell is 1 μM, and if there is twice as much profilin than there is actin, about 30% of the actin would be kept in the monomeric state as a profilin/actin complex which would decrease spontaneous filament nucleation. Recent observations suggesting that the mechanism of action of profilin is more complex are that it inhibits elongation of actin filaments far more at the pointed end than at the barbed end (TILNEY et al. 1983 b), and it inhibits nucleation more strongly than elongation (POLLARD and COOPER 1984). These observations may be explained if it is assumed that profilin can bind to both actin monomers and, with a 10 times lower affinity, to the barbed end of an actin filament. Thus profilin may have certain qualities of an end-binding protein as well.

Perhaps one of the most peculiar activities of actin (and one of undetermined relevance) is its avid and specific binding to a ubiquitous enzyme, DNAse I (LAZARIDES and LINDBERG 1974), from which it can be dissociated only by strong denaturing agents. Like profilin, DNAse I binds to G-actin to form very tight 1:1 complexes (dissociation constant 2 nM), but it will also induce the depolymerization of F-actin (HITCHCOCK et al. 1976), even when the filaments are stabilized by tropomyosin or heavy meromyosin. These properties were exploited in three different ways: 1. as an alternative experimental approach for the specific disruption of actin filaments after microinjection into cells (e.g., WEHLAND and WEBER 1979, GOLDBERG et al. 1980); 2. as a sensitive assay for the determination of the content of monomeric and polymeric actin in cells ("DNAse I inhibition assay"; BLIKSTAD et al. 1978); and 3. in the construction of actin: DNAse I crystals suitable for the analysis of the structure of the actin molecule by crystallography (SUGINO et al. 1979, SUCK et al. 1981).

The other proteins listed in Table 1 all seem to form stable complexes with G-actin, and they also increase the rate of actin depolymerization. However, whether this action requires binding to F-actin has not been demonstrated unequivocally, nor has it been shown that they increase the dissociation rate constant under physiological conditions. Thus it remains uncertain whether these proteins are fundamentally different from profilin in their mechanism of action.

I.3.2. Cross-Linking proteins

Among the most striking characteristics of nonmuscle cells are the distinct supramolecular structures built from actin filaments, namely, three-dimensional networks, and bundles. For both, specific proteins are known that induce and maintain them (Table 2). Studies on the biochemical basis of the phenomenon known as sol-gel transformation led to the isolation and characterization of the factors responsible for the formation and modulation of the gelled state in cytoplasmic extracts (KANE 1975, HARTWIG and

STOSSEL 1975, POLLARD 1976). Certain proteins were found to increase the viscosity of actin solutions and cause the formation of a rigid gel by cross-linking actin filaments into a three-dimensional network. These proteins were therefore called "gelation factors". Later, other factors were identified that induce the lateral aggregation of actin filaments into tight bundles. Gelation factors and bundling proteins are now collectively referred to as

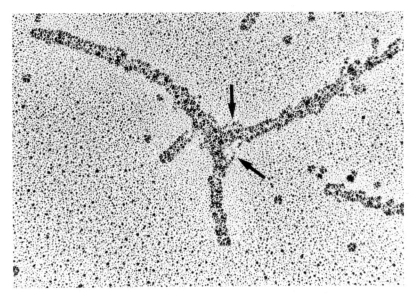

Fig. 6. Rotary-shadowed preparation of actin (1 mg/ml) copolymerized with actin binding protein (1 mg/ml). Arrows indicate strands with the diameter of actin binding protein molecules at an actin filament "branch" point. × 220,000
(Courtesy of J. Hartwig)

"cross-linkers". Their common features are that they possess at least two binding sites for F-actin and, when mixed with actin filaments even at a relatively low concentration, induce gelation or bundling. In some of the cross-linkers, these binding sites are located on a single polypeptide, whereas others are dimers or oligomers of a protein that each contains one F-actin binding site.

The best-characterized of the network-inducing cross-linkers are re-latively large, elongated, flexible molecules consisting of 2 (or 4 in the case of spectrin) subunits. Their flexibility is clearly advantageous in the con-struction of a three-dimensional gel because it allows for substantial freedom in their mode of attachment to actin filaments. Filamin and actin binding protein (ABP) are long, thin rods with dimensions of 160 × 3 nm in the fully extended dimeric state (e.g., TYLER et al. 1980, HARTWIG and

STOSSEL 1981) (Fig. 6). They are immunologically and biochemically very similar and appear to be isoforms of a widespread protein class. Spectrin, a protein originally thought to be restricted to erythrocytes but now known to have counterparts in many cell types (see section IX.3.), can extend to a length of about 200 nm in the tetrameric state (GLENNEY et al. 1982). Spectrin, ABP and filamin bind to F-actin with high affinity, the latter two at a stoichiometry of 1 : 14 to actin monomers at saturation (HARTWIG and STOSSEL 1981). Network formation is induced at much lower ratios (1 : 100–1 : 1,000), while at a concentration approximating saturation, bundles can be induced as well. Thus, inside cells, the concentration of a cross-linker relative to actin may help determine whether networks or bundles are formed.

Whereas the binding of filamin and ABP to actin is calcium-insensitive, other cross-linkers are active only when the calcium concentration is submicromolar. Of these, actinogelin (molecular weight 115,000 d), *Acanthamoeba* gelation protein (2 × 90,000 d), the 90,000 d and 120,000 d *Dictyostelium* actin binding proteins, and the nonmuscle alpha-actinins (2 × 100,000 d) will induce network formation. Other proteins, such as the 30,000 d *Dictyostelium* protein and villin (95,000 d), are actin bundling proteins. Among the calcium-sensitive cross-linkers, the brush border-specific protein, villin, is probably the best-characterized and maybe the most interesting because of its demonstrated dual role as a bundling protein at submicromolar calcium concentrations and a capping protein at micromolar calcium (Fig. 7). Villin has three high affinity (K_d about 5 μM) calcium binding sites, two of which are rapidly exchangeable and one is nonexchangeable (HESTERBERG and WEBER 1983 b). One of the exchangeable and the nonexchangeable site are located on a "core" peptide of 90,000 d molecular weight, whereas the second exchangeable site is located on the 8,500 d "headpiece" fragment. These two fragments are proteolytically separable. Villin core retains the end-blocking capability (GLENNEY and WEBER 1981), whereas the 8,500 d headpiece fragment binds to F-actin—surprisingly, both in the presence and in the absence of calcium (GLENNEY et al. 1981). In the presence of micromolar calcium, villin undergoes a conformational change from a moderately asymmetrical (axial ratio 4.5 : 1) to a more highly asymmetrical molecule (axial ratio 8 : 1). This change seems to be triggered by the headpiece fragment because villin core does not exhibit the strong structural change of the intact villin molecule (HESTERBERG and WEBER 1983 a). So far, this is the best approximate model for how the functional activity of a calcium-sensitive cross-linking molecule might be regulated.

In contrast to gelation factors, many of the bundling proteins are small, globular proteins that bind to actin filaments at regular intervals (see Table 2). They induce the formation of bundles in which all the filaments are of

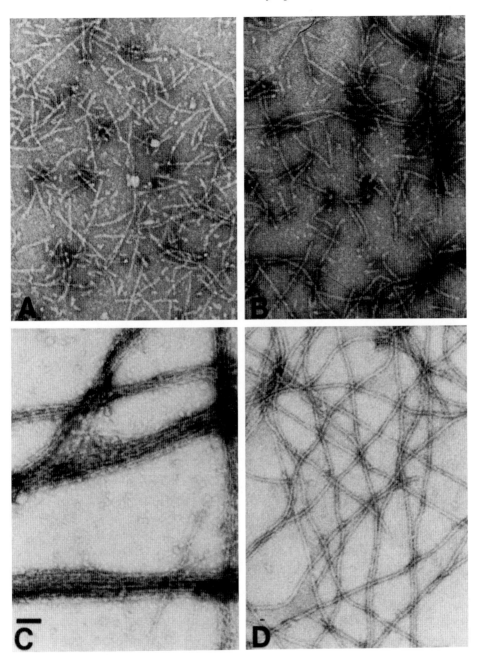

Fig. 7. Electron microscopy (negative staining) of the structures obtained from mixtures of villin, villin core, and actin in the presence and absence of calcium. *A, C* Villin and actin, *B, D* villin core and actin. *A, B* 10 μM free calcium, *C, D* 2 mM EGTA, a calcium chelator. Both villin and villin core fragment actin in the presence of calcium, but only villin induces the formation of actin bundles in its absence. × 84,000. [From GLENNEY and WEBER (1981)]

uniform polarity and all the crossover points of the helices of adjacent actin filaments are in register. Examples of such highly ordered bundles include the stereocilia of chochlear hair cells (TILNEY et al. 1980), and the acrosomal process of certain marine sperm (TILNEY 1975, DE ROSIER et al. 1982). Such bundles are endowed with remarkable tensile strength and yet still retain quite substantial flexibility due to the slight disorder built into the actin filament itself (see section VII.2.1.). Based on the built-in flexibility of actin filaments, in at least one case, the extension of the acrosomal actin bundle of *Limulus* sperm (DE ROSIER et al. 1982), an actin filament bundle can exert a force in the absence of myosin simply by changing the twist of the filaments that compose it.

Several studies of cytoplasmic extracts from a wide variety of cells suggest the presence of other actin cross-linking proteins involved in gel formation. Many of these are high molecular weight, and may be very similar to ABP and filamin (*e.g.*, SCHLOSS and GOLDMAN 1979, WEIHING 1983). In addition to these specific cross-linking molecules, a number of apparently nonspecific cross-linkers and cross-linkers of undetermined function have been identified *in vitro*. These include polylysine, histones, RNAse, nerve growth factor, and several glycolytic enzymes, most prominently aldolase (GRIFFITH and POLLARD 1982). The many (predominantly basic) proteins that can act as cross-linkers may have important features in common with the "specific" cross-linkers. Conversely, it is possible that some of the factors listed in Tables 1–3 are not so specific after all. Since many of them have not yet been extensively characterized, this remains a distinct possibility.

I.3.3. End-Binding Proteins

The proteins listed in Table 3 are characterized by their ability to bind to one of the ends (usually the barbed end) of an actin filament (Fig. 8). As a consequence, they exhibit several or all of the following properties: 1. Accleration of assembly by elimination of the lag phase, usually interpreted as an increase in the rate of nucleus formation by the stabilization of actin oligomers complexed to an end-binding molecule. 2. Inhibition of the overall rate of elongation by preventing the exchange of monomers at the filament end to which they are bound. Blockage of one of the filament's ends is demonstrated by the inhibition of subunit addition to the barbed or the pointed filament end when HMM-decorated filament fragments are used as seeds (see Fig. 4, insert). 3. Reduction of the average length of actin filaments, which lowers the viscosity of F-actin solutions. Length regulation may occur by direct severing (active breaking) of actin filaments, or by binding to the ends of spontaneously broken filaments.

Some end-blocking proteins are calcium-regulated (*e.g.*, gelsolin, villin, fragmin, severin). These proteins require micromolar calcium for the severing or cutting of actin filaments into shorter fragments. Gelsolin, the widely distributed, calcium-sensitive actin severing and end-blocking protein, probably complexes tightly with actin only in the presence of elevated calcium (KURTH and BRYAN 1984). Once formed, this complex is stable even in the absence of calcium, and can bind one more actin molecule to form a ternary complex (BRYAN and KURTH 1984). The activities of these

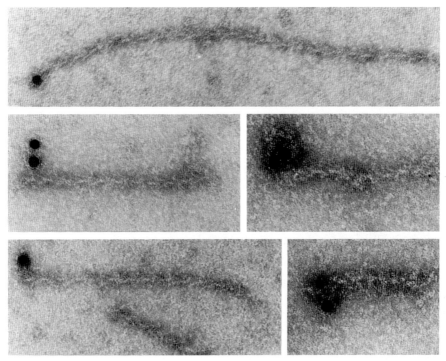

Fig. 8. Demonstration that gelsolin binds to the "barbed" end of actin filaments. Heavy meromyosin-decorated actin filaments are incubated with gelsolin, an antibody against gelsolin, and a secondary antibody coupled to colloidal gold particles. × 90,000. (Courtesy of J. HARTWIG)

complexes remain to be studied in detail. Severing and capping by villin may be controlled by different calcium binding sites on villin because capping seems to require lower calcium concentrations than severing (WALSH et al. 1984). Villin is also an actin-cross-linking (bundling) protein at submicromolar calcium, as described above. The same is true for vinculin, which induces filament bundles in vitro at high vinculin/actin ratios (JOCKUSCH and ISENBERG 1981); its action, however, is calcium-independent.

Of the other calcium-dependent severing/capping proteins, fragmin and cap 42 are particularly intriguing because, by immunological and biochem-

Table 3. *End-binding (capping) proteins*

Protein	Source	Molecular weight (d)	Comments	References
Gelsolin	macrophages, platelets; present in many other cells	90,000	binds to barbed end and shortens F-actin without depolymerization calcium-sensitive	YIN and STOSSEL 1979, 1980, YIN et al. 1981, 1984
90 kd protein	brain	93,000	very similar to, if not identical with, gelsolin	PETRUCCI et al. 1983, NISHIDA et al. 1983, ISENBERG et al. 1983, VERKHOVSKY et al. 1984
Brevin (plasma gelsolin, ADF)	serum	90,000	very similar to gelsolin, but 25 amino acid residues longer; is secreted from cells	HARRIS and SCHWARTZ 1981, THORSTENSSON et al. 1982, HARRIS and WEEDS 1983, WILKINS et al. 1983
Villin	intestinal epith.	95,000	binds to barbed end at micromolar calcium; villin core similar to gelsolin	BRETSCHER and WEBER 1981, GLENNEY et al. 1981, CRAIG and POWELL 1980
Fragmin	*Physarum*	43,000	calcium-sensitive; fragments actin filaments; stimulates ATPase activity of F-actin 15-fold calcium-dependent; binds to barbed	HASEGAWA et al. 1980, HINSSEN 1981 a, b
Severin	*Dictyostelium*	40,000	end; severs actin filaments; forms 1:1 complex with actin	BROWN et al. 1982, YAMAMOTO et al. 1983, GIFFARD et al.1984
Cap 42 (a + b)	*Physarum*	2 × 42,000	calcium-dependent; binds to barbed end; may be distinct from fragmin; subunit (b) phosphorylatable	MARUTA et al. 1983, MARUTA and ISENBERG 1983

Protein	Source	Molecular weight	Properties	References
45,000-molecular weight protein	sea urchin	45,000	calcium-dependent; severs actin filaments	Wang and Spudich 1984
45k-A	sea urchin	45,000	not calcium-sensitive; binds to barbed end; forms 1:1 complex with actin	Hosoya and Mabuchi 1984
Capping proteins	Acanthamoeba	28,000 and 31,000	not calcium-dependent; binds to barbed end	Isenberg et al. 1980, Cooper et al. 1984
Acumentin	macrophages	65,000	not calcium-dependent; binds to pointed end	Southwick and Stossel 1981, Southwick et al. 1982
Beta-actinin	skeletal muscle	34,000 and 37,000	binds to pointed end	Maruyama 1976, Yokota and Maruyama 1983
Cytochalasin-like protein	platelets	65,000 (?)	binds to barbed end	Grumet and Lin 1980
Vinculin	smooth muscle and other cell types	130,000	inhibits rate of elongation, increases critical concentration; phosphorylatable	Jockusch and Isenberg 1981, Wilkins and Lin 1982
Calcium-dependent actin modulator	smooth muscle	85,000	shortens actin filaments (information on other properties very limited)	Hinssen et al. 1984
80k actin modulator	thyroid	80,000	calcium-dependent; nucleates assembly, binds to F-actin (limited information available)	Kobayashi et al. 1983

ical criteria, they share some features with actin itself (MARUTA *et al.* 1984). In fact, they may be regulatory variants of actin that have evolved by duplication and modification of the actin gene. Whether other actin-binding proteins, including those listed in Tables 1 and 2, have actin-related sequences, can not be said with certainty because many are not yet sufficiently characterizied. Some, however, are known in enough detail to conclude that they are unrelated to actin, so modification of actin itself can not be regarded as a general mechanism for the evolution of actin-modulating proteins.

Two of the proteins in this group (beta-actinin from muscle and acumentin from macrophages and leukocytes) bind to the pointed filament end, whereas the others listed in Table 3 bind the barbed end. All of them resemble the cytochalasins in their ability to bind with high specificity to one end of an actin filament. Evidently, it is adavantageous for a cell to regulate actin polymerization via proteins that have a high affinity for filament ends, because the number of ends is low compared to the number of monomers, and because the events at filament ends determine filament length and, ultimately, filament organization (see FÜCHTBAUER *et al.* 1983).

Two additional actin-binding proteins with functions different than those listed in Tables 1–3 include tropomyosin and cofilin. Tropomyosin (BAILEY 1946) has a molecular weigth of about 33,000 d and is a long, rigid molecule consisting of at least 90% alpha-helix. There are two closely related isoforms (alpha and beta) in skeletal muscle. Its 284 amino acid residues form molecules with dimensions of 42×2 nm that associate as coiled-coil heterodimers with the groove of the actin filament helix, spanning approximately seven actin monomers. Tropomyosin stabilizes actin filaments against a variety of depolymerizing stimuli and agents; in skeletal muscle, it is involved in the regulation of contraction. Cofilin, a 21,000 d protein isolated from brain (NISHIDA *et al.* 1984), seems to counteract the binding of tropomyosin to actin filaments by binding to actin filaments in a 1 : 1 molar ratio of cofilin to actin monomer in the filament. It does not depolymerize existing actin filaments, even though it also binds monomeric actin. Cofilin binding to actin dissociates tropomyosin from actin filaments, and it inhibits the interaction of actin with myosin. The function of this unusual actin-binding protein in nonmuscle cells is unknown.

I.3.4. Some General Considerations on Actin Networks

Over the past few years, the consensus has emerged that actin-based networks can be described in terms of FLORY's (1941) network theory which holds that the physical properties of random polymer networks are influenced by two parameters, the number of subunits in the polymer state (P_n) and the number density of cross-linker molecules (C_n). Gelation ensues rather abruptly when the two are present at a critical ratio defined by

$$P_g = \frac{C_n}{P_n} \qquad (1)$$

where P_g is the gel point of the polymer/cross-linker mixture. In practical terms, this relation means that a decrease in average polymer length needs to be compensated by an increase in the number density of cross-linker to maintain the gel point. Indeed, solutions of actin filaments, a capping protein, and a cross-linker were found to exhibit the behavior predicted by this relationship (YIN et al. 1980, NUNNALLY et al. 1981). Network formation in polymer/cross-linker solutions can be effectively regulated by 1. changes in the total amount of subunits in the polymer, 2. changes in the weight average length of the polymer (which is defined as $1/P_g$), and 3. changes in the number density of cross-linkers. Because gel formation is abrupt, only very small changes in one of the three parameters can affect network formation profoundly. It is more than just coincidence that cells have network regulating protein factors in *all* three categories mentioned above: depolymerizing and polymerization-inhibiting proteins affect the total amount of subunits in the polymer, capping proteins can specify the length of the polymers, and gelation factors determine the degree of cross-linking between polymers. Thus, cells seem to have at their disposal the necessary armamentarium to regulate network formation and breakdown. However, the description of cellular networks given here may be somewhat simplistic because in cells, many additional factors will influence network structure and rigidity. For example, actin networks are linked to the plasma membrane (section IX.) and interact with other components of the cytoskeleton (section VII.). Another complication is that networks coexist with bundles made of the same polymer; bundles may even be embedded in networks, and sometimes the same filaments are part of both structures (*i.e.,* filaments emerging from bundles become part of a network). Finally, a different network will be formed when actin polymerizes in the presence of a cross-linking protein, as opposed to a network formed by the addition of a cross-linker to preformed actin filaments. For example, actin assembly in the presence of macrophage ABP leads to a network composed of short, "branched" filaments where ABP molecules occupy the branching points (which, in fact, are points of end-to-side contacts of actin filaments) (HARTWIG et al. 1981, NIEDERMAN et al. 1983). This behavior can be explained by a nucleating as well as a cross-linking activity of ABP which results in the formation of more and shorter filaments connected end-to-side at right angles. Such networks will probably be more rigid than networks composed of long filaments cross-linked after they have formed.

Thinking about cellular gels in terms of polymer networks has greatly advanced our understanding of cytoskeletal organization and function. It should not be overlooked, however, that some important questions still remain to be answered. The molecular basis of the action of many actin-associated proteins is not understood, and the binding sites for these proteins on the actin molecule (and vice versa) are largely unknown. How some of the actin binding proteins are sequestered when not needed is still a mystery. Finally, there seem to be some discrepancies between the *in vitro* properties of several actin-associated proteins and their *in vivo* location. For example, filamin and alpha-actinin induce the formation of three-dimensional networks *in vitro*, but they are prominent components of filament bundles (stress fibers) *in vivo*. Vinculin, which seems to be able to associate with filament bundles along their length in the test tube, is located primarily at the ends of bundles in living cells. Some of these discrepancies may simply be explained by differences in the affinities of these proteins for several available binding sites, but some proteins may actually serve functions in the cell that are different from those identified *in vitro*.

I.4. Myosin

I.4.1. Structure

Myosin is present in only small amounts in most nonmuscle cells, in contrast to actin, one of the most abundant cellular proteins (Table 4), and nonmuscle myosins are more variable than the actins (HARTSHORNE and GORECKA 1980, POLLARD 1981). Presumably, in all cell types, myosins convert chemical energy derived from the cleavage of the gamma-phosphate group of ATP into mechanical energy via a sliding of actin filaments past myosin. However, regarding the mechanism of actin-myosin interaction and myosin function, muscle cells still are the best-studied cell system. Our knowledge of cytoplasmic myosins is sketchy due in part to the small amounts available for isolation and biochemical characterization, and in

Table 4. *Comparison of actin and myosin contents in various cell types (representative values)*

	% of Total Protein		Actin : myosin ratio
	Actin	Myosin	
Skeletal muscle /rabbit)	20	36	6
Smooth muscle (pig)	25	8	34
Acanthamoeba	14	1.5	70
Platelets (human)	10	1	110
Dictyostelium	7	0.5	157

part to the difficulty in identifying myosin molecules in electron micrographs of nonmuscle cells.

Almost without exception, myosins are rather large molecules that exist as a hexameric complex of two non-covalently linked identical heavy chains (molecular weight around 200,000 d), each of which is associated with, as a

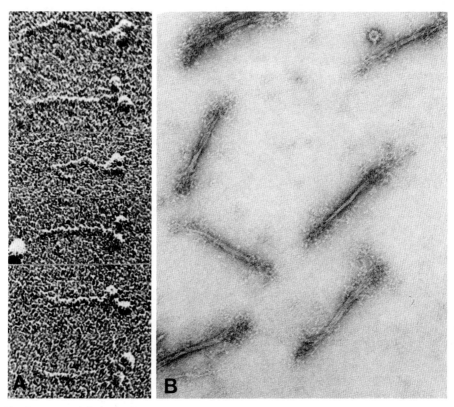

Fig. 9. *A* Unidirectinally shadowed molecules of *Acanthamoeba* myosin II. × 280,000. *B* Negatively stained platelet myosin filaments. These filaments are composed of about 30 myosin molecules and show the typical bare zone in the middle of the filament. × 100,000. (Courtesy of T. Pollard)

rule, one pair of light chains (molecular weight 12–22,000 d). With one known exception, the globular *Acanthamoeba* myosin I (Pollard and Korn 1973 b, Maruta *et al.* 1979), myosin heavy chains are highly asymmetrical molecules, consisting of a large globular head region and a long, alpha-helical, coiled-coil tail region (Fig. 9). Interestingly, even though *Acanthamoeba* myosin I may lack a tail, this globular molecule has domains structurally related to the tail of myosin II (Kiehart *et al.* 1984). The light chains are non-covalently linked to the head regions of the heavy chains. The two major domains of the myosin molecule, head and tail, have distinctly different properties and functions. The head contains the ATPase activity and the actin binding site, whereas the tails have the ability to self-aggregate under appropriate ionic conditions; they form the backbone of the myosin thick filament in striated and smooth muscle. Whether nonmuscle myosins also exist as filaments in the cell, or whether they are present as oligomers or even monomers, is not clear, even though they, too, can form filaments *in vitro* under appropriate ionic conditions (NIEDERMAN and POLLARD 1975). In myosin thick filaments, the heads project from the surface of the assembled core, which is formed by the myosin tails aggregated in an antiparallel fashion. As a result of the antiparallel organization, most myosin filaments are bipolar and consist of a central bare zone flanked by regions where myosin heads protrude as tufts from the myosin core (Fig. 9). The dimensions of myosin filaments formed *in vitro* are variable and depend on a variety of conditions. Striated muscle thick filaments consist of several hundred myosin molecules and are about 1.5 μm long and 18 nm wide (HUXLEY 1963), but they can also be induced to form minifilaments containing only 16–18 myosin molecules (REISLER *et al.* 1980). Platelet or amoeba myosins form much shorter filaments with fewer myosin molecules than in native striated muscle thick filaments.

Myosin can be cleaved by proteolytic enzymes into a number of functional domains. Digestion with trypsin yields two fragments, one, termed heavy meromyosin (HMM), consisting of the heads and part of the tail, while the other, light meromyosin (LMM), represents the distal two thirds of the two rods. Further limited proteolysis of HMM with papain results in two identical S 1 fragments representing the two heads and one S 2 fragment, the rod portion of HMM (LOWEY *et al.* 1969). Both HMM and S 1 are able to bind to F-actin in the absence of ATP in an arrowhead pattern, a property that has been used to detect actin filaments and determine their polarity in nonmuscle cells. The site of proteolytic cleavage between HMM and LMM is interpreted as a region of less ordered structure than the rest of myosin rod. This site is believed to represent a flexible "hinge" between the rod and the HMM portion extending from the thick filament. Another flexible site, termed the "swivel", seems to be located at the juncture between S 1 and S 2. Both sites are considered to play important roles in the power stroke of the myosin molecule during its ATP-dependent interaction with actin. Whereas the heads, S 1, and HMM are soluble, the LMM fragment is rather insoluble and retains the ability to self-aggregate under physiological conditions, indicating that it contains the region responsible for myosin filament formation (HUXLEY 1963, KENDRICK-JONES *et al.* 1971).

The most distinctive feature of the myosin molecule is its unique ATPase activity located in the head region. In the presence of Mg-ATP alone, its acitivity is very low, but it is stimulated by about 200-fold in the presence of actin, or if magnesium is replaced by potassium in the presence of EDTA (SHETERLINE 1983). Whereas the latter ATPase activity is non-physiological and has been used only as a convenient means to identify myosin, the actin-activated ATPase reflects its physiological role and distinguishes myosin from all other cellular ATPases. In cells, the ATPase activity and, therefore, the interaction of actin with myosin is not regulated by the presence or absence of actin, or the presence or absence of ATP, but by specific calcium-dependent regulatory molecules associated with either actin filaments (striated muscle) or myosin (smooth and non-muscle cells) (section I.5.1.).

The head portion of the myosin heavy chain is associated with two classes of light chains, formerly termed "regulatory" and "essential", but now frequently designated simply by their molecular weight. In striated muscle, there are two light chain molecules of molecular weight 18,000 d (LC 18, former regulatory light chains) that are phosphorylatable by myosin light chain kinases, whereas the "essential", light chains, LC 22 und LC 16, are non-phosphorylatable. Each myosin head has one of the LC 18 and either the LC 22 or the LC 16 light chain. The complement of light chains and their molecular weights in smooth and nonmuscle cells is variable and species-dependent, but all have in common that at least one of the light chains is phosphorylatable, a property that plays a major role in myosin assembly and function.

Light chains have several properties in common with other proteins known to play a role in calcium regulation and contraction, including troponin C and calmodulin. For example, light chains share substantial sequence homology with both of these molecules (KRETSINGER 1980). The precise function of the light chains in striated muscle is not understood, since neither LC 18 removal nor its phosphorylation appear to influence actomyosin ATPase activity (ADELSTEIN and EISENBERG 1980). Similarly, LC 22 and LC 16 are not essential either for the ATPase activity of striated muscle myosin (WAGNER and GINIGER 1980). On the other hand, in both smooth and nonmuscle cells, phosphorylation of the LC 18-equivalent is a major regulator of myosin filament formation and actin myosin interaction, as described in the next two sections.

I.4.2. Myosin Filament Formation

All myosins can self-associate *in vitro* to form filamentous structures that resemble those isolated directly from cells. The mechanism of association is believed to be equivalent to a condensation-polymerization scheme. The ability to form filaments resides in the properties of the myosin rod, but the process of assembly and the final structure of the filaments depends on the source of myosin. Striated muscle myosin will self-assemble *in vitro* into structures resembling native filaments under physiological conditions of ionic strength, pH, ATP, and magnesium concentration (HUXLEY 1963). Structural features of these *in vitro* filaments include the bare zone and bipolarity, but they are usually longer than *in vivo* filaments. In contrast, the precise morphology of myosin filaments in smooth muscle, and even more so in nonmuscle cells, and the way in which the filaments are assembled, is not established. Smooth muscle myosin filaments are unstable in the presence of ATP, and may not be well preserved during processing for electron microscopy (SOMLYO 1980). In both smooth muscle (SUZUKI *et al.* 1978, KENDRICK-JONES and SCHOLEY 1981, KENDRICK-JONES *et al.* 1983) and nonmuscle myosins (SCHOLEY *et al.* 1980, CRAIG *et al.* 1983, BROSCHAT *et al.* 1983), filament formation (as well as activation of the ATPase activity: see next section) are greatly promoted by the phosphorylation of one of the light chains. The molecular mechanism by which light chain phosphorylation promotes filament assembly is not understood. The filaments formed *in vitro* from smooth muscle myosin (*e.g.,* HINSSEN *et al.* 1978) usually lack a bare zone in the middle. Instead, antiparallel myosin molecules are present all along the filament, their polarities alternatining between neighboring longitudinal rows of myosin molecules. Nonmuscle myosin, such as the myosin II from *Acanthamoeba* or platelet myosin, will form short, bipolar filaments consisting of 12–18 molecules under conditions of low salt, and larger aggregates in high magnesium (NIEDERMANN and POLLARD 1975, POLLARD 1982) (Fig. 9). Their state of aggregation in the cell is not known, but under the conditions prevalent in cells, myosin should be almost completely polymerized. However, myosin filaments are very difficult to identify in electron micrographs of thin sections. If filaments are present at all, they may be very short. Another reason why myosin filaments may be difficult to detect is that myosin not engaged in contractile acitivity may be disassembled in the cell. Dephosphorylation of the light chains may lead to disassembly of myosin filaments (SUZUKI *et al.* 1978, MARTIN *et al.* 1981).

In some cell types, the tail portion of the heavy chain is itself an important site of regulation of nonmuscle myosin assembly. *Acanthamoeba* myosin II, for example, can be phosphorylated to the extent of 3 moles of phosphate per mole of heavy chain (COLLINS and KORN 1980, COTE *et al.* 1981). Both dephosphorylated and phosphorylated forms of myosin can form filaments *in vitro*, but the filaments formed from desphosphorylated myosin are more stable (KUCZMARSKI and SPUDICH 1980) and larger (COLLINS *et al.* 1982 a). Heavy chain dephosphorylation also increases the actin-activated ATPase activity of myosin (see next section) and therefore also is an important site of regulation of contraction.

I.5. Actin-Myosin Interaction

I.5.1. Regulation of Actin-Myosin Interaction

The mechanism of regulation of actin-myosin interaction during contraction is different for skeletal muscle than for smooth or nonmuscle cells. A complete discussion of skeletal muscle regulation, however fascinating, is beyond the scope of this monograph (for recent overviews, see HUXLEY 1980, SQUIRE 1981, HIGHSMITH and COOKE 1983, SHETERLINE 1983). However, it needs to be considered briefly to compare and contrast it with the smooth and nonmuscle systems.

The interaction of myosin with F-actin to generate force for the relative sliding movement of one past the other is triggered by calcium ions and regulated via the tropomyosin/troponin system associated with actin filaments. The rod-shaped tropomyosin molecules are associated head-to-tail with each other to form molecule chains that lie in the grooves on either side of the actin filament. Troponin is a 70,000 d complex of three subunits, termed Tn T (for tropomyosin-binding), Tn I (for inhibitory), and Tn C (for calcium-binding). Tn T binds to tropomyosin close to one end of the tropomyosin molecule. Tn I also binds to tropomyosin, while Tn C binds to Tn I and Tn T, but not to tropomyosin. In the resting state, tropomyosin is located in a position along the actin filament where it sterically hinders access of the myosin heads to their binding sites on actin. Upon activation, troponin is believed to evoke a movement of the tropomyosin rod deeper into the groove of the actin helix, thereby exposing the myosin binding sites on F-actin (TAYLOR and AMOS 1981). Activation is initiated by a rise in the intracellular free calcium concentration released from the sarcoplasmic reticulum. Calcium will bind to Tn C, which has four calcium binding sites, but which are responsible for calcium regulation of Tn C function is not quite clear. Upon binding of calcium, Tn C will relieve the inhibition of actin-myosin interaction imposed by the Tn I/Tn T/tropomyosin complex. Myosin heads will then interact cyclically with actin filaments. The result of this cyclical interaction which, in skeletal muscle, occurs at a frequency of 5–10 cycles/second, is a relative movement between myosin and actin filaments. This sliding movement is the structural basis for force production. While the details of the crossbridge cycle remain to be elucidated, the principle has remained essentially unchanged for the past 30 years (HUXLEY 1957). The crossbridge cycle involves the attachment of myosin heads to actin, a conformational change in the myosin molecule, and the subsequent detachment of the head from actin, only to bind to a new site on the actin filament a few nanomemters away from the first site. This mechanical process is interlaced with the chemical process of ATP hydrolysis by the myosin ATPase. The precise site of the conformational change in the myosin molecule has not yet been identified unequivocally, but it is believed to involve a rotation of the myosin head (or a substantial portion thereof) relative to the rest of the molecule (for recent discussions of the crossbridge cycle, see SHETERLINE 1983, HUXLEY et al. 1983). Removal of calcium by pumps located in the sarcoplasmic reticulum membrane will revert the sequence of events that initiated contraction.

This model for the regulation of skeletal muscle contraction is widely accepted, even though some of the details are still disputed (e.g., SOBIESZEK and SMALL 1981, POULSON and LOWY 1983). The scheme of regulation involves a calcium-triggered release of the inhibition of contraction imposed by tropomyosin. In contrast, calcium influx directly activates contraction in smooth and nonmuscle cells. Activation involves calcium/calmodulin-mediated stimulation of kinases that phosphorylate the myosin light chains and/or heavy chains, depending on species. In many mammalian cell types, for example, the actin-activated ATPase activity of myosin is enhanced only when LC 20 is phosphorylated by a specific light chain kinase (e.g., ADELSTEIN and CONTI 1975, YERNA et al. 1979, SAGARA et al. 1982). In both smooth muscle (SOBIESZEK 1977) and nonmsucle cells (ADELSTEIN and CONTI 1975, TROTTER and ADELSTEIN 1979), myosin light chain phosphorylation is mediated by calmodulin, the troponin C-like molecule present in essentially all cells (MEANS and DEDMAN 1980, CAMPBELL 1983). Calcium-calmodulin will activate a kinase highly specific for myosin light chains. The kinases comprise a family of proteins that have diverged early in evolution and therefore may vary strikingly in size in different species (summaries in HARTSHORNE 1982, NUNNALLY and STULL 1984). The kinase is fully activated only when calcium-calmodulin is bound to it. At least 3, or all 4, of the calcium binding sites on calmodulin need to be occupied in order to activate the kinase (BLUMENTHAL and STULL 1980). The activated kinase will phosphorylate each of the phosphorylatable light chains to the extent of one mole of phosphate per mole of light chain. However, it is not known whether one or both of the light chains need be phosphorylated for activation of myosin in vivo. Deactivation of the contractile machinery not only involves the removal of calcium, but also the dephosphorylation of light chains. This ist believed to be accomplished by another specific enzyme, myosin light chain phosphatase, of which two forms have been isolated from smooth muscle (PATO and ADELSTEIN 1980). Myosin light chain phosphatase seems to be active both in the presence and absence of calcium, but its acitivity is considerably lower than that of the light chain kinase so that it will effectively dephosphorylate the light chains only in the absence of calcium. Interestingly, the myosin light chain kinase itself is subject to regulation by phosphorylation, mediated by a cyclic AMP-dependent kinease (ADELSTEIN et al. 1978, NISHIKAWA et al. 1984). Phosphorylation of light chain kinase results in a decreased affinity for calmodulin and, therefore, in decreased kinease activity (for a review, see ADELSTEIN et al. 1981).

The scheme briefly outlined here is known as the "phosphorylation/de-phosphorylation theory" of regulation. It is widely accepted as a model for regulation of contraction in smooth and nonmuscle cells, and the experimental evidence in its support is considerable. However, certain aspects are not fully explained by the theory (reviewed in HARTSHORNE 1982), suggesting that additional factors are involved.

In contrast to mammalian cells, no evidence has been obtained for phosphorylation of the light chains in certain amoebae. Rather, their myosins are regulated by heavy chain phosphorylation. The three phosphorylation sites of *Acanthamoeba* myosin II heavy chains are located close to the tip of the tail (COLLINS *et al.* 1982 b) and are phosphorylated by a specific myosin heavy chain kinase. In this as well as other species, *e.g.,* *Dictyostelium* (KUCZMARKI and SPUDICH 1980, PAGH *et al.* 1984), only the dephosphorylated form of the molecule has the highest ATPase activity. Phosphorylation of myosin tails regulates the activity of myosin by affecting the formation of myosin filaments, rather than the heads of the myosin molecules only (KUZNICKI *et al.* 1983). This conclusion is corroborated by proteolytic removal of a small peptide carrying the three phosphorylation sites (KUZNICKI *et al.* 1985), and by specific monoclonal antibodies that bind to the tip of the heavy chain tail where the sites are located (KIEHART and POLLARD 1984). In both experiments, the ATPase activity as well as the ability to form filaments are inhibited, suggesting that intact filaments are required for full expression of the ATPase activity and the ability to induce contraction.

I.5.2. Tropomyosin, Troponin, and Alpha Actinin in Smooth and Nonmuscle Cells

Tropomyosin has been known to exist in smooth and nonmuscle cells (Fig. 10) for some time (reviewed in HARTSHORNE 1982, KORN 1982). Although they are generally similar to their skeletal muscle counterparts (*e.g.,* antibodies against skeletal muscle tropomyosin will crossreact with nonmuscle tropomyosin), the nonmuscle tropomyosins are shorter by 37 amino acid residues, which allows them to span only 6, instead of 7, actin monomers in a filament. The function of tropomyosin in smooth and nonmuscle cells is not fully understood. It is evidently not involved in actin-linked calcium regulation of actin-myosin interaction, but it seems to help protect actin filaments against spontaneous or actin-depolymerizing factor-induced fragmentation (WEGNER 1982, BERNSTEIN and BAMBURG 1982). Tropomyosin may also enhance the activation of myosin ATPase. However, because nonmuscle tropomyosins are shorter than their muscle counterparts, they are less effective in stabilizing filaments.

Of the three troponin subunits, only a troponin C-like molecule has apparently been found in several nonmuscle sources (*e.g.,* Fine *et al.* 1975,

KUO and COFFEE 1976). Its function is obscure. Since it seems to be present in only minute amounts, and since troponin I and troponin T are not found in cells other than striated muscle, a functional actin-linked regulatory system analogous to that of skeletal muscle may not be operating in smooth muscle and nonmuscle cells. In addition, most other calcium-linked

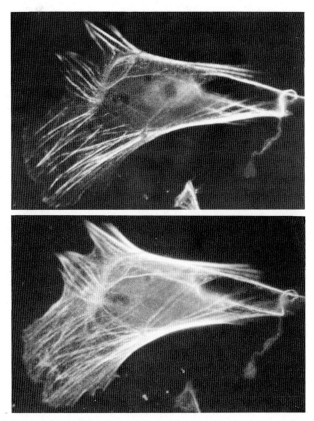

Fig. 10. Chicken embryo fibroblasts double-labeled with antibodies against alpha-actinin (top) and tropomyosin (bottom). × 1,000. (Courtesy of K. FUJIWARA)

processes in smooth and nonmuscle cells are regulated by troponin C's close relative, calmodulin.

Alpha actinin, the major (but not the only) protein of the Z-line in striated muscle to which actin filaments are anchored, is a prominent component of smooth and nonmuscle cells. In the former, alpha actinin is a major constituent of the so-called dense bodies which are scattered both within the cytoplasm and attached to the plasma membrane (reviewed in SOMLYO 1980). They are believed to be Z-line analogs that aid in the transmission of tension throughout the cell. A prominent location of alpha-actinin in many nonmuscle cells is within stress fibers (Fig. 10), where it is found both at the ends and as periodic patches distributed along their length (*e.g.,* LAZARIDES 1976, FERAMISCO and BLOSE 1980). Other studies found alpha actinin in association with the cell membrane (CRAIG and PARDO 1979, FUJIWARA and POLLARD 1978). By immunofluorescence microscopy, little alpha actinin appears to be associated with network regions in smooth and nonmuscle cells, even though, as mentioned earlier, *in vitro* it seems to be able to cross-link actin filaments into a network.

I.5.3. Myosin and Its Interaction with Actin: Outlook

Two fundamental aspects of the contractile mechanism in muscle and nonmuscle cells are generally regarded as identical: first, the sliding filament model, originally put forth as a logical and satisfying explanation of skeletal muscle contraction, can clearly be extended also to smooth and nonmuscle cells. Second, in all cells, the contractile activity is regulated by changes in the concentration of calcium ions in the cytoplasm. Beyond these two basic similarities, major differences are apparent, and large gaps still exist in our knowledge of details in the regulation and function of "primitive" contractile systems. Even though in all cell types the principal regulator is the concentration of calcium ions which either promotes or inhibits actin-myosin interactions, the molecular vehicles of calcium regulation are different for striated vs. smooth or nonmuscle cells: it is actin-linked, via tropomyosin-troponin, in the former and myosin-linked, via calmodulin and the light chain kinase/phosphatase system, in the latter. The two systems are evolutionarily related, as evidenced by the homologies between calmodulin and troponin C, but the details of that relationship need to be elucidated.

Despite the fact that current concepts of smooth and nonmuscle regulation overwhelmingly favour the myosin phosphorylation mechanism, many features of this regulatory system are not understood. First, relatively little is known about nonmuscle myosins, which are just beginning to be characterized. Clearly, they are much more diverse in structure, peptide composition, and possibly also function than the actins. Second, the state of myosin in the cell (*i.e.,* monomer vs. polymer), and the possibility of length changes of myosin filaments associated with the activation of the contractile machinery, need to be assessed. If assembly of myosin filaments from dispersed subunits is required prior to initiation of contraction, the contractile machinery of these cell types would be more flexible and versatile, but slower in its reaction to activating stimuli. Third, the precise role of some of the muscle accessory proteins (alpha-actinin, tropomyosin, troponin C) in nonmuscle cells is not understood. Fourth, the kinetics of the crossbridge cycle need to be evaluated. The cross-bridge cycle has been found to be slower in nonmuscle actomyosin than in striated muscle, but whether this is attributable solely to the lower ATPase activity of nonmuscle myosins, or whether other factors are involved, needs to be determined. Finally, it would be of interest to determine whether myosin in nonmuscle cells can perform other functions apart from force generation. The two heads of a myosin molecule can bind to two different actin filaments, thereby cross-linking them (TRINICK and OFFER 1979). Thus myosin might be able to act as a cross-linker as well as a force-generating molecule. Whether this is of relevance for certain cellular activities remains to be shown.

I.6. Aspects of Contractile Protein Distribution and Function

Even though the organization and function of contractile proteins in various cellular activities will be dealt with in more detail later, this section will discuss, however sketchily, some dynamic features of contractile proteins. The similarities between the contractile mechanisms of striated and nonmuscle cells have been emphasized repeatedly. However, beyond the basic fact that in both muscle and nonmuscle cells, force is generated by the sliding of actin filaments past myosin, there are fundamental differences between the two systems in terms of the subcellular and supramolecular organization of the contractile proteins. Both skeletal and smooth muscle cells are extremely specialized for generating a force in one dimension only, shortening of the sarcomere in the former, and of the entire cell in the latter. Nonmuscle cells also generate linear forces, but, in addition, the contractile proteins are sometimes arranged in more complex three-dimensional lattices so that a force generated within the lattice will result in complicated three-dimensional deformations. Furthermore, the contractile machinery of nonmuscle cells can be remodelled rapidly both in space and in time, whereas the paracrystalline array of skeletal muscle remains virtually unchanged. Thus, a fundamental difference between skeletal and nonmuscle systems resides in the degree of *plasticity* of the contractile apparatus. A second, eminently important aspect ist that in nonmuscle cells, the contractile system may serve other purposes aside from generating a contractile force. There are many examples where contractile proteins play an essential *structural* role. Such a function is generally associated with actin, and is brought about, and regulated, by the proteins that interact with it. These two aspects—structural roles of contractile proteins, and plasticity of contractile protein organization—shall now be discussed briefly.

Structural role of contractile proteins. Many cells possess a cortical cytoplasmic zone characterized by a gelatinous consistency. This ectoplasmic gel layer comprises a variety of filamentous assemblies with actin as the chief component whose purpose is not only to contract the cell, or part of it, but also to provide mechanical strength and rigidity in the absence of contractile activity. Within the cortical gel, two distinct differentiations can be distinguished: networks and bundles. As discussed in a previous section (I.3.2.), both require specific associated proteins, some of which will lead to bundle formation, while others will induce networks. Bundles are involved in the formation of rigid cellular extensions whose shape, size, and location varies with their presumptive function. For example, filopodia are long, needle-shaped extensions that can be waved around by the cell and are thought to have an exploratory function (ALBRECHT-BUEHLER 1976, TOSNEY and WESSELS 1983). Microvilli are short, finger-like extensions of uncertain function that cover the surfaces of many cells, often being concentrated, as in many epithelia, on one side of the cell. Both filopodia and microvilli are anisotropic structures, that is, the filaments that support them are all highly aligned. In contrast, networks are isotropic, that is, the polymers from which they are constructed seem to have no preferred orientation. While this may be true for networks formed from actin and a cross-linker *in vitro* and for certain cellular structures, some networks found in cells are in fact more highly organized. For example, the organization of filaments in the leading edge of certain motile cells was long considered to be a paradigm for a cortical, isotropic network. Now it seems that as the procedures used to visualize these structures become increasingly sophisticated, the degree of organization in the texture of the network increases proportionally. Whole mount preparations visualized by negative staining or high voltage electron microscopy show an open network structure, but the filaments clearly have a preferred orientation, giving the network a weave-like texture (HOGLUND *et al.* 1980). It still is a quasi three-dimensional network, but it is not truely isotropic. Order might be imposed by a combination of two factors: the presence of actin binding protein molecules, which can induce the formation of a perpendicular branching network, and the association of the network with the plasma membrane, which might provide both nucleation sites and anchorage points (SMALL 1981).

Networks and bundles are interconvertible. This ist not only suggested by their coexistence in the cortices of many cells, and by the fact that the same filament may be part of both structures (Fig. 11); it is also indicated by the dynamics of the cortices of living cells, where bundles may appear within network-containing regions and melt back into them without creating discontinuities in cortical organization (*e.g.,* EDDS 1975). Interconversion can also be demonstrated *in vitro* where a stress or shear applied to an isotropic actin network will lead to the appearance of anisotropic (birefringent) domains. Stress-induced anisotropy as a mechanism of bundle formation may also operate *in vivo*. If supported by the intercalation of appropriate bundling proteins, it is a self-reinforcing process.

Plasticity of contractile protein organization. Networks and bundles are

filamentous assemblies that provide mechanical strength and resist de-
formation. Network formation (gelation) of F-actin and a cross-linker
depends on the cross-linker concentration and the average filament length;
a critical ratio of the two allows for incipient gelation at a rather sharply
defined gel point (see section I.3.4.). Because gelation is abrupt and can be

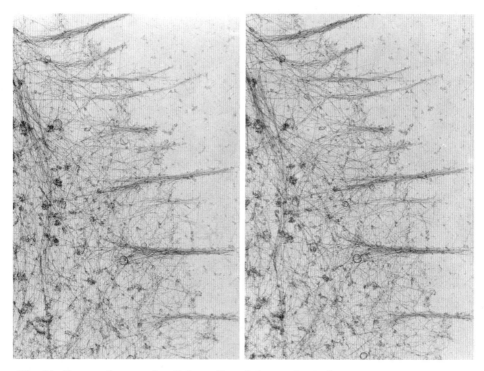

Fig. 11. Stereomicrograph of the cell periphery of a Triton X-100 extracted rat
carcinoma cell, demonstrating the splaying of filopodial actin filaments into the
peripheral three-dimensional network. × 9,800. (From the work of J. GIBBINS and
M. SCHLIWA)

induced by only slight changes in the state of its components, it provides an
ideal point of control for large changes in cytoplasmic consistency.
However, plasticity of gels can only be maintained if not only the formation
of gels, but also their breakdown, is under precise cellular control.
Theoretically, a gel can be broken down by two possible mechanisms: by
decreasing the density of cross-linkers, and by decreasing the average length
of actin filaments (without reducing the total filament mass), which is
equivalent to raising the critical cross-linker concentration required for
gelation. In practice, i.e., in a cell, the latter mechanism is the more likely

candidate: conditions which would allow cross-linker molecules to come off actin filaments without promoting a change in the state of the filaments have so far not been identified; on the other hand, several protein factors are known that reduce actin filament length. Therefore, networks probably are broken down by cutting the fibers between cross-linked vertices.

Formation and breakdown of networks, *i.e.,* cross-linking and severing of actin filaments, may well be the molecular basis for the phenomenon of *gel-sol-transformation.* Known for many decades, gel-sol-transformations are thought to play an essential role in many motile acitivities of cells, particularly those involved in the generation of cytoplasmic streaming. A prime example of a cell type whose locomotion is believed to be based on gel-sol-transformations is the giant freeliving protozoan, *Amoeba proteus.* Its cytoplasm is divided into two microscopically distinguishable regions, a central fluid endoplasm containing the bulk of the cell organelles, and a transparent, stationary ectoplasm, the cell cortex. The cortical gel-like layer undergoes a transition from a gelled state to a more fluid sol state at the cell posterior, while the endoplasm streaming into an extending pseudopod is seen to solidify into gelled ectoplasm near its tip as the cell is moving forward (MAST 1926; see summary by TAYLOR and CONDEELIS 1979). Gelation-solation also plays a role in shuttle streaming of slime mold plasmodia (sumarized in Kamiya 1981), and the crawling movement of many other metazoan cells (OSTER 1984). Gel-sol-transformations are a good example for dramatic, large-scale consistency changes of the cytoplasm essentially made possible by the versatility of cytoplasmic actin and its associated proteins.

Plasticity of contractile protein organization may be further increased by interlacing sol-gel-transformation and contraction. Exactly how the state of gelation (*i.e.,* the three-dimensional organization of actin filaments) and contractile activity (*i.e.,* actin-myosin interaction) are coordinated in the generation of cellular motility is not known. *A priori,* however, several conditions need to be fulfilled to allow for myosin-mediated force generation, and some assumptions can be made about the outcome of such contractile acitivity: 1. By analogy to the skeletal muscle paradigm, actin filaments must be present in an antiparallel organization. Contraction will not be possible in organelles where this prerequisite is not met, such as microvilli and filopodia, where all actin filaments are of uniform polarity. Consistent with this postulate, myosin has not been found in these structures. 2. Again by analogy, myosin must be present as a bipolar filament. Though such filaments have been demonstrated in the cytoplasm of only a few nonmuscle cells, they are likely to exist in many of them, given the known *in vitro* properties of cytoplasmic myosins. 3. If the network, or bundle, of actin filaments is highly cross-linked, myosin is likely to be ineffective in producing a sliding motion, and, consequently, a full contractile response. 4. Conversely, a highly solated state in which there are many, but short, actin filaments insufficiently cross-linked due to a lack of gelation factors, myosin-mediated sliding will be largely inconsequential because of a lack of coherence in the actin-based network. 5. On the basis of the latter two premises, it is reasonable to assume that modulation of the contractile response occurs at or around the gel point of an actin filament network, ranging from partial solation where actin filaments still show limited interactions, to a gel state which will allow for some flexibility in the arrangement of actin filaments. Generally speaking, strong solation and gelation antagonize contraction, while partial solation (or gelation) promotes it.

Considerations of the possible way by which gel-sol-transformations might be related to contractile activity led to the formulation of the *solation-contraction coupling hypothesis* (CONDEELIS and TAYLOR 1977, HELLEWELL and TAYLOR 1979, CONDEELIS 1981, TAYLOR and FECHHEIMER 1982). The hypothesis postulates that a gel constructed from actin filaments and a cross-linker has to be weakened in order to allow myosin molecules interspersed within the gel to exert a contractile force. Weakening of the gel, *i.e.,* partial solation, may be achieved by the action of one of the calcium-activated filament severing proteins such as gelsolin, villin, or severin, which reduce average filament length. In this model, calcium ions would serve as a trigger for both gel weakening and initiation of contraction. An experimental test of this hypothesis has been conducted by STENDAHL and STOSSEL (1980) and TAYLOR and FECHHEIMER (1982). The first two investigators used reconstituted mixtures of actin, myosin, filamin (a cross-linker) and gelsolin (a severing protein), while the latter employed cytoplasmic extracts from *Dictyostelium*. The reconstituted mixture or the extract were placed in glass capillaries and were allowed to gel. One end of the capillary was exposed to a solution containing micromolar calcium. Contraction induced by the exposure to this calcium solution was directional and invariably occurred from the site of high calcium to the site of low calcium in both experiments. The two groups differ in their interpretation as to the mechanism by which contraction is related to the state of gelation in this experiment. STENDHAL and STOSSEL (1980) showed that increasing quantities of cross-linker amplified the rate of contraction and concluded that cross-linking decreases the concentration of myosin required for contraction (in effect, they proposed a gelation-contraction model). TAYLOR and FECHHEIMER (1982) observed contraction within the extract only in regions of high calcium and concluded that the gel needs to be weakened for contraction to occur. Both groups agree, however, that a certain degree of cross-linking is required for amplification of contraction and for transmission of the force away from the site of contraction. The details of the interplay between gel-sol-transformation and contraction may well be very complex and may depend on a number of critical factors that can produce either antagonisitic or synergistic effects. .

In summary, plasticity of contractile protein organization and versatility of contractile protein function are secured at several levels: at the level of assembly/disassembly of the two major components, actin and myosin; at the level of the supramolecular organization of actin; and at the level of actin-myosin interaction. The high degree of variablility in the combination of these three factors provides for the astounding flexibility of contractile protein organization and function in nonmuscle cells.

II. Microtubules

II.1. Historical Aspects

Among the three major filamentous components of the cytoskeleton, microtubules have so far received the most attention, primarily for the following reasons: they were the first cytoskeletal component to be described as ubiquitous in eukaryotic cells; they form extremely complex and esthetically pleasing patterns in microtubular organelles; and they are involved in the construction of the mitotic spindle, a cell organelle that has fascinated generations of cell biologists. The history of research on microtubules is marked by discontinuous jumps where periods of rapid progress—initiated, in most cases, by the advent of new techniques—alternate with periods of relatively slow progress. Although the exponential (or shall we say explosive?) phase of microtubule research began with the introduction of glutaraldehyde as a fixative for electron microscopy (SABATINI et al. 1963), microtubule-containing structures or organelles had been described decades earlier. Similarly, the onset of what were to become the fields of microtubule physiology and biochemistry can be traced to the pharmacological experiments of PERNICE (1889).

KÖLLIKER (1849) was probably the first to study a microtubule-containing cytoplasmic organelle. He described the heliozoan, *Actinophrys sol*, and noted the presence of an "Achsenfaden" (axial rod) in the axopodia extending from the cell body. This axial rod, as we now know, consists of a bundle of microtubules arranged in an intricate pattern of two interlocking coils (ROTH et al. 1970). Even earlier, REMAK (1844) had noted a fibrillar substructure in neurons. However, because of the rather harsh fixatives used at that time it is more likely that he had visualized the more stable neurofilaments, rather than the microtubules. In 1875, RANVIER and, in more detail later MEVES (1911), described a bundle of minute filamentous components encircling the nucleus of erythrocytes of cold-blooded vertebrates (Fig. 12). This organelle, the marginal band, is known today to

consist of a bundle of microtubules. MEVES' study marks the first occasion
that a microtubule-containing structure was proposed to be involved in the
maintenance of cell shape, which today is still viewed as one of the most
likely functions of cytoplasmic microtubules. Fibrillar structures of the
spindle were described first by FLEMMING (1879). His discovery aroused

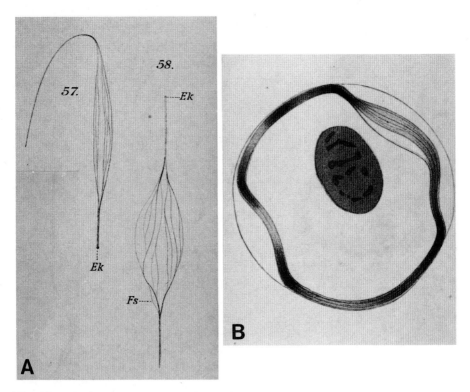

Fig. 12. *A* Isolated sperm tails of the chaffinch in which separation into several
subfibers (11 in "58.") was induced by maceration. Stained with gentian violet.
[From BALLOWITZ (1888).] *B* Amphibian erythrocyte stained with gentian violet,
showing a peripheral ring of fibrils, the marginal band of microtubules. [From
MEVES (1911)]

much speculation on spindle dynamics and the nature of these fibers, but
many decades passed before these fibers were finally accepted as true
components of the spindle apparatus. Last but not least, the fibrillar
substructure of cilia und flagella was disclosed as early as 1888 by
BALLOWITZ, who observed them to fray into several (up to 11) fine filaments
(Fig. 12). (A remarkable oberservation, considering the limited optical
resources of that time; the successful visualization of single microtubules is,
therefore, almost 100 years old!)

Even though physiological and biochemical analyses on microtubules could begin only after their nature and identity was established, the roots of these fields are connected with the name of the Sicilian anatomist PERNICE (1889). He observed an increase in the frequency of mitotic figures in the gastrointestinal tract of two dogs that he had given an extract of *colchicum* (containing colchicine as its active ingredient). He interpreted his images as a stimulatory action of this extract on mitosis, a misinterpretation that was not to be rectified until 1936 (LUDFORD, 1936). At around this time colchicine began to be used to induce polyploidy in plants (BLAKESLEE 1937). Two physical treatments, low temperature and high pressure, were also known early on to affect certain cellular strucutres containing microtubules. In 1890, HERTWIG noted an arrest of mitosis by low temperature, while MARSLAND (1938) began a series of studies on the effect of high hydrostatic pressure on dividing sea urchin eggs. These as well as polarization optical studies of the mitotic apparatus (SCHMIDT 1937) suggested early on that the spindle fibers are labile structures that appear to be in a dynamic equilibrium with the sub-components of which they are made.

As soon as workable electron microscopes became available, filamentous structures in cilia were revealed; their number was estimated to be 11 per cilium (RUSKA 1939). During the first 20 years of electron microscopy following this observation, increasingly detailed micrographs of the stable tubular elements in cilia and flagella were provided. MANTON and CLARKE (1952) were the first to reconstruct, in a remarkably accurate way, the structure of a cilium. However, it was not before improved thin sectioning techniques were developed (FAWCETT and PORTER 1954) that the filaments of the cilium were shown to look like tiny tubes. Although several studies were able to demonstrate tubular structures in the cytoplasm (*e.g.,* protozoa: ROTH 1958, nerve cells: PALAY 1956, toadfish erythrocytes: FAWCETT 1959), their widespread occurrence was established in the early 60ies, when they were also given the name microtubules (SLAUTTERBACK 1963). This year marks the beginning of the remarkable development of research on microtubules.

The biochemical properties of the protein that makes up microtubules, later named tubulin (MOHRI 1968), were beginning to be unraveled in the mid60ies with the first attempts at purifying microtubules and microtubule proteins. The first isolation of a microtubule-rich structure, however, dates back to 1952, when MAZIA and DAN (1952) reported the successful isolation of the mitotic apparatus of sea urchin eggs—even though this experiment did not lead to the identification of the protein that makes up the spindle fibers. This was accomplished later by the development of a radioactively labelled colchicine analog (BORISY and TAYLOR 1967a, b), which also allowed the quantification of the "colchicine-binding protein" in various tissues. In 1972, the stage was set for biochemical studies on microtubules when WEISENBERG (1972) identified the conditions for reversible *in vitro* polymerization of microtubules, providing the essential prerequisites for functional studies on tubulin assembly. The latest revolution in the field of

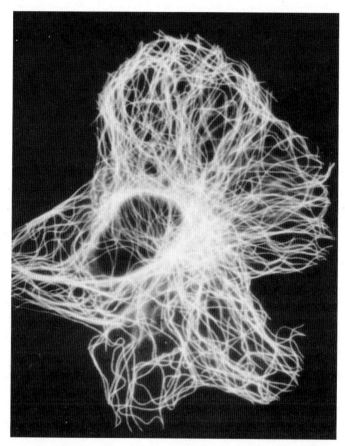

Fig. 13. Microtubule organization in a chicken embryo fibroblast as seen by immunofluorescence microscopy with an antibody against tubulin. × 1,100

microtubule research (and the study of other cytoskeletal elements) was the development of imunocytochemical probes for tubulin and actin (WEBER *et al.* 1975, LAZARIDES and WEBER 1974). Immunofluorescence microscopy is now an indispensable tool in research on the cytoskeleton because it allows the survey of the distribution of an antigen in large numbers of cells (Fig. 13).

More than in any other subfield of cell motility, the statement that "the literature is voluminous and widely scattered" applies to the subject of microtubules. The list of just the review publications includes a monograph (DUSTIN 1984), several collections of review articles and original research papers (SOIFER 1975, BORGERS and DE BRABANDER 1975, ROBERTS and HYAMS 1979, DE BRABANDER and DE MEY 1980, SAKAI *et al.* 1982, and other conference summaries on topics that include microtubules), and countless overview papers, of which only some of the more recent general reviews will be listed here (ROBERTS 1974, STEPHENS and EDDS 1976, KIRSCHNER 1978, RAFF 1979, BRINKLEY *et al.* 1980, MCKEITHAN and ROSENBAUM 1984, SOIFER and MACK 1984). Reviews on more specialized topics will be introduced where appropriate.

II.2. Microtubule Structure

Microtubules are hollow cylinders of approximately 25 nm outer diameter, 15 nm inner diameter, and indefinite length. Their general morphology is similar in all cell types studied, although, as we shall see, they often vary in the details of their construction. The wall of the microtubule cylinder is made up of linear elements, termed protofilaments (Fig. 14), which are formed from microtubule subunits aligned in rows (LEDBETTER and PORTER 1963, TILNEY *et al.* 1973). In their early study, LEDBETTER and PORTER were able to visualize protofilaments and determine that there are 13 per closed tubule due to the fortuitous selection of their object of study, the tannin-rich juniper root tip. Later, tannic acid was (re-)discovered as an agent that helps visualize the protofilamentous nature of microtubules (TILNEY *et al.* 1973). Its use as an additive to glutaraldehyde was instrumental in establishing that indeed 13 is the usual number of protofilaments in microtubules from a wide range of cell types. Why the prevailing number of protofilaments is 13 is unclear. *A priori*, one might expect them to contain 12, not 13, protofilaments, based on the observation that microtubules form superstructures with sixfold symmetry. Later, microtubules made of 11 to 16 protofilaments were described for a number of cell types fixed with tannic acid and glutaraldehyde (*e.g.,* BURTON and HINKLEY 1974, BURTON *et al.* 1975, SAITO and HAMA 1982). The functional significance of constructing microtubules from more or less than 13 protofilaments has eluded researchers to-date. In at least one protozoan, microtubules with different numbers of protofilaments are spatially segregated into different cellular domains; cytoplasmic microtubules have 13 protofilaments, while intranuclear microtubules contain from 13 to 16 (EICHENLAUB-RITTER and TUCKER 1984). Microtubules containing different numbers of protofilaments may also respond differently to microtubule-disrupting agents (CHALFIE and THOMSON 1982). These examples suggest that the number of protofilaments is functionally significant, and that cells have precise spatial and perhaps temporal control over this aspect of microtubule structure. When microtubules known to

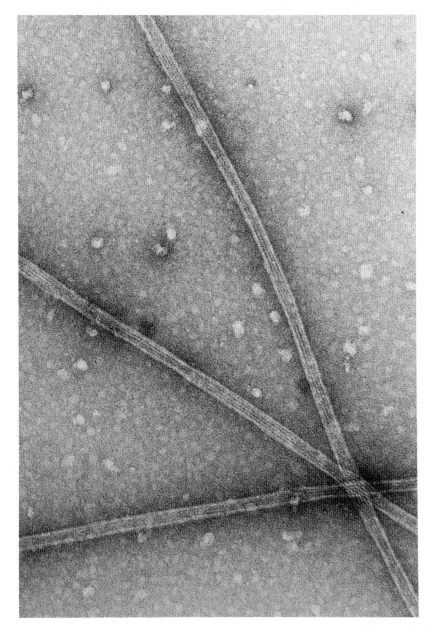

Fig. 14. Negatively stained bovine brain microtubules repolymerized from diethyl
aminoethyl cellulose-purified bovine brain tubulin in the presence of 1 µg/ml taxol.
× 160,000

have 13 protofilaments *in vivo* are depolymerized and repolymerized two or more times *in vitro*, the number of protofilaments per microtubule ranges between 13 and 16, but the most common number is 14 (PIERSON *et al.* 1978). Why this is so remains obscure, but it might suggest that the *in vitro* preparations lack some kind of a control factor (a nucleation center?) that, in living cells, assures a constant structure of microtubules (SCHEELE *et. al.* 1982; EVANS *et al.* 1985). An alternative possibility is that the conditions for polymerization *in vitro* are less than optimal.

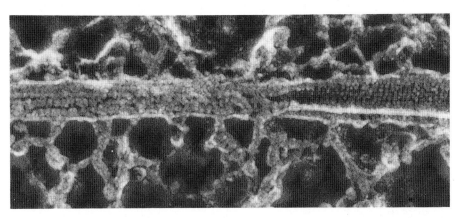

Fig. 15. Purified microtubules fractured and deep-etched after quick-freezing. The microtublule shown here is fractured open and reveals oblique striations that form a three-start helix on its inner surface. The outer surface displays longitudinal bands of bumps, corresponding to the protofilaments. × 350,000. Courtesy of John Heuser

Optical diffraction studies of the surface lattice of negatively stained microtubule preparations clearly reveal two additional aspects of microtubule substructure that are of importance for an understanding of microtubule assembly. First, protofilaments appear to be composed of stacks of 4 nm subunits (GRIMSTONE and KLUG 1966, AMOS and KLUG 1974, ERICKSON 1974); secondly, protofilaments are staggered in such a way that the 4 nm subunits form two helical families, a 10-start right-handed helix and a 3-start helix originally thought to be left-handed (AMOS 1975), but more recently proposed to be right-handed (SCHULTHEISS and MANDELKOW 1983) (Fig. 15). Each of the 4 nm globular subunits arranged along the protofilaments probably corresponds to one of the two monomeric subunits of the tubulin dimer. The existence of an 8 nm layer line in the diffraction pattern, though weak, suggests that two slightly different subunits alternate along the protofilament length. This image is consistent with the fact the functional subunit in microtubule assembly is a heterodimer composed of one each of alpha and beta tubulin. The arrangement of subunits in the microtubule wall suggests that each subunit has at least 4 binding sites, two within the protofilaments and one each to subunits in neighboring protofilaments. Intraprotofilament bonding appears to be much stronger than the latter, as suggested by frequent images of microtubules with frayed ends revealing individual protofilaments.

II.3. Tubulin

The microtubule cylinder is constructed from identical subunits of a protein called tubulin. Originally identified in various tissues as a colchicine-binding protein (BORISY and TAYLOR 1967 a, b, WEISENBERG *et al.* 1968), it was shown to be a globular molecule of approximately 110,000 d native molecular weight. The tubulin molecule itself is composed of two very similar proteins, alpha and beta tubulin. On polyacrylamide gels, most tubulins migrate as two distinct polypeptide species of 54,000 (beta) and 56,000 (alpha) daltons. The dissociation constant of the heterodimer is very low (7.4 10^{-7} M for brain tubulin) (DETRICH and WILLIAMS 1978). The complete amino acid sequences for both tubulins have recently been established by sequencing of proteolytic fragments of porcine brain tubulin (PONSTINGL *et al.* 1981, KRAUHS *et al.* 1981) and chick brain tubulin mRNA (VALENZUELA *et al.* 1981), respectively. One of the remarkable features of the amino acid sequence is that the C-terminal portions of both subunits have an unusually high proportion of acidic residues (19 out of 40 residues in brain alpha tubulin are glutamate or aspartate). Both subunits of tubulin from most species are acidic, with an isoelectric point near 5.5. The native molecular weight of each tubulin is about 50,000 d based on the sequence data, showing that both alpha and beta tubulin migrate anomalously in SDS gels.

Like actin and histones, tubulin is a highly conserved protein. Earlier partial amino acid sequencing (LUDUENA and WOODWARD 1973) revealed a higher degree of homology between alpha tubulins or beta tubulins, respectively, from distant species (*e.g.,* sea urchin and chicken), than between alpha and beta tubulin from the same species. Although the evolutionary conservation is high, there is evidence for substantial microheterogeneity of both tubulin subunits. In the amoeboflagellate *Naegleria*, for example, there are at least two forms of tubulin, one specific for cytoplasmic microtubules, and one specific for flagellar microtubules. When this organism changes from the amoeboid form (in medium rich in nutrients) to the flagellate form (in nutrient poor medium), it does not use any of the cytoplasmic tubulin for the construction of flagellae (FULTON and SIMPSON 1976). The two forms of tubulin are immunologically different (FULTON and KOWIT 1975). By a variety of techniques, including isoelectric focusing and peptide mapping, microheterogeneity has also been detected in neurons (GOZES and SWEADNER 1981, GEORGE et al. 1981, SULLIVAN and WILSON 1984), sea urchins (BIBRING et al. 1976, WILSON et al. 1984), flagellar A, B, and central pair subfibers (STEPHENS 1975, 1978), the fungus Aspergillus (SHEIR-NASS et al. 1978, MORRIS et al. 1979), and the algae Chlamydomonas (LEFEBVRE et al. 1980) and Polytomella (MCKEITHAN et al. 1981, 1983). A survey of brain tubulins from six vertebrates shows the extent of heterogeneity to be similar in all these organisms, with at least six isotubulins in the alpha subunit and 12 in the beta subunit (Field et al. 1984). In some instances, the biochemical and immunological differences between the tubulins correlate with different subcellular locations of the microtubules they build, and therefore seem to imply different functions. This is particularly obvious in the case of Neagleria (cytoplasmic versus flagellar microtubules). In many instances, however, the differences in the function(s) of the microtubules made from different forms of tubulin in a single cell are not obvious (for a more complete discussion, see MCKEITHAN and ROSENBAUM 1984).

Microheterogeneity of tubulins can arise from either multiple tubulin genes or posttranslational modification, or a combination of both. DNA hybridization analysis has shown that in many species tubulin genes comprise a gene family whose members may be dispersed over many chromosomes; alpha and beta tubulin are usually not found in tandem (CLEVELAND et al. 1980). As many as 13 genes for each alpha and beta tubulin are present in sea urchin DNA (CLEVELAND et al. 1980), while Chlamydomonas and Aspergillus have just two for each (BRUNKE et al. 1982, SILFLOW and ROSENBAUM 1981, WEATHERBEE and MORRIS 1983). There are 4 genes for each subunit in the chicken and Drosophila (LOPATA et al. 1983, SANCHEZ et al. 1980), and ten in the human (CLEVELAND et al. 1980), most of which appear to be processed pseudogenes (WILDE et al. 1982). The sequences of the different isotypes of tubulin differ more or less subtly; for example, the four alpha tubulins in porcine brain show only minor differences in their primary sequence (PONSTINGL et al. 1981), while the beta tubulins of the chicken may diverge substantially (by 8.9%) in their primary sequence (SULLIVAN and CLEVELAND 1984) and in the construction of their messenger RNAs (LOPATA et al. 1983). The functional siginificance of multiple gene copies for tubulin is not clear. RAFF (1984) discusses two possibilities: 1. Although all tubulins are similar and can copolymerize into functional microtubules, each tubulin may possess slightly different properties, allowing for a slightly different mode of function. 2. The gene families arose from the need for differential regulation, i.e., to allow control of the timing, location, or amount of tubulin synthesis. The two possibilities are not mutually exclusive and may operate in different cell systems to different extents (see CLEVELAND 1983, COWAN and DUDLEY, 1984, for reviews).

Posttranslational modification can create further variation among the tubulins. For example, the chicken has only four beta tubulin genes, but by biochemical criteria has about 12 different beta tubulins (FIELD *et al.* 1984). Three major forms of posttranslational modification of tubulin have been identified so far; there is indirect evidence for other forms of modification whose nature remains obscure to-date. First, alpha tubulin can be modified by C-terminal tyrosination. Curiously, the C-terminal tyrosine residue is present when alpha tubulin is first produced (VALENZUELA *et al.* 1980), but later it is removed from most of the subunits under the action of a specific carboxypeptidase (KUMAR and FLAVIN 1981). Alpha tubulin can be retyrosylated by another specific enzyme, tubulin-tyrosine ligase (*e.g.,* RAYBIN and FLAVIN 1975), which requires ATP. Tyrosinated and non-tyrosinated tubulins do not differ in their ability to polymerize into microtubules (RAYBIN and FLAVIN 1977). A specific function for tubulin tyrosination has not yet been identified but there appear to be changes in the ability of tubulin to become tyrosinated during the cell cycle (NATH *et al.* 1978). Increased tubulin tyrosination also accompanies chemotactic stimulation of leukocytes (NATH *et al.* 1982), *i.e.,* a dramatic change in the activity of this cell type. In addition, tyrosinated and non-tyrosinated tubulins may have different subcellular distributions (GUNDERSON *et al.* 1984). Different staining patterns of microtubules within the same cell were also detected with a set of monoclonal antibodies (THOMPSON *et al.* 1984), but the specific nature of the antigenic sites detected by these antibodies is not known. The second major form of posttranslational modification, phosphorylation, seems to occur only on the beta subunit (EIPPER 1974). It, too, appears to vary with the cell cycle (PIRAS and PIRAS 1975), again suggesting some functional significance. Last, tubulin can also be glycosylated (see discussion by LUDUENA 1979), but the level of glycosylation seems to be low.

Another possible source for microheterogeneity of tubulins, namely, differential processing of mRNAs transcribed from a single gene (*e.g.,* modified initiation, termination, or splicing) has not been demonstrated for tubulin so far.

II.4. Interactions with Other Molecules

Tubulin specifically interacts with other molecules from the four following categories: ions, nucleotides, microtubule-associated proteins, and drugs (natural products and synthetic substances). All of these have been intensely studied because they are able to affect the polymerization of microtubules. While the effects of colchicine had been known for about a century, the importance of calcium ions, GTP, and proteins coassembling with microtubules was not revealed until conditions for the polymerization of microtubules *in vitro* were developed (WEISENBERG 1972).

II.4.1. Ions

The two ions of greatest importance for microtubule assembly in the living cell and *in vitro* are magnesium and calcium. Millimolar magnesium is required for assembly, while even micromolar calcium is generally regarded as inhibitory. The tubulin dimer itself appears to have one high affinity and several low affinity binding sites for calcium (SOLOMON 1977), and many binding sites for magnesium. Since WEISENBERG's (1972) demonstration that assembly of microtubules in crude extracts of brain tissue is inhibited by micromolar calcium levels, a possible regulatory role for calcium has captured the imagination of many investigators. Whether micromolar or millimolar calcium is required to inhibit assembly *in vitro* (OLMSTED and BORISY 1975, ROSENFELD *et al.* 1976) was settled when it was discovered that crude extracts contain a factor that mediates the inhibitory action of calcium, and that this factor is lost after further purification of microtubule protein. This protein factor has been identified as calmodulin. It potentiates the inhibitory action of calcium on microtubule assembly *in vitro* (NISHIDA and SAKAI 1977, MARCUM *et al.* 1978), and, when added to preformed microtubules, induces disassembly if sufficient calcium (above micromolar) is present. On the basis of immunological criteria, calmodulin can associate with at least certain subsets of microtubules in living cells (WELSH *et al.* 1978, DE MEY *et al.* 1980), indicating a specific interaction with certain microtubular components. Anticalmodulin drugs reduce the calcium sensitivity of cellular microtubules in an *in vitro* model (SCHLIWA *et al.* 1981), confirming the notion that calmodulin is associated with cellular microtubules (see also DEERY *et al.* 1984). Studies on living cells also show that experimental elevation of intracellular calcium to concentrations within the micromolar range by microinjection (KIEHART 1981, KEITH *et al.* 1983) of calcium or addition of the calcium ionophore A 23187 (SCHLIWA 1976) will induce microtubule disassembly. With the demonstration of calcium-sequestering membrane systems associated with microtubular structures such as the mitotic spindle (HEPLER 1980, PAWELETZ 1981, PETZELT 1984), the concept emerged that calcium ions may serve to regulate microtubule assembly locally in living cells. However, the mechanism of calcium/calmodulin-induced microtubule disassembly is not fully understood. Initially, tubulin was proposed to be the molecule with which calmodulin interacts (KUMAGAI and NISHIDA 1979, KUMAGAI *et al.* 1982), but other studies suggest that microtubule-associated proteins (REBHUN *et al.* 1980, SOBUE *et al.* 1981), specifically the "tau" proteins and MAP 2 (LEE and WOLF 1984 a, b), mediate calcium/calmodulin-induced microtubule disassembly. According to these studies, calmodulin exerts its effect on microtubule assembly through the formation of a calcium-calmodulin-MAP complex, but the specific mechanism of action of this inhibitory complex is not understood (for a more complete discussion, see MARGOLIS 1983).

II.4.2. Nucleotides

Tubulin has two bound guanine nucleotides per dimer, one of which is exchangeable with unbound GTP (WEISENBERG et al. 1968). GTP at the exchangeable (or E) site, located on the beta subunit, rapidly exchanges with free GTP and is hydrolyzed during polymerization. The GTP at the nonexchangeable (or N) site is not hydrolyzed and persists as such in polymerized microtubules. E-site GDP ist either replaced by GTP or rephosphorylated by the action of a nucleoside diphosphokinase, which uses ATP as a substrate (JACOBS 1975). Interestingly, even though GTP at the E-site is normally hydrolyzed during assembly, hydrolysis is not absolutely required for microtubule assembly. Tubulin prepared in the presence of nonhydrolyzable analogs of GTP is apparently able to polymerize into microtubules (LOCKWOOD et al. 1975, PENNINGROTH et al. 1976). The possible importance of GTP hydrolysis for microtubule assembly will be discussed in more detail in section IV. More complete accounts of the complex interactions between nucleotides and tubulin are given by JACOBS (1979) and KIRSCHNER (1978).

Even though ATP hydrolysis does not take place during microtubule assembly, ATP may affect microtubules indirectly. ATP increases the rate of nucleation (ZABRECKI and COLE 1982) and treadmilling (MARGOLIS and WILSON 1979) in vitro, and it promotes microtubule disassembly in detergent-lysed cells (BERSHADSKY and GELFAND 1981, 1983). Microtubule stability and the rate of microtubule assembly may also be affected by ATP-dependent phosphorylation of microtubule-associated proteins (see next section).

II.4.3. Microtubule-Associated Proteins

A set of proteins copurify with tubulin stoichiometrically when microtubules are prepared by several cycles of temperature-dependent assembly and disassembly in vitro. These proteins are collectively referred to as microtubule-associated proteins or MAPs; the mixture of MAPs and tubulin is known as microtubule protein. In brain tissue, the first source of microtubule protein to be studied in great detail, two major classes of MAPs were identified: A group of high molecular weight proteins with two major components, termed MAP 1 and MAP 2, with molecular weights of 330 and 280 kd, respectively (SLOBODA et al. 1975, VALLEE and BORISY 1978), and a group of proteins termed tau factor resolved in one-dimensional polyacrylamide gels into several bands between 55 and 70 kd (WEINGARTEN et al. 1975, PENNINGROTH et al. 1976). Both groups of proteins bind to microtubules and stimulate their polymerization in vitro. They reduce the critical concentration of tubulin for assembly and increase the rate of microtubule polymerization (MURPHY et al. 1977). Contrary to initial suggestions, they are not absolutely required for assembly; pure tubulin from which MAPs are removed by phosphocellulose column chromatography can under certain conditions polymerize into microtubules (*e.g.,* BRYAN 1976, DENTLER *et al.* 1975, HERZOG and WEBER 1977). The binding of high molecular weight MAPs to tubulin seems to be regulated by the level of phosphorylation of MAPs, which is mediated by both cAMP-dependent and Ca/calmodulin-dependent kinases (BURNS and ISLAM 1982, SELDEN and POLLARD 1983, THEURKAUF and VALLEE 1983, SCHULMAN 1984). In fact, MAP 2 is a major substrate for phosphorylation in brain tissue extracts (SLOBODA *et al.* 1975, THEURKAUF and VALLEE 1983). The cAMP-dependent kinase, which is directly associated with MAP 2 in brain tissue (VALLEE *et al.* 1981) (it is, in effect, a microtubule-associated protein-associated protein, a "MAPAP"), accounts for the phosphorylation of about half of the 20–22 phosphorylation sites on MAP 2. Phosphorylation reduces the rate and extent of microtubule assembly *in vitro* by decreasing the affinity of MAP 2 for microtubules (BURNS *et al.* 1984), but its effect on the function of MAPs and/or microtubules *in vivo* has not been demonstrated. Tau protein also is phosphorylated and dephosphorylated *in vivo* (LINDWALL and COLE 1984). In cultured sympathetic neurons, a group of closely related polypeptides about the same size as tau may be made from one protein that is phosphorylated to different extents (BLACK and KURDYLA 1983). The role tau phosphorylation might play in the control and regulation of microtubules in not known.

MAPs, in particular the high molecular weight species of brain tissue, were studied in great detail during the past 10 years (summary in VALLEE 1984), for several reasons. First, high molecular weight MAPs are long, flexible molecules (VOTER and ERICKSON 1982) that form regularly spaced projections from the microtubule surface (MURPHY and BORISY 1975, BINDER and ROSENBAUM 1978, SLOBODA and ROSENBAUM 1979, KIM *et al.* 1979) (Fig. 16). These projections appear similar to those seen on microtubules in electron micrographs of intact cells, particularly neurons (see WUERKER and KIRKPATRICK 1972 for a review). The projections were believed to be responsible for presumptive physical interactions among microtubules, and between microtubules and other structural components of the cytoplasm. Second, electron microscopy of axonal processes had revealed crossbridges resembling MAP projections between microtubules and organelles, such as synaptic vesicles and mitochondria, believed to be transported rapidly along the axon (*e.g.,* SMITH 1971, RAINE *et al.* 1971, HIROKAWA 1982). Therefore, MAP-like proteins were considered possible candidates for force-producing elements in intracellular organelle transport. Third, since MAPs are able to stimulate the polymerization of, and to stabilize polymerized microtubules, they were believed to be involved in the regulation of microtubule polymerization *in vivo*. Fourth, because proteins similar to brain MAPs were assumed to exist in other cell types, a study of

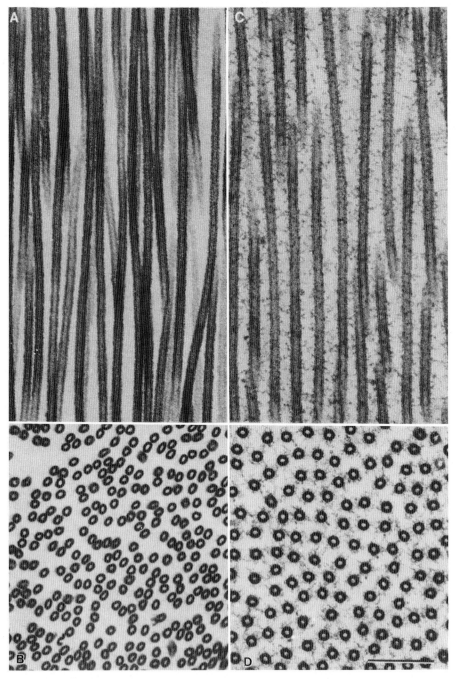

Fig. 16. Longitudinal and transverse sections of *in vitro* assembled flagellar tubulin in the absence (A, B) and presence (C, D) of brain MAPs. The microtubules polymerized in the presence of MAPs have wispy projections and are more regularly spaced. [From BINDER and ROSENBAUM (1978)]

MAPs of a readily available tissue providing high yields of these proteins was hoped to facilitate the identification and characterization of similar proteins in other cell types.

However, only some of these points could be supported by experimental evidence:

1. Procedures that measure the interaction between linear polymers, such as falling ball viscometry, did not demonstrate any significant interactions between microtubules containing MAPs (RUNGE *et al.* 1981). Though complex microtubular organelles with corssbridges or links between microtubules exist, notably in protozoa (*e.g.,* TILNEY and PORTER 1965, TUCKER 1968), no study has identified a cytoplasmic (nonciliary) protein that serves such a function *in vivo*. High molecular weight proteins very similar to MAPs may, however, act as linkers between intermediate filaments and microtubules, and between actin filaments and microtubules (see section VII.3.).

2. An involvement of MAPs in the generation of the motive force for intracellular organelle movements was much debated in the late 70ies but could not be demonstrated convincingly. Early studies suggested the presence of a dynein-like ATPase activity in brain tissue (BURNS and POLLARD 1974), suspected to be associated with microtubules. An ATPase activity was considered to be a prerequisite for MAPs to function in microtubule-associated movements. Initial studies which showed ATPase activity copurifying with brain microtubule protein through several cycles of assembly/disassembly looked promising. However, it was not possible to assign this ATPase activity to any of the known MAPs. Finally, MURPHY *et al.* (1983 a, b) showed that the ATPase activity of brain microtubule preperations is an impurity identical with F 1-ATPase from mitochondria. MAP 1 and MAP 2 themselves have no ATPase activity. Furthermore, MAP 2 can not be involved in axonal organelle transport on the basis of its topology: immunolocalization shows this protein to be restricted to the dendritic portions of neurons (MATUS *et al.* 1981); it is absent from axons.

3. There ist no direct evidence for a regulatory role of MAPs in microtubule assembly *in vivo*. Although the accessory proteins shift the equilibrium towards assembly *in vitro*, similar effects can be produced by other, nonspecific, cationic proteins (ERICKSON and VOTER 1976). It is possible that the demonstration of a specific function for MAPs in the regulation of microtubule assembly in the living cell has to await the development of sensitive techniques able to probe and manipulate assembly conditions inside cells.

4. The last point, widespread distribution of at least some of the neuronal MAPs, is in part supported by experimental evidence. Initial immunocyto-chemical studies suggested a rather widespread distribution of neuronal MAP-related antigens (SHERLINE and SCHIAVONE 1977, CONNOLLY *et al.*

1977, 1978). Later studies showed the cellular and subcellular distribution of MAPs to be more complex than initially believed. First, MAP 2 seems to be present only in differentiated neurons (IZANT and MCINTOSH 1980), where it is specifically enriched in dendrites and perikarya (MATUS et al. 1981). Only minute amounts of MAP 2 may be present in HeLa cells (WEATHERBEE et al. 1982) and other cell types. Second, MAP 1, which has substantial structural similarity with MAP 2 on the basis of peptide mapping and immunological criteria (HERRMANN et al. 1984), seems to be more widely distributed in nervous tissue although, again, the cellular and subcellular levels of MAP 1 expression seem to vary (BLOOM et al. 1984 a, HUBER and MATUS 1984). The story is complicated further by the finding that MAP 1 comprises a family of polypeptides (BLOOM et al. 1984 a). Its principal component, MAP 1 A, seems to be widely distributed in non-neuronal cell types, where it associates with both interphase and spindle microtubules (BLOOM et al. 1984 b). Antibodies raised against MAP 1 stain not only microtubular structures, but also centrosomes (SHERLINE and MASCARDO 1982, SATO et al. 1983) and spots within nuclei (IZANT et al. 1982, BENNETT and DAVIS 1981). In addition, MAP 1 is structurally similar to ankyrin, a protein of the red blood cell cytoskeleton (see section IX.2.2.) (BENNETT and DAVIS 1981). These studies point to a rather widespread distribution of immunoreactive forms of high molecular weight proteins related to neuronal MAP 1.

In general, functional studies of MAPs from sources other that brain have been difficult. In most nonneuronal cells, the concentration of tubulin is too low to allow the isolation and characterization of MAPs by the same protocols used with brain tissue, i.e., temperature-dependent cycles of assembly and disassembly. Furthermore, isolation of microtubule protein requires rather large quantities of starting material, a requirement often difficult to meet with cultured cells. These drawbacks were overcome by using 1. coassembly with exogenous carrier tubulin (SOLOMON et al. 1979), 2. suspension cultures of rapidly growing cells (BULINKSI and BORISY 1980), and 3. the newly discovered microtubule-stabilizing drug taxol to increase the yield of microtubule protein (SCHIFF et al. 1979, VALLEE 1982). On the basis of these and similar studies, high and low molecular weight MAPs of both widespread and specific distribution were discovered in nonneuronal cells (BULINSKI and BORISY 1979, VALLEE and BLOOM 1983, DUERR et al. 1981, ZIEVE and SOLOMON 1983). Several of these proteins have been shown to promote microtubule assembly in vitro (BULINSKI and BORISY 1980, PALLAS and SOLOMON 1982), but other specific functions were not yet identified. Some are phosphoproteins, and their ability to bind to micro-tubules depends on the state of phosporylation (PALLAS and SOLOMON 1982), a feature that these proteins share with MAP 2 (MURTI and FLAVIN 1983, BURNS et al. 1984).

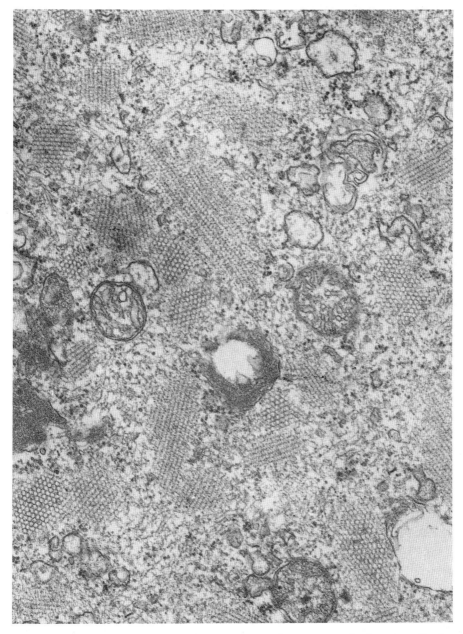

Fig. 17. Paracrystals of tubulin in the cytoplasm of a rat sympathetic neuron *in vitro* treated with 10 µM vincristine for one hour. × 40,000. (Courtesy of P. Dustin)

The search for MAPs associated with specific cellular activities has led to the discovery of microtubule-binding proteins with ATPase activity (PRATT 1980, SCHOLEY *et al.* 1984, HOLLENBECK and CANDE 1984). These MAPs can cross-link microtubules into large aggregates in an ATP-dependent manner. They are tentatively referred to as "cytoplasmic dynein" because they share one unique feature with this major microtubule-associated protein from cilia an flagella, namely, the associated ATPase activity (see section II.8.3.).

II.4.4. Drugs

A number of synthetic and natural compounds specifically interact with tubulin and/or microtubules. Since essentially all of them interfere with microtubule assembly, they are collectively referred to as "antimicrotubular agents". While this is not the place to review their actions extensively (see DUSTIN 1984, for a more complete discussion), some of the most widely used agents should be discussed because they were important in the dissection of the properties and functions of microtubules.

II.4.4.1. Colchicine and Its Analogues

Colchicine is by far the best known of all microtubule drugs. An alkaloid isolated from the meadow saffron, *Colchicum autumnale*, colchicine binds specifically and with high affinity to tubulin to form a 1 : 1 complex (WILSON *et al.* 1974) that dissociates very slowly. This property made it a uselful tool for determining the amount of tubulin ("colchicine-binding protein") in various cell types (*e.g.,* BORISY and TAYLOR 1967 a, b). Colchicine binding induces a conformational change in tubulin (DETRICH *et al.* 1982, ANDREU and TIMASHEFF 1984). The colchicine-tubulin dimer is able to bind to microtubule ends (MARGOLIS and WILSON 1977) where it inhibits further addition of unliganded tubulin subunits (substoichiometric "poisoning" of microtubule assembly). Colchicine-tubulin dimers can also incorporate into microtubules to some extent, destabilizing the polymer (STERNLICHT and RINGEL 1979), or they may form nonmicrotubular tubulin-colchicine aggregates (SALTARELLI and PANTALONI 1983, ANDREU *et al.* 1983). Colcemid, a close relative of colchicine, has very similar effects, but its action is more immediate and more readily reversible. Colcemid, therefore, has been more frequently employed in experiments on living cells.

II.4.4.2. Vinblastine and Vincristine

The plant products vinblastine and vincristine, isolated from the periwinkle, *Vinca rosea*, bind tubulin at a different site than colchicine. There are two binding sites per tubulin dimer, one of high and one of moderate affinity (BHATTACHARRYA and WOLFF 1976). At high (millimolar) concentration, both drugs induce the formation of paracrystalline structures composed of microtubule protein (Fig. 17) which can be used as starting material for the isolation of tubulin. Other actions of the *Vinca* alkaloids include the formation of macrotubules (tubules with a diameter greater than 30 nm) and the induction of helical protofilamentous arrays (*e.g.,* HIMES *et al.* 1976). In most cells, the effects of the *Vinca* alkaloids are slowly reversible or irreversible.

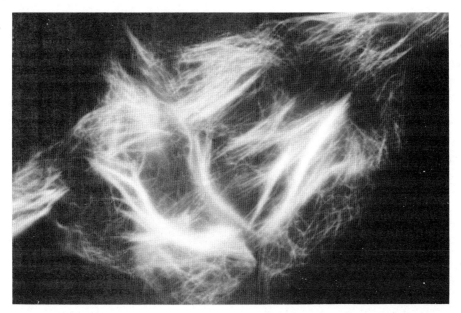

Fig. 18. Chicken embryo fibroblast treated with 10 µM taxol for 2.5 hours. Note criss-crossing bundles of microtubules. × 800

II.4.4.3. Nocodazole

Nocodazole is a benzimidazole derivative that recently has gained popularity because it can bind tubulin rapidly an with high specificity (DE BRABANDER *et al.* 1976). It inhibits microtubule assembly and induces disassembly of existing microtubules. Because its action is even more readily reversible than that of colcemid, it is the drug of choice in studies demanding rapid action and fast recovery (*e.g.,* SNYDER *et al.* 1982).

II.4.4.4. Taxol

Taxol is the latest addition to the arsenal of microtubule-binding compounds, and one whose arrival has been anxiously awaited. In contrast to the drugs discussed above, it binds to the microtubule polymer, rather than the subunit, and it stabilizes instead of depolymerizes microtubules (SCHIFF *et al.* 1979, SCHIFF and HORWITZ 1980). Taxol shifts the euqilibrium between tubulin and microtubules in favor of the polymer by decreasing the critical concentration for assembly. It can also bind to preformed micro-tubules, which are thereby stabilized against depolymerizing conditions, such as elevated calcium or depolymerizing drugs. When applied to living cells, taxol induces the formation of unusual microtubular arrays (*e.g.,* DE BRABANDER *et al.* 1981, ANTIN *et al.* 1981) (Fig. 18). The drug has been a valuable tool in biochemical studies of tubulin, in structural studies of the cytoskeleton, and in functional anylses of the role of microtubules in living cells (see HORWITZ *et al.* 1982, GREEN and GOLDMAN 1983).

II.4.4.5. Other Microtubule Poisons

Compounds such as griseofulvin, podophyllotoxin, maytansin, isopropyl-N-phenylcarbamate, halothane, and many more have played important roles in studies of the biochemistry and function of microtubules, but space considerations do not allow their discussion here. For more complete imformation, see DUSTIN (1984), BISWAS *et al.* (1981), and GUNNING and HARDHAM (1982).

II.5. Mechanism of Microtubule Polymerization

Microtubule assembly proceeds in two distinct steps, nucleation and elongation. In *in vitro* assembly, there is an initial lag period during which no polymer forms in a solution of tubulin under polymerizing conditions. This lag is interpreted as a nucelation phase during which stable oligomers form onto which subunits can add. This interpretation is supported by the observation that short microtubule segments or "seeds" eliminate the initial lag phase. Elongation of microtubules then proceeds by the direct addition of subunits to the growing polymer. Consistent with a simple condensation polymerization scheme (OOSAWA and KASAI 1963), the rate of elongation is directly proportional to the concentration of tubulin and the number concentration of microtubule ends (JOHNSON and BORISY 1977). *In vitro*, elongation of the polymer continues to equilibrium, *i.e.,* until the free monomer reaches a *critical concentration* below which net assembly can not take place. At this stage, the "plateau" or "steady state" phase, subunits continue to exchange between the monomer pool and polymer, if sufficient GTP is available. It is during this phase that "treadmilling" may occur *in vitro*; this process and its possible implications for the regulation of assembly *in vivo* will be discussed in section IV. In analogy to the "monomer activation step" of actin assembly, CARLIER (1983) has proposed that a conformational change of tubulin precedes microtubule assembly. Whether this occurs as a discrete third step remains to be shown.

Subunits add to and are lost from both ends of microtubules (KARR *et al.* 1980 a, b, BERGEN and BORISY 1980), but at different rates at each end. If flagellar axonemes or basal bodies are used as seeds, subunits are added more rapidly at the distal end, *i.e.,* the *in vivo* site of assembly, than at the proximal end (ALLEN and BORISY 1974, DENTLER *et al.* 1974, BINDER and ROSENBAUM 1978) (Fig. 19). Polar growth is also shown by single micro-

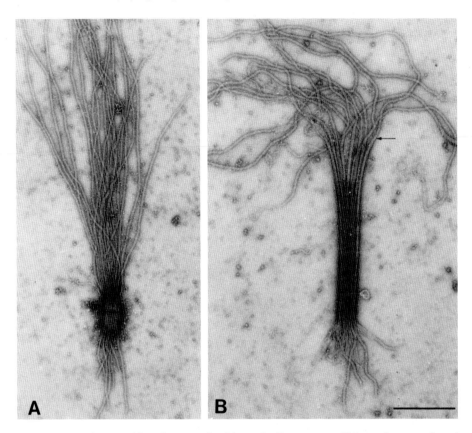

Fig. 19. Biased assembly of outer doublet tubulin onto a *Chlamydomonas* basal body (*A*) and axoneme (*B*). In both cases, longer microtubules have grown from the distal ends of the initiating structures (representing the plus ends). Arrow indicates an outer doublet with microtubules emanating from both A-and B-subfibers. × 30,000. [From BINDER and ROSENBAUM (1978)]

tubules (OLMSTED *et al.* 1974, SUMMERS and KIRSCHNER 1979). Thus microtubules have a functional polarity in their ability to add and subtract subunits from their ends. This is consistent with the structural polarity of the microtubule derived from the unipolar orientation of tubulin dimers within the microtubule lattice. The possibility that subunits can leave and enter the surface lattice laterally along the length of microtubules has been proposed (*e.g.,* INOUE and SATO 1967, SALMON *et al.* 1984, MANDELKOW and MANDELKOW 1985), but not yet proven conclusively.

The discovery of a nucleation phase for microtubule assembly *in vitro* suggests that an oligomeric intermediate must be formed during the initial stages of assembly. Time-resolved studies using negative-staining electron microscopy reveal a variety of intermediate structures in the early stages of assembly, including discs, rings, double rings, spirals, and straight or curved protofilament ribbons composed of a variable number of protofilaments (*e.g.,* ERICKSON 1974, OLMSTED *et al.* 1974, KIRSCHNER *et al.* 1975). The ring-shaped aggregates are the prevailing form of oligomers in depolymerized tubulin (*e.g.,* at 0° C), whereas stacked rings, sheets, and ribbons are typical of the nucleation phase; both disappear during elongation. KIRSCHNER (1978) discusses in detail various models proposed to explain how one form of oligomer converts into another and then into complete microtubules. Even though rings are not obligatory intermediates in assembly (BRYAN 1976), most models ascribe a major role to rings in microtubule nucleation. Rings are proposed to un-roll and associate side-by-side to form small sheets that are viewed as the assembly unit for microtubule nucleation. In this model, rings represent coiled protofilaments, an idea also supported by analysis of X-ray patterns obtained by synchroton radiation of solutions of tubulin rings (MANDELKOW *et al.* 1983). The sheets, in turn, could roll up to form a short segment of a closed microtubule which elongates by endwise addition of subunits and/or incorporation of ring-derived oligomers (PANTALONI *et al.* 1981).

A more refined model of assembly is based on a technique that circumvents some of the difficulties inherent in electron microscopic studies, yet provides good time resolution and structural information. On the basis of an analysis of microtubule assembly by synchroton X-ray scattering of microtubule protein in solution, BORDAS *et al.* (1983) suggest that during the nucleation phase, both disassembly and assembly processes take place. Rings appear to break down into fragments, which break down further into subunits, while assembly appears to occur by lateral association of protofilament fragments to form nuclei for polymerization. As in other models, nucleating centers are proposed to be derived from breakdown products of rings. Elongation of microtubule seeds shows the behavior expected for endwise addition of single subunits, but the incorporation of tubulin in the form of oligomers is not excluded. Indeed, oligomeric tubulin species with (BURNS and ISLAM 1984, CARLIER *et al.* 1984 d) or without (KRAVIT *et al.* 1984) associated MAPs exist in solutions of cycled micro-tubule protein, and may incorporate into polymerizing microtubules. Depolymerization induced by lowering the temperature causes loss of subunits or small fragments from the ends, and below 10° C rings re-form.

How does the assembly-promoting action of MAPs relate to the pathways of microtubule assembly outlined here? MAP 2 promotes the formation of ring-shaped oligomers even in the cold and is likely to be associated with ring aggregates in *in vitro* preparations of microtubules (VALLEE and BORISY 1978, BURNS *et al.* 1984, CARLIER *et al.* 1984 d). To reconcile the assembly-promoting effect of high molecular weight MAPs with schemes for microtubule polymerization *in vitro*, VALLEE has (1984) proposed a model in which MAPs associate with several dimers via the extended microtubule-binding domain of the MAP molecule to form a MAP-oligomer complex. The organization of this complex is such that it can easily incorporate into both the ring oligomer and the microtubule wall. During polymerization, the MAP-oligomer complexes could incorporate in such a way that the MAP projections will form regular patterns along assembled microtubules, as previously observed (see AMOS 1977, KIM *et al.* 1979). However, experimental evidence supporting the proposed scheme of MAPs in promoting microtubule assembly is not available at present.

Whether microtubule assembly in the cell uses any of the mechanisms identified *in vitro* is unknown. There is no conclusive evidence for the existence of ring oligomers or other intermediates of polymerization in microtubule assembly in cells. Most cell types organize microtubule arrays from specific nucleation centers or organizing sites (described in section II.7.), which seem to obviate the need for rings and other assembly intermediates because they might act in a manner analogous to "seeds" or microtubule fragments; other cells seem to possess microtubules that form free in the cytoplasm transiently.

II.6. Microtubule Polarity

Microtubules possess an intrinsic structural and functional polarity which has been revealed by determining the orientation of dimers in the microtubule lattice, or by measuring the rate constants for addition of subunits to microtubule ends. Both these approaches are tedious and time-consuming. Because of the potential importance of microtubule polarity for many cellular activities, including chromosome movement in mitosis, intracellular transport, and cell locomotion, attempts were made to reveal the polarity of cellular microtubules in a more simple way. These efforts produced two techniques which now can be used as conveniently as heavy meromyosin decoration of actin filaments. In the first, dynein molecules isolated from flagellar axonemes are used to decorate microtubules. The direction of the slight tilt of the bound dynein reveals microtubule polarity and is visible in both transverse and longitudinal sections (HAIMO *et al.* 1979). The second technique takes advantage of the ability of tubulin to polymerize onto the walls of preexisting microtubules under rather unusual polymerization conditions (BURTON and HIMES 1978), thereby forming curved protofilament sheets that run lengthwise along microtubules. When viewed in cross-section, they appear as curved appendages or "hooks"

(MANDELKOW and MANDELKOW 1979) whose handedness reliably reveals microtubule polarity (HEIDEMANN and MCINTOSH 1980, EUTENEUER and MCINTOSH 1981 a, b, MCINTOSH and EUTENEUER 1984) (Fig. 20). There is now strong evidence that cells strictly control microtubule polarity. First, in the cells in which microtubule polarity has been determined, microtubules are more than 90% uniform im polarity with respect to the cell center and/or the microtubule-organizing center, their fast-growing or "plus-ends" being distal. Examples include the microtubules of mitotic spindles of both animal (EUTENEUER and MCINTOSH 1981 b) and plant cells (EUTENEUER et al. 1982), axons (e.g., BURTON and PAIGE 1981, HEIDEMANN et al. 1981), pigment cells (EUTENEUER and MCINTOSH 1981 a, MCNIVEN et al. 1984), and heliozoan axopodia (EUTENEUER and MCINTOSH 1981 a). This uniformity of microtubule organization in terms of their instrinsic polarity is remarkable and indicates an evolutionary conservation (plants and animals, protozoa to mammals) of the mechanisms that control microtubule organization in cells. Second, there is one reported case, fish pigment cells, where a change in a physiological parameter is accompanied by a change in microtubule polarity (MCNIVEN et al. 1984). If cell processes of isolated melanophores are severed from the cell body, pigment granules will initially aggregate to the former base of the severed cell process. Eventually, however, granules reverse the direction of transport and aggregate in the center of the "minicell" created by severing. This reversal in granule transport is accompanied by a reversal in the polarity of at least some of the microtubules which now are focused on the new center of the minicell with their minus-ends, as determined by the "hook" method. While the precise role of microtubule polarity in force generation and directionality of organelle movements is not known, this experiment is an excellent indication that cells care about microtubule polarity and can change polarity patterns according to physiological needs.

II.7. Nucleation Sites and Organizing Centers

Microtubule assembly in cells is under strict temporal and spatial control. As a cell traverses the cell cycle, for example, the interphase network of microtubules is reorganized into a ephemeral structure, the mitotic spindle, which, in turn, is replaced again by interphase arrays in the two daughter cells. At no time is the number and distribution of microtubules a simple function of the concentration of tubulin available for polymerization. In fact, in many cells, only about one half of the tubulin is polymerized even though the tubulin concentration probably exceeds the critical concentration for assembly. The highly precise, ordered coming and going of microtubules suggests that the cell possesses an effective system for the control of the distribution of microtubules. This possibility was realized

Fig. 20. Microtubule polarity in an axonemal bundle of the heliozoan *Actinosphaerium* as revealed by the hook-decoration method. All microtubules are of uniform polarity. The hooks are curving clockwise, so this axoneme is viewed from the tip towards the base. × 80,000. (Courtesy of U. Euteneuer)

as early as 1966 by PORTER, who wrote: "it is evident that the microtubules follow a programme of changes with time which involves their disappearance relative to one or several initiation sites and their reformation relative to others". The "initiation sites", or nucleation centers, in cells fall into two categories, microtubular and nonmicrotubular. Evidently, microtubules grow by addition of subunits to their end(s); this would be a microtubular initiation site. According to this definition, basal bodies are microtubular initiation sites for the growth of axonemal microtubules. There are, however, nonmicrotubular sites from which the growth of cellular microtubules is initiatied; these sites and their structural and biochemical properties shall be discussed briefly in this section. PICKETT-HEAPS (1969) has coined the term microtubule-organizing centers (MTOCs) to discribe this morphologically heterogeneous class of cell organelles. In many instances, an organizing center is identified on the basis of morphological evidence only, such as the finding of microtubules inserting into, or originating from, a certain cellular site. However, there are a few organizing centers whose properties have been studied in some detail. This is not the place to review the extreme diversity of MTOCs since this has been done frequently and elegantly before (PORTER 1966, TILNEY 1971, TUCKER 1979, DUSTIN 1984). Only some of the basic characteristics of cellular organizing sites—at least as far as they are understood to-date—will be outlined.

II.7.1. Structure

Initially, MTOCs were identified as discrete cellular sites from which microtubules originate. The first of these sites to be studied in some detail were the centrosomes of mammalian cells (DE HARVEN and BERNARD 1956) which comprise the centriole pair and a surrounding amorphous cloud of electron-dense structures, the pericentriolar material (ROBBINS et al. 1968, reviewed in Wheatley 1982). Microtubules insert into the dense material without necessarily making contact with the centrioles, suggesting that the nucleating capacity resides with the pericentriolar material rather than the centrioles themselves. This is most convincingly demonstrated by experimental analyses (see next section), and by animal cell lines that lack centrioles, but not pericentriolar material (DEBEC et al. 1982). Further, the "fuzzy", electron-dense material is the common theme for MTOCs from other sources, not the presence of centrioles. Cells that possess neither centrioles nor pericentriolar material are unable to develop an organized array of microtubules (KARSENTI et al. 1984). The extreme morphological diversity of MTOCs of different cell types is illustrated in Fig. 21. The number of microtubules radiating from such sites may range from several tens (e.g., leukocytes, PRYZWANSKY et al. 1983) to several thousand (e.g., pigment cells, SCHLIWA et al. 1979). Although many cells possess discrete

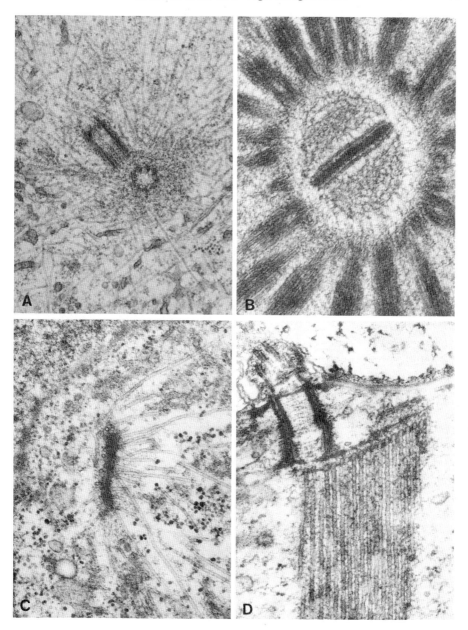

Fig. 21. Structural diversity of microtubule-organizing centers of animal cells. *A* Centrosome of a cultured PtK1 cell. × 22,000. Courtesy of K. McDonald. *B* Centroblast of a centrohelidian heliozoan. × 38,000. From Bardele (1977). *C* Dense bar of *Acanthamoeba*. × 26,000. From Bowers and Korn (1968). *D* Perforated sheet at the top of a cytopharyngeal microtubule bundle of the ciliate *Nassula*. × 46,000. [From Tucker (1970)]

and sometimes elaborate MTOCs, some cell types may organize microtubules from less conspicuous structures. A case in point is the specialized
sensory neuron of *Caenorhabditis elegans*. Serial thin sectioning has shown
that in the short microtubules dispersed along the length of the axon, one of
the ends appears "filled" with electron dense-material (CHALFIE and
THOMSON 1979). There is no evidence that these filled ends are nucleation
sites for individual microtubules, but it is interesting to note that they are
always proximal to the cell body, suggesting uniform polarity of the
microtubules.

II.7.2. Demonstration of Nucleating Activity

Microtubule growth from presumptive nucleation centers has been
demonstrated in many *in vitro* experiments. SNYDER and MCINTOSH (1975)
and MCGILL and BRINKLEY (1975) reported microtubule nucleation with
exogenous brain tubulin in lysed mammalian cells from centrosomes and, to
a lesser extent, from kinetochores in mitotic cells. GOULD and BORISY (1977)
gave convincing experimental demonstration that the pericentriolar
material rather than the centrioles is responsible for the growth of aster-like
microtubule arrays. These authors separated the pericentriolar material
from the centrioles and showed that while centrioles were able to grow
microtubules in an axonemal pattern, only the pericentriolar material
initiated microtubules in an aster-like fashion. There is evidence from
studies of microtubule nucleation using isolated spindle pole bodies of yeast
cells or centrosomes of CHO cells that an MTOC can specify the number of
microtubules growing from it (BYERS *et al.* 1978, HYAMS and BORISY 1978).
For example, interphase centrosomes of CHO cells initiate approximately
50 microtubules when incubated with bovine brain tubulin (MITCHISON and
KIRSCHNER 1984a) (Fig. 22). The centrosome can also determine the
direction of microtubule growth, at least initially. For example, in mitoses
of early mouse embryos, which do not contain centrioles (SZOLLOSI *et al.*
1972), the centrosome is a flattened band of pericentriolar material from
which microtubules emerge laterally (CALARCO-GILLAM *et al.* 1983). The
material assembles in this form only during mitosis and disperses between
cleavages. This and other observations on the specific morphologies of
organizing sites (see, for example, TUCKER 1984, PAWELETZ *et al.* 1984)
indicate the centrosome or, more specifically, its active ingredient, the
pericentriolar material, to be a bearer of morphological information, as well
as of chemical instructions for the initiation of microtubule growth (see
discussion by MAZIA 1984).

The nucleation capacity of centrosomes appears to be cell-cycle-
dependent, more microtubules being initiated from mitotic than from
interphase centrosomes (SNYDER and MCINTOSH 1975, KURIYAMA and
BORISY 1981). Changes in the nucleating capacity of MTOCs have also been

reported in other instances. For example, only *activated* sea urchin eggs polymerize microtubules in an astral-like array (WEISENBERG and ROSENFELD 1975), and only fish melanophores with *dispersed* pigment granules initiate large numbers of microtubules from their organizing center in lysed cells (SCHLIWA *et al.* 1979) (Fig. 23). In the melanophore, the

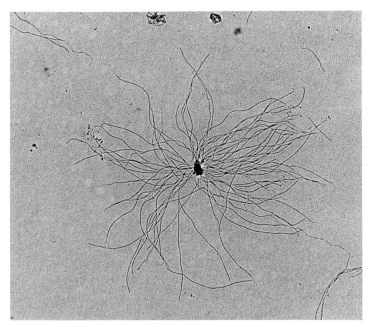

Fig. 22. Centrosome isolated from a Chinese hamster ovary cell, incubated with 2.5 mg/ml brain tubulin and visualized by rotary shadowing. The centrosome (which includes a centriole pair) initiated the growth of about 50 microtubules in a radial pattern. × 5,600. (Courtesy of T. MITCHISON and M. KIRSCHNER)

dramatic change in the nucleation capacity correlates with a similarly striking change in the morphological appearance of the MTOC, namely, an almost complete disappearance of the electron-dense cloud so abundant in the center of cells with dispersed pigment. These observations underscore the importance of the dense material in microtubule nucleation and indicate that the nucleating capacity of MTOCs appears to be under precise temporal and spatial cellular control.

II.7.3. Biochemistry of MTOCs

Knowledge of the composition of MTOCs and the molecular basis for microtubule nucleation is still very fragmentary. Some major polypeptide components of isolated basal bodies were identified by gel electrophoretic

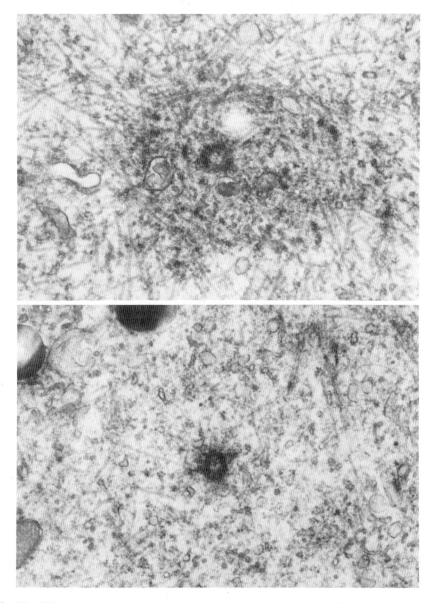

Fig. 23. Cell center of an angelfish melanophore with pigment dispersed (top) and pigment aggregated (bottom). Numerous electron-dense aggregates from which microtubules emerge surround the centrioles in dispersed cells, but are virtually absent from aggregated cells. × 38,000

and immunological analyses (ANDERSON and FLOYD 1981, LIN *et al*. 1982, TURKSEN *et al*. 1982), but no specific functions have yet been assigned to any of these proteins. Other studies took advantage of the existence of autoantibodies that specifically stain centrosomes and/or pericentriolar material (*e.g.,* MAUNOURY 1978, BRENNER and BRINKLEY 1982, TUFFANELLI *et al*. 1983) to follow the behavior of centrosomes during different cellular activities. However, these antibodies have yet to be used to dissect the chemical composition of centrosomes and to identify specific functions of centrosomal components. In one study, a set of high molecular weight proteins were isolated from the centriolar rootlets of the green alga, *Polytomella*. These proteins increase the efficiency of microtubule initiation *in vitro* (STEARNS and BROWN 1979). So far this seems to be the only example of proteins with a "microtubule-initiation-promoting" capacity.

One feature of centrosomes that attracted much attention was the possibility that they contain RNA essential for some of their acitivities. When *Tetrahymena* basal bodies are injected into unfertilized eggs of *Xenopus laevis*, they elicit aster formation. If the basal bodies are pretreated with RNAse, however, the potential to elicit aster formation is obliterated (HEIDEMANN *et al*. 1977). Since basal bodies of *Tetrahymena* are not known to have such a high microtubule-nucleating capacity *per se*, the generation of large astral arrays inside frog oocytes may be viewed as the congregation and RNA-dependent activation of microtubule-initiating sites present in the oocyte itself; in other words, basal bodies would serve to "organize the organizing sites". This is an intriguing possibility that merits further study. Additional evidence for a functional role of RNA in centrosomal activity was provided by BERNS *et al*. (1977) who used acridine orange to sensitize centrosomes for subsequent laser microirradiation, which interferes with centrosomal activity. The existence of RNA in centrosomes is strongly supported by cytochemical evidence, but little is known about its chemistry and precise function (see summary by MCINTOSH 1983).

II.8. Microtubular Machineries: Cilia and Flagella

Although this monograph is concerned with chiefly the disposition and function of cytoplasmic filamentous components, we need to consider briefly some structural and functional aspects of cilia and flagella, mainly for two reasons. First, they are the best-studied and best-understood of all microtubule-based organelles; and second, their mode of operation has served as a paradigm for the mechanism of certain microtubule-dependent processes in the cytoplasm, such as spindle elongation and intracellular transport.

Cilia and flagella generate the force that propels cells through a liquid (in most instances, water), or that moves liquid past stationary cells. Force is exerted by bending movements in a whip-like fashion (cilia) or sinusoidal waves (flagella). Their structures are basically identical, including a variety of highly specific subcomponents such as links, spokes, and sidearms interconnecting an evolutionarily conserved array of microtubules. Nine outer doublet microtubules are composed of one complete or A-microtubule and an incomplete or B-microtubule that shares 3 protofilaments with the A-subfiber. The ring of outer doublet microtubules surrounds two singlet central pair microtubules. This "9 + 2" structure is remarkably constant among plants and animals alike, even though departures from this scheme are known, notably among insect sperm.

II.8.1. Structure

The detailed structure of the "9 + 2" pattern of the axoneme with all its appendages and cross-links is now fairly well understood, thanks to the use of a variety of electron microscopic techniques (for a review of the history of our understanding of axonemal structure, see GIBBONS, 1981). It comprises six major components: 1. outer doublet and central pair microtubules. 2. radial spokes projecting from each A-microtubule towards the 3. central sheath that envelopes the central pair microtubules, 4. spoke heads in which the spokes terminate, 5. bridges between the A-microtubules of the outer doublets, called nexin links, and 6. dynein arms attached to the A-subfiber, extending as curved hooks towards the B-subfiber of the neighbouring doublet. These appendages and links are spaced at regular intervals along the length of the axoneme. The spokes, either doublets (e.g., Chlamydomonas) or triplets (e.g., Tetrahymena), occur at intervals of 96 nm, as are the nexin links. The components of the central sheath, which consists of two rows of projections from the two central microtubules, have an axial periodicity of 8 nm. Dynein arms are spaced 24 nm apart so that the inner and outer rows of arms are staggered with respect to each other. The fact that these periodicities are multiples of 8 nm is consistent with the 8 nm spacing of tubulin dimers in the microtubule wall and suggests that the axial spacing between adjacent binding sites on axonemal microtubules is 8 nm.

Despite its relatively simple architecture, the axoneme is a biochemically complex organelle consisting of more than 250 polypeptides. The identification of the polypeptide composition of some of the axonemal accessory structures was made possible through the use of flagellar mutants, in combination with two-dimensional polyacrylamide gel electrophoresis (reviewed in LUCK 1984). Radial spokes, for example, are made of 17 polypeptides (PIPERNO et al. 1981), while the spoke head consists of 5 polypeptides. Central pair microtubules contain 19 polypeptides besides the two tubulins. A similar complexity was revealed for the dynein arms.

II.8.2. Function

AFZELIUS (1959) first proposed that the dynein arms attached to the A-tubule might be involved in a sliding motion to produce ciliary and flagellar beating. Subsequent studies confirmed this proposal and revealed the molecular details of the dynein molecule and how it interacts with the B-subfiber to produce sliding (reviewed in WARNER and MITCHELL 1980). Dynein is an ATPase that requires magnesium or calcium ions for its activity and can be inhibited by EDTA (BROKAW 1972), vanadate ions (GIBBONS et al. 1978), and erythro-9-[3-(2-hydroxynonyl]adenine (EHNA) (BOUCHARD et al. 1981). An essential role for the dynein ATPase in axonemal motility is suggested by three lines of evidence: 1. in demembranated axonemes reactivated by ATP, the beat frequency is proportional to the number of dynein arms present in the axoneme; 2. mutants of cilia and flagella which lack dynein arms are unable to beat ("paralyzed"); and 3. antibodies prepared against flagellar dynein inhibit both its ATPase acitivity and the ability of demembranated sperm to be reactivated upon the addition of ATP (see review by WARNER and MITCHELL 1980). That dynein generates force by inducing coordinated sliding of outer doublets past each other was implicated first in thin sections of cilia (SATIR 1965), and then shown directly by dark-field microscopy of demembranated, reactivated axonemes in which interconnecting links were partially digested with trypsin (SUMMERS and GIBBONS 1971). Addition of ATP to these axonemes causes outer doublets to slide apart such that the dynein arms on the A-tubule exert a tip-directed force on the B-tubule of the neighboring doublet (SALE and SATIR 1977). These experiments also provide evidence that the axonemal links (spokes and nexin) convert sliding into bending because, disruption of the links by enzyme digestion, prevents bending.

The axoneme with its highly specific microtubule-microtubule interactions mediated by dynein has served as a paradigm for other microtubule-based cellular events. Sliding of microtubules past each other medieated by "bridges" was proposed to be involved in spindle elongation during mitosis (McINTOSH et al. 1969), cell elongation (WARREN 1974), and movements of organelles along microtubules (CLARK and ROSENBAUM 1982). Probably the best-documented case for force generation by a cytoplasmic microtubular organelle, mediated by the sliding of microtubules past each other, is found in certain parasitic protozoa habitating the hind gut of termites. These organelles, called axostyles, run through the cell body along the cell's long axis and propagate bends that propel the organism (LANGFORD and INOUE 1979, WOODRUM and LINCK 1980). Bend propagation appears to be mediated by crossbridges between the rows of microtubules that make up the organelle (McINTOSH 1973), an acitivity that can be reactivated in isolated axostyles in an ATP-dependent fashion (MOOSEKER and TILNEY 1973, BLOODGOOD 1975).

II.8.3. Ciliary and Cytoplasmic Dyneins

Considerable attention has been focused on the dynein molecule since it is the only microtubule-associated, force-generating protein characterized in some detail. Over the years, the structural and molecular complexity of this molecule has increased proportional to the degree of sophistication of the procedures applied to study it. The view of its structure has changed from that of a hook-shaped appendage (by conventional thin-sectioning) to that of a rod composed of three globular subunits (by negative staining) to that of a complex of globular heads and slender stalks arranged in a configuration whose details depend on the species studied and the method of preparation used (quick-freezing and deep-etching of intact axonemes: GOODENOUGH and HEUSER 1982, 1984; scanning transmission electron microscopy of unstained, isolated dynein: JOHNSON and HALL 1983; see HYAMS 1984). However, models for the rearrangement of these components during force generation in outer doublet sliding have yet to be reconciled with details of a proposed mechanism of the mechanochemical cycle of dynein (SATIR et al. 1981, AVIOLO et al. 1984).

The apparent structural complexity of axonemal dynein is surpassed only by its biochemical complexity. Despite species-specific differences, the dyneins that have been studied in some detail have two main features in common: first, the presence of (two or three) magnesium or calcium-activatable and EDTA-inhibitable ATPase activities with sedimentation coefficients ranging from 10 to 30 S; second, a complex polypeptide composition that characteristically includes one to several high molecular weight proteins in the range of 300 to 350 kd. Each of these can be equated with a single structural head and one ATPase. The entire complexes may have a molecular weight of 1–2 million daltons (GIBBONS 1981, LUCK 1984). In *Chlamydomonas*, for example, the outer and inner arms are complexes of distinctly different sets of high molecular weight proteins, and each arm appears to have three different ATPase activities. Dynein isolated from flagellar axonemes will rebind to outer doublet microtubules at its original sites on the A-microtubule, but it can also attach to the B-subfiber in an ATP-dependent fashion, presumably to the sites with which it interacts during normal beating (MITCHELL and WARNER 1981). Axonemal dynein will also bind to cytoplasmic microtubules with a lateral spacing of 24 nm and will induce ATP-dependent crossbridging into large bundles (HAIMO et al. 1979). This observation ties in with reports indicating the existence of magnesium-activated cytoplasmic dynein ATPases (PRATT 1980) that can bind to, and crossbridge, cytoplasmic microtubules (PRATT et al. 1980, HISANAGA and SAKAI 1983, SCHOLEY et al. 1984, HOLLENBECK et al. 1984). While at this point it can not be excluded with absolute certainty that the dyneins identified in these studies represent precursors of axonemal dyneins (PIPERNO 1984, ASAI and WILSON 1985), it is possible that these molecules are geniune cytoplasmic microtubule-associated enzymes long suspected to be involved in a variety of microtubule-dependent cellular acitivities (see review by PRATT 1984).

II.9. The Distribution and Function of Microtubules in Cells: A Summary

In this introductory overview of microtubules, primarily biochemical aspects of microtubules have been dealt with. Ultimately, however, the knowledge acquired in *in vitro* studies should help explain the diverse aspects of microtubule distribution, function, and regulation in eukaryotic cells. One should not be misled to view microtubular organization of a mammalian tissue culture cell as being representative for microtubule organization in other eukaryotic cells. The extreme morphological and functional diversity of microtubule arrays in protozoa, for example, has only begun to be explored. While other chapters will deal with the specific role of microtubules, alone or in conjunction with other cytoskeletal elements, in the many acitivities of the cytoskeleton, a note on microtubule control and function in the living cell shall close this introductory survey.

Control of microtubular organization in cells can be subdivided into the three major categories nucleation, polymerization, and interaction. There can be little doubt that in order for a cell to achieve and maintain control over the distribution of microtubules, it requires a mechanism that specifies *where* and *when* microtubules are to be formed (nucleation). This requirement was clearly recognized 20 years ago by PORTER (1966). Cellular structures that might do this job have been identified and termed MTOCs, but precisely how they function is unknown. Regarding polymerization, possible regulators for the control of the extent of microtubule assembly suggested by *in vitro* studies are calcium ions and MAPs, but whether the cell uses these, alone ore in combination with other, unknown, factors, is unclear. Finally, the ways by which microtubules become distributed in the cell once they are formed, and the patterns they may form, are poorly understood. In a cultured cell, the distribution of microtubules may simply be the consequence of the ways by which microtubules are able to interact with other components of the cytoskeleton (reviewed in section VII.)—if one is inclined to call these forms of interactions "simple". But there are other, highly elaborate patterns dictated primarily by the interactions of microtubules with other microtubules. Prime examples are heliozoan axopodia, the feeding basket of *Nassula*, or the axostyle of certain parasitic protozoa. Such interactions are mediated by links (*i.e.,* MAPs) we know very little about; in fact, they must be quite different from the much-studied MAPs of mammalian cells because the latter do not act as microtubule-microtubule linkers.

The functions that cytoplasmic mictrotubules have been associated with are manifold, but generally fall into 4 categories: development and maintenance of an asymmetrical cell shape, intracellular movements (encompassing such diverse processes as axonal transport, color change, exo- and endocytosis), mitosis, and cell translocation. Microtubule involvement in these cellular events is inferred on the basis of two main lines of evidence which are, in fact, two sides of the same coin: the close proximity of microtubules with the event in question, and the inhibition or modification of these events by experimentally induced changes in the state of microtubule assembly. These are crude and indirect criteria. The specific role of microtubules in most of these phenomena said to be "microtubule-dependent" remains to be determined.

So, surprisingly, despite a wealth of detailed knowledge about many aspects of microtubule structure and biochemistry, little is known about what microtubules really do in cells and how they do it. There is a need for an approach that might be called "Intracellular Biochemistry" which will allow investigators to test the knowledge acquired in *in vitro* studies in a more meaningful environment, the living cell. First steps towards the implementation of this new brand of cell biology are being taken now (using *e.g.,* fluorescence analog cytochemistry, microinjection, photo-bleaching recovery, genetic dissection), and hopefully the future will see more of this.

III. Intermediate Filaments

III.1. Historical Aspects

Filaments 8–11 nm in diameter (intermediate filaments) were established as a separate class of major fibrous cellular structures only relatively recently. Recognized 15 years ago as unique filaments different from F-actin in muscle cells (ISHIKAWA *et al.* 1968, KELLY *et al.* 1968, RASH *et al.* 1970; but see also HENSON-STIENNON 1965), they were soon discovered to be a prominent component of many other cell types, both fibroblastic and epithelial (GOLDMAN and FOLLET 1969, ISHIKAWA *et al.* 1969). In addition, filaments of approximately 10 nm diameter were long known to be present in nerve cells (*e.g.,* BUNGE *et al.* 1961), glial cells (*e.g.,* GRAY 1959), and epidermal cells (*e.g.,* WEISS and FERRIS 1954, BODY 1960), but they were not yet thought of as different members of one family of filaments. Some of the earlier literature was also clouded by confusion over the issue whether these filaments might be breakdown products and/or a different supramolecular form of other cytoplasmic components, such as microtubules (*e.g.,* WISNIEWSKI *et al.* 1968, ISHIKAWA *et al.* 1968, DANIELS 1973, DE BRABANDER *et al.* 1975). This issue was laid to rest in the midseventies when the intermediate filament proteins were firmly established as a distinct group. Since then, increasing interest in the nature, distribution, molecular composition, and function of these filaments has led to the rapid accumulation of a staggering amount of information. Indeed, the popularity of this research field has made it one of the most rapidly devolping areas in cell

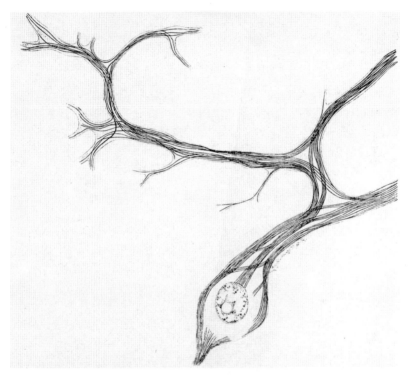

Fig. 24. Purkinje cell of the dog stained with molybdenum-toluidine blue, showing a fibrillar substructure. [After BETHE (1911)]

biology of the past ten years. However, it should not be overlooked that some of the most fundamental questions concerning intermediate filaments are still unresolved: First, and most important, what is the precise function of these filaments in cells? How is their assembly controlled within the cell? In what way are the different intermediate filament classes related? How do they interact with other cellular components? Some observations pertinent to these questions will be addressed in this chapter.

Intermediate filament-containing cellular structures have been known for more than 150 years, largely due to their extraordinary stability and their high affinity for certain histological stains (Fig. 24). Historically, the most important cell in this respect is the neuron, the study of which contributed to the establishment of the "fibrillar theory" of protoplasm at the turn of the century. Not surprisingly, the fibrillar substructure of nerve fibers was observed as early as 1828 by REMAK. Through the work of DEITERS (1865), ARNOLD (1865), and especially SCHULTZE (1868), among many others, the

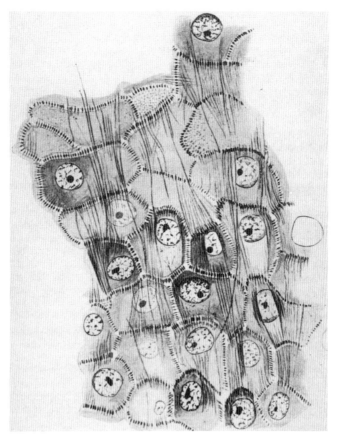

Fig. 25. Epithelium of the jaw of a newborm cat stained with iron hematoxylene, showing the transcellular organization of filament bundles. [From HEIDENHAIN (1911)]

concept of the fibrillar structure of nerve cells was developed, and became widely acknowledged after the introduction of the novel metal impregnation techniques of RAMON y CAJAL (1899) and BIELSCHOWSKY (1903, 1907). These procedures revealed with remarkable detail an intricate, fibrillar network in the cell processes and the cell body of neurons. While the cellular substrate for these impregnation procedures was unknown at that time, the modified reduced silver staining procedure introduced in 1936 by

BODIAN was recently shown to stain the proteins that make up neurofilaments (the so-called neurofilament triplet) in polyacrylamide gels (GAMBETTI et al. 1981). This observation elegantly bridges the gap between the classical neuroanatomical studies initiated by CAJAL, and more recent protein-chemical analyses of neuronal intermediate filaments. In the first half of this century it was much disputed whether fibers demonstrated only in fixed tissue by metal impregnation really existed in living nerve cells. But, in 1936, WEISS and WANG demonstrated well developed birefringent fibrils in living cultured ganglion cells, and BEAR et al. (1937) quantitated the fiber birefringence in squid nerve axons.

The only other tissue in which a fibrillar substructure based on intermediate filaments was observed early on is the epidermis, though there were occasional reports of fibrils in connective tissue cells (UNNA 1895, SCHNEIDER 1908). Following the discovery of a pervasive system of "plasma fibrils" in the vertebrate epidermis by RANVIER (1882), it became apparent that the fibrillar bundles form a transcellular network encompassing many cells, individual bundles being linked by dense "nodules" at the cell boundaries (RANVIER 1882, KROMAYER 1892) (Fig. 25). As we now know, the structures observed by these investigators are bundles of cytokeratin fibrils joined by desmosomes at the cell membrane. These bundles were later given the name tonofibrils (HEIDENHAIN 1911) in recognition of their presumptive role as resistive fibers that strengthen the coherence of epidermal tissues. Interest in the subcellular distribution and function of these fibers then subsided temporarily. Subsequent studies focused on the process of keratinization, chiefly because of the commercial importance of wool, the product of morphologically transformed tonofibril-containing cells.

Recent work on intermediate filaments has been summarized by LAZARIDES (1980, 1982). ZACKROFF et al. (1981), ANDERTON et al. (1981), and OSBORN and WEBER (1982). As in the previous two chapters, summaries of more specialized subtropics will be mentioned where appropriate.

III.2. Classification of Intermediate Filaments: Molecular Composition and Tissue Distribution

The term intermediate filaments originated from fine structural studies of muscle cells, which revealed filaments intermediate in size between the thin (6 nm) actin filaments and the thick (16 nm) myosin filaments (HENSON-STIENNON 1965, ISHIKAWA et al. 1968). A nondiscriminative, essentially generic term, it later was applied to filaments in other cells with a diameter of around 10 nm (range 8–12 nm). In addition to their rather uniform diameter and smooth surface, all intermediate filaments are generally insoluble in neutral buffers over a broad range of ionic strengths, and in nonionic detergents. They are soluble, to varying degrees, in denaturing agents (*e.g.,* 8 M urea), low ionic strength solutions (*e.g.,* 5 mM total salt), low pH (*e.g.,* pH 2), or ionic detergents (*e.g.,* sarkosyl), depending on the type of filament. Though intermediate filaments appear to be a rather homogeneous class of filaments morphologically and by solubility properties, they can be subdivided into 5 classes on the basis of biochemical and immunological criteria. In vertebrates, these 5 classes of intermediate filaments are: 1. vimentin (MW 57,000 d), characteristic of cells of mesenchymal origin; 2. desmin (MW 53,000 d), found in sarcomeric and smooth muscle cells; 3. cytokeratin (about 20 different polypeptides in the range between 40,000 and 70,000 d), found in epithelial cells, both keratinizing and non-keratinizing; 4. the "neurofilament triplet" (three polypeptides of 200,000 d, 150,000 d, and 68,000 d) present in neurons; and 5. glial fibrillar acidic protein (GFAP; MW 50,000 d), present in glial cells and astrocytes. This classification is satisfying because it parallels established histological criteria. While initial biochemical and immunological studies tended to emphasize differences in protein composition among these 5 classes, growing evidence demonstrates quite striking similarities in amino acid sequence, antigenicity, assembly properties, subunit structure, and gene organization. This suggests that all intermediate filaments belong to a chemically heterogeneous, yet evolutionarily related, homologous group of proteins.

III.2.1. Vimentin

Vimentin (FRANKE *et al.* 1978 b) was first identified in chicken fibroblasts as the major subunit of a population of filaments (BROWN *et al.* 1976) that remain after microtubules and most of the actin filaments are removed by detergent and high salt treatment. By immunofluorescence microscopy, vimentin was shown to be present not only in cells of mesenchymal origin, but also in a wide variety of cultured cells and permanent cell lines, both normal and transformed, derived from other tissues (TRAUB *et al.* 1983, and references therein). In these cells, vimentin is expressed in addition to the intermediate filament type characteristic of the tissue of origin (see OSBORN *et al.* 1982 for a summary). Thus, epithelial PtK 2 cells (a rat kangaroo line), and HeLa cells (derived from a human cervical carcinoma) express cytokeratin and vimentin (FRANKE *et al.* 1979 a, b); BHK 21 cells contain desmin and vimentin (GARD *et al.* 1979), and certain glioma cell lines possess GFAP and vimentin (PATEAU *et al.* 1979). Most cells *in situ* express only one intermediate filament type. Exceptions to this rule include some glial cells, aortic smooth muscle cells, and neurons (see OSBORN and WEBER 1983 for a summary). In all these cases, as in cultured cells, vimentin is the second partner in addition to the tissue-specific intermediate filament protein. The reason for the appearance of vimentin in certain tissues and many cultured cells is not known, but it is possible that its expression is correlated with a less differentiated state. By immunofluorescence and electron microscopy, vimentin intermediate filaments occur as single filaments or gently curving, loose bundles that have a more or less radial organization, with a higher concentration of filaments in the perinuclear area (Fig. 26). Double labelling experiments with either actin or tubulin antibodies and vimentin antibodies show that the patterns of distribution are different, though vimentin is often closely associated with microtubules (*e.g.*, FRANKE *et al.*

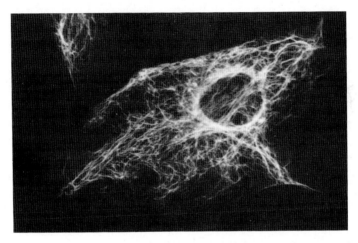

Fig. 26. Cultured PtK cell stained with an antibody against vimentin. × 700. (Courtesy of W. FRANKE)

1978 c, GEIGER and SINGER 1982; see also section VII.3.3.). On the basis of the cross-reactivity of vimentin antibodies with cells from all five vertebrate classes, vimentin can be considered an evolutionarily conserved protein. On the other hand, the existence of specific antibodies with restricted cross-reactivity, and the occurrence of minor differences in molecular weight and isoelectric point between vimentins from various sources suggest some degree of divergence (GRANGER and LAZARIDES 1979, GARD and LAZARIDES 1980). The genomes of various vertebrate cell types are known to have a single copy of a gene for vimentin (QUAX *et al.* 1983, 1984), but this gene seems to produce at least two mRNA species which differ in the presence of a 3′ nontranslated region of uncertain function (ZEHNER and PATERSON 1983).

III.2.2. Desmin

The major subunit of intermediate filaments of striated and smooth muscle cells has been termed desmin (LAZARIDES and HUBBARD 1976). [The synonymous term skeletin (SMALL and SOBIESZEK 1977) is less frequently used.] The isolation and characterization of desmin was facilitated by the finding that intermediate filaments in smooth muscle are insoluble in high ionic strength buffers (COOKE 1976), in which contractile proteins are soluble. The extraction of actin and myosin leaves behind a residue consisting primarily of an interconnected network of intermediate filaments linking dense bodies (the presumptive Z-line analogs of smooth muscle cells) and membrane-bound dense plaques (UEHARA *et al.* 1971, COOKE 1976). In mature striated (cardiac and skeletal) muscle desmin is concentrated at the Z-lines (Fig. 27), in regions of cell-cell contact, and, at least in cardiac muscle, in intercalated discs (LAZARIDES and HUBBARD 1976, LAZARIDES and BALZER 1978, LAZARIDES *et al.* 1982). Using a procedure which shears muscle fibers transversely to produce sheets of Z-discs which are then viewed *en face* by immunofluorescence microscopy, GRANGER and LAZARIDES (1978) have shown that desmin is located at the periphery of each Z-disc, while alpha-actinin is present within each disc and actin is found throughout the Z plane (Fig. 28). Since the major function of the Z-line appears to be to provide a point of anchorage for the actin filaments of the sarcomere, it has been hypothesized (GRANGER and LAZARIDES 1978, LAZARIDES 1980) that the peripheral desmin-containing domain conveys mechanical stability to muscle fibers by integrating and coordinating the activities of these fibers during successive contraction-relaxation cycles. In this way desmin would constitute a truely "skeletal" component.

While in mature cardiac and skeletal muscle desmin seems to be associated primarily with the Z-lines, its organization in less mature and developing muscle cells is quite different. Here desmin exists in the form of a pervasive, more or less uniformly distributed cytoplasmic network (BENNETT *et al.* 1978, 1979, LAZARIDES 1978). By immunofluorescence microscopy, desmin is found in only a subpopulation of myoblasts, but in all myotubes, suggesting that its synthesis is restricted to postmitotic, prefusion myoblasts (GARD and LAZARIDES 1980). In early post-fusion myotubes, desmin exists in the form of cytoplasmic intermediate filaments that have little or no apparent association with forming Z-discs, which are clearly visualized with antibodies against alpha-actinin. Later, desmin begins to associate with the already well-developed Z-discs; at this time they can no longer be aggregated into coiled filament masses by colcemid

---

Fig. 27. Colocalization of alpha-actinin (*A*) and desmin (*B*) in 7-day-old myotubes. Alpha-actinin localizes at the Z-lines, which are already in register. Desmin is found in the same location, but, in addition, it forms dense networks in cell processes. The corresponding phase contrast micrograph is shown in *C*. × 940. (Courtesy of E. LAZARIDES)

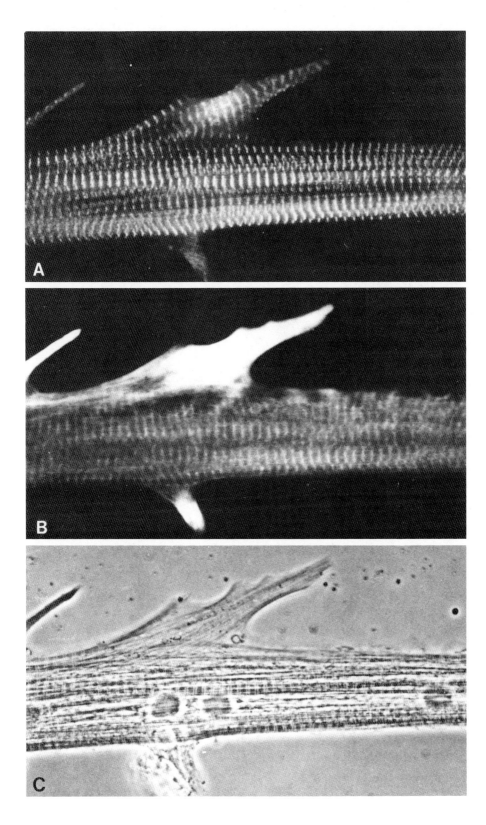

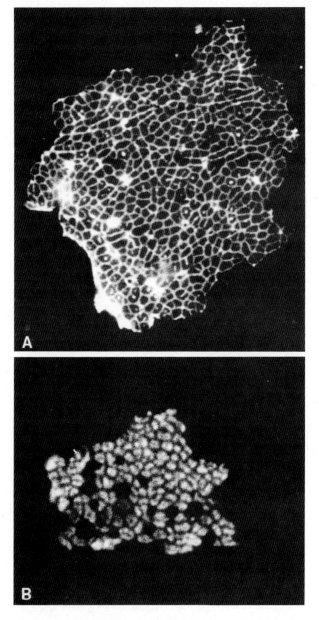

Fig. 28. Isolated Z-disc sheets stained with antibodies against desmin (*A*) and alpha-actinin (*B*). × 4,200. (Courtesy of E. Lazarides)

(BENNETT *et al.* 1979, GARD and LAZARIDES 1980). This physiological change seems to reflect a molecular change in desmin that accompanies its association with the Z-disc. Interestingly, desmin's association with the Z-discs is coordinated with lateral association of the discs into the characteristic Z-lines of adult muscle. Thus, desmin organization in myogenesis progresses through a series of developmentally regulated steps in its subcellular distribution and, presumably, in its physical-chemical properties. How this progression is regulated is presently poorly understood, but it is possible that certain intermediate-filament-associated proteins (see section III.5.) or post-translational modifications such as phosphorylation (see section III.6.) may be important in this intriguing developmental process. The possible coexistence of vimentin and desmin at the periphery of Z-discs (GRANGER and LAZARIDES 1978, 1979, LAZARIDES 1980, LAZARIDES *et al.* 1982) was not confirmed by others (BENNETT *et al.* 1979, HOLTZER *et al.* 1982, OSBORN *et al.* 1982). Thus, whether vimentin is retained during myogenesis and forms a major structural component of the Z-disc along with desmin, can not be said with certainty. There are, however, certain vascular smooth muscle types that contain both desmin and vimentin (SCHMID *et al.* 1981, TRAVO *et al.* 1982, QUINLAN and FRANKE 1982). This is one of the rare cases where two different intermediate filament types are known to be coexpressed in differentiated vertebrate cells.

Desmins from various vertebrate species and muscle tissues are related, but not identical, as indicated by immunological and biochemical criteria (LAZARIDES and HUBBARD 1976, LAZARIDES and BALZER 1978, GEISLER and WEBER 1980). The gene for desmin is present in a single copy (QUAX *et al.* 1984, CAPETANAKI *et al.* 1984), analogous to vimentin.

III.2.3. Cytokeratins

Whereas intermediate filaments in mesenchymal and muscle cells are composed of identical subunits of a single polypeptide chain, the cytokeratin filaments of epithelial cells, both ectodermal (*e.g.,* epidermis) and endodermal (*e.g.,* oesophagus), are constructed from several different subunits in the molecular weight range between 40,000 and 70,000 d. The polypeptide patterns of cytokeratins may vary considerably between tissues and even from cell type to cell type within the same tissue. Perhaps the best-known of the cytokeratin (or prekeratin) filaments are the tonofilaments or tonofibrils of vertebrate epidermal cells. These cells contain tight bundles of 8–10 nm filaments frequently associated with desmosomes at the cell periphery. Cultured epithelial cells, such as PtK, show a network of curly filament bundles (Figs. 29 and 30). Cytokeratins may account for 20–30% of the total cellular protein in some cell types. In keratinizing tissue, the cytokeratin filaments undergo molecular modification (*e.g.,* formation of disulfide linkages, interaction with keratohyalin granules) as the cells move from the basal layer to the outer, dead-cell layer of the stratum corneum, a process that takes 20–50 days in many mammals. As mentioned, however, cytokeratin filaments are also present in nonkeratinizing cells of epithelial

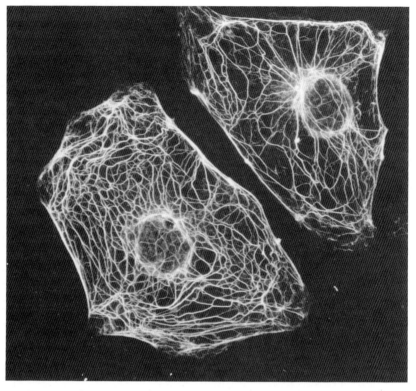

Fig. 29. Cultured PtK cells stained with an antibody against cytokeratin, showing the network of filament bundles. × 400. (Courtesy of W. FRANKE)

origin (SUN and GREEN 1977, 1978, SUN 1979, FRANKE et al. 1978 a, b, 1979 a).

Biochemical analysis, principally by two-dimensional gel electrophoresis, shows that the cytokeratins comprise a complex family of about 20 different polypeptides ranging from 40,000 to 70,000 d in molecular weight and from 5 to 8.5 in isoelectric point (e.g., FUCHS and GREEN 1979, MILSTONE and MCGUIRE 1981, FRANKE et al. 1981 b, MOLL et al. 1982, WU et al. 1982, ANDERTON 1983). In addition, each cytokeratin polypeptide may have a set of isoelectric variants derived usually by phosphorylation (SUN and GREEN 1978 b, FRANKE et al. 1981 b, STEINERT et al. 1982, GILMARTIN 1983). As reviewed in detail by MOLL et al. (1982), the members of this

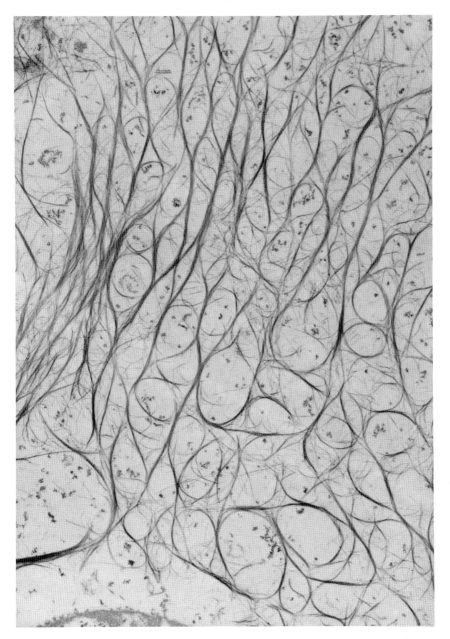

Fig. 30. Cytokeratin filament bundles in a PtK cell extracted with Triton X-100. × 18,000. (Courtesy of U. EUTENEUER)

family of polypeptides are differently expressed in different tissues and cell types; never are all of these polypeptides expressed simultaneously by any one epithelial cell. Cytoskeletal preparations of human epithelial tissues extracted with detergent and high salt buffers exhibit complex and tissue-specific patterns of cytokeratin polypeptides. For example, larger and more basic polypeptides are typical of stratified epithelia, while cytokeratins of intermediate size and electrical charge are found in diverse simple epithelia. Epithelial cells can be subclassified further according to the specific complement of cytokeratin polypeptides they normally express. Some polypeptides have a restricted distribution, while others may be found in many kinds of epithelial tissues. The differential expression of cytokeratin genes, reflected in these polypeptide patterns, also occurs within one keratinocyte under altered environmental conditions. Thus, keratinocytes of different origin, normally expressing different sets of cytokeratins, will synthesize similar sets of cytokeratins when cultured under the same conditions (SUN and GREEN 1977, FUCHS and GREEN 1980).

Most members of the cytokeratin family fall into one of two classes, the first composed of acidic (also termed type I) polypeptides, the second of primarily basic or neutral (also termed type II) polypeptides (FUCHS et al. 1981, KIM et al. 1983, MAGIN et al. 1983, ROOP et al. 1983, FRANKE et al. 1983). Within the acidic (type I) class, two different subtypes may be distinguished by their carboxyterminal sequences (JORCANO et al. 1984). At least one representative of the "basic" and one of the "acidic" family is expressed in all epithelial cells and tissues examined (KIM et al. 1983), and a combination of at least one of each is required for the construction of cytokeratin filaments in vitro (STEINERT et al. 1976, FRANKE et al. 1983). Although two cytokeratins, one from each subfamily, are sufficient to form 8–10 nm filaments by copolymerization, most cytokeratin filaments isolated from epithelial cells have a more complex polypeptide composition (MOLL et al. 1982). Apparently, all of the cytokeratins can be incorporated into filaments, even in heterologous assembly. For example, kidney cells expressing a characteristic set of epithelial keratins, when microinjected with epidermal keratin mRNA, will translate it and incorporate the foreign keratin into the same intermediate filament system, constructing a novel, hybrid keratin filament (FRANKE et al. 1984).

Although cytokeratins are heterogeneous in gel electrophoretic analyses, immunological and molecular criteria imply substantial amino acid sequence homology. Mammalian keratins from a variety of species are immunologically related; antibodies against either human (SUN and GREEN 1978 a) or bovine epidermal keratins (FRANKE et al. 1978 a, b, 1980, 1981 a) cross-react with a variety of epithelial cells from fish to man. GIGI et al. (1982) have isolated a monoclonal antibody against bovine epidermal prekeratin which cross-reacts broadly and probably recognizes a cytokeratin determinant present in many kinds of cytokeratin polypeptides, both basic and acidic. This supports the concept that the basic and acidic subfamilies of polypeptides are immunologically related. Both polypeptide fingerprints and amino acid composition of keratins of very different sizes indicate common structural features (STEINERT and IDLER 1975, FUCHS and GREEN 1978). Their genomic organization (LEHNERT et al. 1984) also supports the hypothesis that the different cytokeratins are derived from a common ancestral gene. As a rule, the greater the size difference between any two keratins, the smaller the degree of homology in structure. This conclusion is based on recent partial and/or complete amino acid sequencing data obtained through direct sequencing, or by deduction from the cDNA sequences (HANUKOGLU and FUCHS 1982, 1983, STEINERT 1983). These studies reveal a distant relation between the two major keratin subfamilies, and an even more distant relation with other intermediate filament subunits (see section III.3.). Therefore, the cytokeratins display a gradient of relatedness that roughly parallels their molecular size, with substantial differences between the basic and acidic subclasses. The latter two form the foundation for the cytokeratin filament substructure in all vertebrate epidermal keratinocytes from fish to man, and presumably also in other epithelial cells that express this class of intermediate filaments (FUCHS and MARCHUK 1983).

III.2.4. Neurofilaments

The cytoskeleton of all vertebrate and some invertebrate neurons comprises microtubules and a class of intermediate filaments known as neurofilaments. Known for over 100 years to be a component of neurites and cell bodies of neurons, their subunit composition was not unraveled until about 10 years ago. Structurally, neurofilaments are very similar to the 10 nm intermediate filaments of other cell types. They are a prominent component particularly of axons, where they extend parallel to the long axis. In many cases, they are interconnected with each other and with microtubules by wispy side arms or bridges visible in the electron microscope (Fig. 31 a). Neurofilaments can be either randomly dispersed in the cross-sectional area of an axon, or they may occur in clusters or bundles. Usually they extend uninterrupted for long distances, but an apparent discontinuity may exist at the node of RANVIER, as suggested by an analysis of serial thin sections (TSUKITA and ISHIKAWA 1981).

Initial biochemical analyses of neuronal tissue erroneously identified a protein of approximately 50,000 d as a major subunit of neurofilaments (e.g., DAVISON and WINSLOW 1974, YEN et al. 1976). This protein was later shown to be the major subunit polypeptide of glial filaments (e.g., LIEM et al. 1978) which were a prominent contaminant in the earlier preparations of

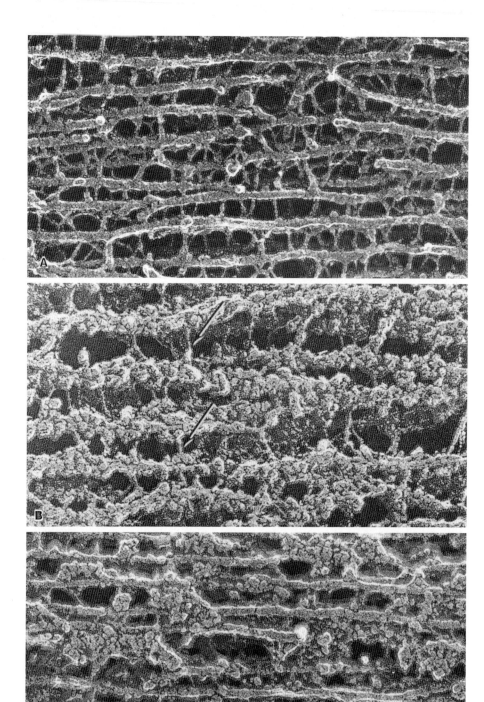

neuronal tissue. Several experimental approaches have shown neurofilaments to comprise a complex of three polypeptides with molecular weights of approximately 200,000 d, 150,000 d, and 68,000 d, referred to as the "neurofilament triplet" (HOFFMAN and LASEK 1975). First, the major nontubulin polypeptides in peripheral nerves that are substantially free of contaminating structures from glial cells are the triplet proteins, with no or little material in the 50,000 d range (MICKO and SCHLAEPFER 1978, SCHLAEPFER and FREEMAN 1978). Second, in transected nerves, the disappearance of identifiable neurofilaments due to an increase in the concentration of calcium, coincides with the disappearance of the neurofilament triplet proteins (SCHLAEPFER and MICKO 1979). Third, antisera against filament preparations from peripheral nerves stain neuronal structures specifically, but not glial tissue, and bind to neuronal 10 nm filaments in immunoelectron microscopy (e.g., SCHLAEPFER and LYNCH 1977, ANDERTON et al. 1980, YEN and FIELDS 1981, SHAW and WEBER 1981, SHARP et al. 1982). Finally, studies of axonal transport showed that the slowest component consists of 5 major polypeptides, namely, the two tubulins, and the neurofilament triplet. HOFFMAN and LASEK (1975) suggested that this component of axoplasmic transport represents the movement of the microtubules and neurofilaments of the neuronal cytoskeleton itself. Although comigration of proteins does not prove that they are components of the same structure, it is strong suggestive evidence.

Many studies of the neurofilament triplet polypeptides isolated by a variety of procedures (see WILLIAMS and RUNGE 1983 for a summary) strongly suggest that the neurofilament triplet polypeptides are distinct, but evolutionarily related proteins. Antibodies of both broad (e.g., LIEM et al. 1978, YEN and FIELDS 1981, SHAW et al. 1984) and restricted cross-reacitvity (e.g., LEE et al. 1982, DEBUS et al. 1982, SHAW et al. 1984) have been prepared, indicating the presence of both unique and shared antigenic determinants on each polypeptide. The triplet proteins are not proteolytic fragments of a precursor because each is translated from different mRNAs, at least in vitro (CZOSNEK et al. 1980). Neurofilament polypeptides from

---

Fig. 31. Neurofilament organization and antibody labeling of neurofilament polypeptides in triton X-100 extracted spinal cord neurons prepared by quick-freezing, deep-etching, and rotary shadowing. A Neurofilament-rich portion of the cytoplasm. Note the presence of numerous filamentous crossbridges. B Similar preparation, descorated with antibodies against the 68,000 d polypeptide. Neurofilament cores, but not the crossbridges, are heavily labeled (arrows). C Similar preparation, labeled with an antibody against the 200,000 d polypeptide. Globular masses of antibody complexes are present in places where one would expect to see crossbridges. The neurofilament cores tend to be smooth. × 206,000. (Courtesy of N. HIROKAWA)

reptiles, birds, and mammals are clearly closely related (*e.g.,* SHAW *et al.* 1984), although polyacrylamide gels consistently show slight differences in the molecular weights of homologous proteins (CHIN *et al.* 1980, DAVISON and JONES 1980). Also, certain species may possess only two of the triplet proteins, with the third (usually the high molecular weight component) absent or greatly reduced (SCHEKET and LASEK 1980, TAPSCOTT *et al.* 1981). The latter observation suggests that the high molecular weight protein may have the status of an associated protein which is not absolutely required for neurofilament formation. Indeed, immunoelectron microscopic localization of the three neurofilament proteins indicates that the 68,000 d protein forms the core of the filament, while the 200,000 d protein occurs periodically along the periphery of individual filaments either as cross-bridges or in a helical fashion (WILLARD and SIMON 1981, SHARP *et al.* 1982). In addition, recent studies on the repolymerization of neurofilaments from dissociated triplet proteins (GEISLER and WEBER 1981, LIEM and HUTCHINSON 1982) show that only the 68,000 d subunit is competent to self-assemble into morphologically normal neurofilaments, while the 150,000 d and the 200,000 d polypeptides are able to copolymerize only if the 68,000 d subunit is present. This may appear somewhat surprising in the light of the fact that even the middle and high molecular weight neurofilament proteins share structural similarities with the 68,000 d subunit, such as the presence of a rod-shaped alpha-helical domain flanked by globular head- and tailpiece-like regions. The only difference is the presence of carboxyterminal extensions of unique amino acid composition; this extension is longer for the 200,000 d subunit than for the 150,000 d subunit (GEISLER *et al.* 1983, 1984, 1985 b). Thus the two high molecular weight polypeptides may perform two functions. Via the alpha-helical domain they may interact with the neurofilament core and contribute to the formation of the neurofilament backbone, while the tailpiece extensions protrude from the filament wall and may engage in other macromolecular associations.

Invertebrate neurofilaments so far appear to be significantly different from vertebrate neurofilaments. The two best-studied systems are the giant nerve fiber of the marine fan worm, *Myxicola,* and the squid giant axon and brain. Neurofilaments of the former are composed of two polypeptides with molecular weights of 150,000 d and 160,000 d (GILBERT *et al.* 1975, LASEK *et al.* 1979), while the latter include peptides of 50,000 d and 200,000 d (LASEK *et al.* 1979, ROSLANSKY *et al.* 1980). In the squid, both subunits copolymerize *in vitro* to form 10 nm filaments (ZACKROFF and GOLDMAN 1980) that seem to have a helical substructure (KRISHNAN *et al.* 1979).

III.2.5. Glial Filaments

Many glial cells have a massive system of filaments approximately 8 nm in diameter that course the cytoplasm in gently curving, packed bundles

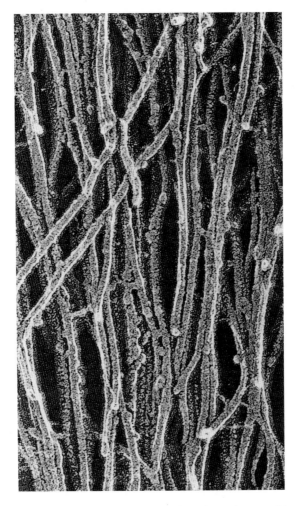

Fig. 32. Glial filaments prepared by quick-freezing, deep-etching, and rotary shadowing. In contrast to neurofilaments, glial filament bundles are devoid of crossbridges. × 257,000. (Courtesy of N. HIROKAWA)

(Fig. 32). The subunit of these filaments was identified as an acidic protein of approximately 50,000 d molecular weight (ENG *et al.* 1971, DAHL 1976, GOLDMAN *et al.* 1978), termed glial fibrillary acidic protein (GFAP). This protein shows species-specific heterogeneity, and even differs among different glial cells of the same species. For example, human brain GFAP has a subunit molecular weight of 49,000 d (GOLDMAN *et al.* 1978), while bovine GFAP appears to be substantially larger, about 54,000 d (RUEGER *et al.* 1979). Using a novel isolation procedure to purify GFAP from mouse optic nerve astrocytes, TSUKITA *et al.* (1981) identified two polypeptides of

45,000 d and 55,000 d in an approximate molar ratio of 1 : 1. By immunofluorescence microscopy, antibodies against GFAP react with glial cells *in situ* and *in vitro* (*e.g.,* DAHL and BIGNAMI 1973, ANTANITUS *et al.* 1975, ANDERTON *et al.* 1980), but not with neuronal processes, thus establishing GFAP filaments as distinctly glial-cell-specific intermediate filaments.

III.2.6. Cells Without Intermediate Filaments?

Among vertebrates, a few cell types appear not to have any of the 5 intermediate filament types just discussed. For example, intermediate filaments are generally not found in mammalian germ cells. During early development of the mouse, the first intermediate filament proteins appear at the morula stage, *i.e.,* when tissue differentiation first becomes apparent (*e.g.,* JACKSON *et al.* 1980, BRULET *et al.* 1980); these filaments belong to the cytokeratin class. In contrast, amphibian oocytes and eggs have, in addition to 6 nm actin filaments, 10 nm filaments and express cytokeratin polypeptides (FRANZ *et al.* 1983, GALL *et al.* 1983) as well as vimentin (GODSAVE *et al.* 1984). Their presence at such early stages has been related to the much faster rate of development of amphibian embryos, which may require precleavage storage and continued synthesis of these proteins in preparation for the onset of epithelial and mesenchymal cell differentiation. An absence of morphologically identifiable intermediate filaments has also been claimed for late stages of spermatogenesis in vertebrates, even though an immunoreactive form of vimentin stains a band-like region in the human spermatozoon (VIRTANEN *et al.* 1984). In certain insect spermatozoa, a keratin-like protein has been claimed to be present in accessory filamentous structures (BACCETTI *et al.* 1984). In general, however, reports on the presence or absence of intermediate filament-like proteins in invertebrates are too few to allow firm conclusions. Assessment of their distribution in lower eukaryotes and determination of their relatedness to vertebrate intermediate filament proteins is a challenging task for the future and may help solve the mystery of the evolutionary origin and relatedness of intermediate filament proteins.

III.2.7. Intermediate Filaments as Markers for Tumor Characterization

The classification of tumor tissues on the basis fo the type of intermediate filaments they contain is an important biomedical application that emerged from the discovery of the cell type-specific expression of intermediate filament proteins. In tumor therapy, the choice of the optimum therapeutic scheme often depends critically on the correct identification of the histological origin of the tumor. While a variety of diagnostic tools are available and currently in use, some of them are time-consuming and/or provide insufficient information on the histiotype of the tumor. Practical experience with several hundred human tumors has confirmed that immunocytochemical tests (in most cases immunofluorescence microscopy) with antibodies against the major intermediate filament classes are a convenient, time-saving, powerful and reliable new tool in tumor diagnosis. Intermediate filament antibodies can distinguish the major tumor groups, and, in some cases, such as cytokeratin-containing carcinomas, may allow even further subclassification based on the expression of different subsets of cytokeratin polypeptides. For overviews on the use of intermediate filament typing in pathology, see RAMAEKERS et al. (1982) and OSBORN and WEBER (1983).

III.3. Structural, Physical, and Chemical Similarities Between Intermediate Filament Proteins

As suggested in the above discussion, the intermediate filament proteins appear to be less conserved through evolution than the microtubule protein tubulin, or the microfilament protein actin. Nevertheless, different intermediate filament proteins have in common their rather uniform diameter and unusual stability, and share principles of structural organization that probably reflect biochemical homologies. Indeed, over the past 5 years the assumption that intermediate filament proteins comprise a family of homologous polypeptides has been substantiated by a large body of biophysical, proteinbiochemical, and immunological evidence.

All intermediate filament subclasses are far more stable under physiological conditions than the other two major fibrous structures, microtubules and actin filaments. Virtually all intermediate filament types require a denaturant (6–8 M urea, guanidine-HCl, or an ionic detergent) for disassembly. Most of them also disassemble in solutions of low ionic strength (see summary by STEINERT et al. 1982). Upon return to more physiological conditions, intermediate filaments can re-assemble into bona fide 10 nm filaments in vitro. None of the intermediate filament types require specific accessory proteins, cofactors, specific ions, or high-energy phosphates in order to assemble into filaments, features that again distinguish them from the other two major cytoskeletal elements.

Intermediate filaments isolated from a variety of different cell types have been shown to be composed of bundles of protofilaments approximately 3 nm in diameter that are twisted around each other in a helical fashion (STEINERT 1978, STEINERT et al. 1980, FRANKE et al. 1982, STEVEN et al. 1982, AEBI et al. 1983). In addition, vertebrate desmin, cytokeratin, and neurofilaments (HENDERSON et al. 1982, MILAM and ERICKSON 1982, FRANKE et al. 1982) and invertebrate neurofilaments (WAIS-STEIDER et al. 1983) reveal an axially repeated substructure with a periodicity of between 19 and 25 nm and show extended alpha-helical coiled-coil domains by X-ray diffraction (STEINERT 1978, STEINERT et al. 1980). Thus, intermediate filaments from as widely separated sources as annelids and man share structural features which suggest that the primary sequences of the subunits are similar. This suggestion is supported by the existence of monoclonal antibodies that cross-react not only with all homopolymeric intermediate filaments (i.e., desmin, vimentin, the 68,000 d neurofilament protein, and GFA), but also certain members of the cytokeratin family (PRUSS et al. 1981, DELLAGI et al. 1982). Striking similarities of intermediate filament structural domains are now confirmed by extensive amino acid sequencing studies. The information obtained by either direct sequencing or deduction of the amino acid sequence from that of cDNAs points to varying degree of relatedness among the various intermediate filament types (for overviews, see OSBORN and WEBER 1983, FUCHS and HANUKOGLU 1983). The extent to which various intermediate filament proteins have been sequenced varies. For example, the entire molecular structure of desmin, vimentin, the 68,000 d neurofilament protein, and GFAP is known (GEISLER and WEBER 1981, 1982 a, b, GEISLER et al. 1983, 1985, LEWIS et al. 1984); other intermediate filament polypeptides are currently sequenced and more complete information can be expected in the near future. These studies reveal that porcine and chicken desmin show at least 90% sequence homology, whereas porcine desmin and vimentin have "only" about 70% identity (GEISLER and WEBER 1981, QUAX et al. 1984). The small neurofilament subunit and GFAP also have substantial, though lower, homologies with desmin. Thus these four classes of intermediate filament proteins are closely related. However, as expected, tissue specificity is clearly reflected in the sequence data and far overrides species divergence. Because of its complexity, the cytokeratin family of proteins was expected to be less closely related to the other intermediate filament proteins. About 30% sequence homology was detected between a 50,000 d cytokeratin and the four other intermediate filament proteins (HANUKOGLU and FUCHS 1982, STEINERT et al. 1983). The homology between two of the cytokeratins is similarly low (HANUKOGLU and FUCHS 1983), indicating a certain degree of divergence even within this subclass.

Of particular interest is the finding that the sequence homolgy corresponds to a common substructure of all intermediate filament proteins. Each protomer consists of a central rod-like portion rich in alpha-helical domains, flanked by amino-terminal and carboxyterminal nonhelical regions termed headpiece and tailpiece, respectively, which are of unequal size. In desmin, for example, the 8,000 d headpiece and the 5,500 d tailpiece flank a 39,000 d middle domain (GEISLER et al. 1982). As expected, the greatest degree of homology between the different intermediate filament proteins exists in the alpha-helical middle portion, even though many amino acid substitutions, all compatible with the alpha-helical disposition, have occurred. Both the head and the tailpiece are involved in linking the subunits to form filaments and interprotofibril associations. Limited proteolysis, which digests primarily the nonhelical regions, induces filament disassembly (GEISLER *et al.* 1982a, STEINERT *et al.* 1983, SAUK *et al.* 1984).

A final point which should be considered here briefly concerns the evolutionary origin of the intermediate filament polypeptide family. So far, there are few reports of intermediate filament-related polypeptides in invertebrates. Examples include two proteins in *Drosophila* of 40,000 d and 46,000 d, that have certain antigenic determinants in common with vertebrate desmin and vimentin (FALKNER *et al.* 1981), and that form intermediate filament-like structures (WALTER and BIESSMANN 1984); the neurofilament proteins of the fanworm and the squid; and a 49,000 d protein of the ciliate *Tetrahymena* that resembles intermediate filament proteins of mammalian cells by amino acid composition and solubility properties (NUMATA and WATANABE 1982). FUCHS and MARCHUK (1983) found sequences in *Drosophila* and yeast distantly related to keratin polypeptides, but whether these organisms contain keratin filaments needs to be established by more rigorous criteria. FUCHS and HANUKOGLU (1983) speculate that intermediate filament proteins might have originated from a single gene. A common origin is supported by the finding of a remarkable degree of conservation in the structure of the genes, including the position of the introns, for two distantly related intermediate filament proteins, a 50,000 d human cytokeratin, and hamster vimentin (QUAX *et al.* 1983, MARCHUK *et al.* 1984). The hypothetical ancestral gene might have duplicated twice to lead to the formation of three genes, namely, the two major types of cytokeratin proteins, and the precursor of the other four (homopolymeric) intermediate filament classes. More recent gene duplications may have given rise to the multiplicity of related sequences within these three "primordial" classes. Whether this hypothetical scheme can be considered valid has to await further sequence analyses.

III.4. Co-Existence and Co-Assembly of Different Intermediate Filament Subunits

As outlined in the preceeding sections, each of the 5 intermediate filament types is characteristic of a set of tissues or cell types and can serve as reliable markers for tissue typing. However, there are well-documented cases where cells express two different intermediate filament subunits simultaneously. In all instances where co-expression has been observed, one of the partners invariably is vimentin, and the other the intermediate filament subunit characteristic for the cell's origin. Coexpression is common in cultured cells.

Well-known examples include PtK cells, an epithelial rat kangaroo line, and HeLa cells, which are derived from a human cervical carcinoma. Both these cell lines express cytokeratin and vimentin. The two types of intermediate filament proteins form separate networks, as demonstrated by double immunofluorescence and immunoelectron microscopy (*e.g.,* FRANKE *et al.* 1979 a, b, OSBORN *et al.* 1980, HENDERSON and WEBER 1981). Cytokeratin forms a wavy network of tight bundles, while vimentin is organized into wispy filaments with a more or less radial organization (Figs. 26 and 29). Only the vimentin filaments collapse towards the perinuclear area upon colcemid treatment (OSBORN *et al.* 1980), while the system of wavy cytokeratin bundles remains extended. Thus the two filament systems appear to be largely independent of one another. *In situ*, coexpression of vimentin and cytokeratin seems to be a rare event (*e.g.,* RAMAEKERS *et al.* 1983, LANE *et al.* 1983). On the other hand, there are several documented cases for the coexistence of vimentin with one of the other three major intermediate filament proteins. For example, certain glial cells (SCHNITZER *et al.* 1981, YEN and FIELDS 1981), some developing neurons (TAPSCOTT *et al.* 1981), and vascular smooth muscle cells (FRANK and WARREN 1981, SCHMID *et al.* 1982, TRAVO *et al.* 1982) are characterized by the simultaneous presence of vimentin with either GFAP, or neurofilaments, or desmin. Reports that vimentin and desmin coexist at the Z-disc of adult striated muscle (GRANGER and LAZARIDES 1979, GARD and LAZARIDES 1979), are not confirmed by other studies (BENNETT *et al.* 1979, HOLTZER *et al.* 1982). While there is ample evidence for the coexistence of vimentin and desmin in developing skeletal muscle (BENNETT *et al.* 1979, GRANGER and LAZARIDES 1979, TOKUYASU *et al.* 1984), vimentin seems to disappear gradually as the muscle matures. The confusion about this point may be resolved in the near future since TOKUYASU *et al.* (1984) detected in mature muscle cells a hitherto unidentified protein similar to vimentin in molecular weight, but distinct in antigenic properties.

In some instances of coexpression it is not clear whether the two intermediate filament subunits exist in different intermediate filament systems, as in the afore-mentioned cultured cells, or whether they are incorporated into the same filament. *A priori*, copolymerization is a reasonable possibility given the substantial structural and sequence homologies of intermediate filament proteins. Indeed, the existence of heteropolymers of vimentin and desmin has been demonstrated by chemical cross-linking for vascular smooth muscle tissue (QUINLAN and FRANKE 1982). Likewise, copolymers of vimentin and GFAP exist in certain glioma cell lines (QUINLAN and FRANKE 1983, SHARP *et al.* 1982), while vimentin and desmin copolymers are believed to be present in the established cell line, baby hamster kidney (BHK) cells (TUSZYNSKI *et al.* 1979, GARD *et al.* 1979), and possibly also other fibroblastic cell lines (IP *et al.* 1983). In agreement with these findings, several studies demonstrated the potential for copolymerization between vimentin and either desmin, GFAP, or one of the cytokeratin polypeptides *in vitro* (STEINERT *et al.* 1981, 1982). Thus the structural homologies between many, but possibly not all, of the intermediate filament proteins are sufficient to allow polymerization into structures that have the characteristics expected for 10 nm filaments. The physiological significance of copolymerization of different intermediate filament subunits remains to be established.

III.5. Intermediate Filament-Associated Proteins

Both the actin filament and microtubule systems are characterized by proteins that profoundly affect their assembly properties, their distribution in the cell, and, presumably, their function(s). In fact, the various roles of actin-based cellular structures is largely determined not by actin itself, but by its numerous associated proteins. *A priori*, one might therefore expect intermediate filaments to have similar accessory proteins, and indeed such proteins have been found. It should be emphasized, however, that many are insufficiently characterized and their importance for intermediate filament function is poorly understood. This is a young research topic in which many discoveries can be expected in the near future.

Among the best-characterized intermediate filament-associated proteins are synemin and paranemin, two high molecular weight proteins known to bind to desmin and vimentin filaments. Synemin and paranemin were originally discovered as proteins that copurify and associate with desmin isolated from chicken smooth or skeletal muscle (GRANGER and LAZARIDES 1980, BRECKLER and LAZARIDES 1982), but they both also bind to vimentin filaments. Synemin has an apparent molecular weight of 230,000 d and is present at a ration of 1:50–100 relative to desmin, while peranemin (280,000 d) is 5–10 times more abundant than synemin in embryonic muscle. The soluble form of synemin is a tetramer that can be highly phosporylated and can bind to desmin filaments *in vitro* (SANDOVAL *et al.* 1983). The precise functions of synemin and paranemin are unknown. The former is localized at regular intervals with a periodicity of between 180 and 230 nm along native vimentin filaments of chicken erythrocytes (GRANGER and LAZARIDES 1982) (Fig. 33). Synemin-containing crossbridges between individual vimentin filaments and the lateral registration of synemin in filament bundles, suggest that synemin mediates filament-filament interactions, but such a function has not yet been confirmed experimentally. The expression of synemin and paranemin changes during chicken development (PRICE *et al.* 1983), suggesting that both serve specific functions, but what these functions are is presently undetermined.

Another intermediate filament-associated protein that is now well characterized has been isolated from mammalian epidermis (DALE 1977, DALE *et al.* 1978) and has been given the name filaggrin (STEINERT *et al.* 1981 b). This protein interacts with keratin filaments to induce the formation of highly organized, tight bundles, or macrofibrils. Filaggrin actually comprises a family of basic, histidine-rich proteins whose molecular weights and amino acid composition vary from species to species

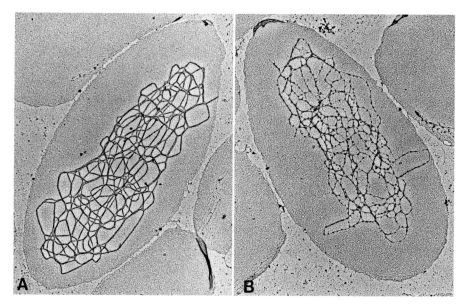

Fig. 33. Patches of the plasma membrane of chicken erythrocytes adhering to a glass substratum, prepared by low-angle rotary shadowing. *A* Staining with a vimentin antibody of a reticular filament system. *B* Staining with a synemim antibody, which decorates the filaments at regular intervals. × 6,000. [From GRANGER and LAZARIDES (1982)]

(mouse: 26,000 d, rat: 49,000 d, human: 35,000 d). Additional heterogeneity is introduced by the existence of several isoelectric variants of each of the members of the protein family (HARDING and SCOTT 1983, LYNLEY and DALE 1983). Filaggrin is synthesized as a high molecular weight precursor (approximately 250,000 d) that may be highly phosphorylated (LONSDALE-ECCLES *et al.* 1983, MEEK *et al.* 1983). The precursor does not interact with keratin filaments. Even though filaggrin is an intermediate filament-associated protein specific for epithelial cells, it interacts *in vitro* with nonepithelial intermediate filaments to form large aggregates. Filaggrin may therefore bind to a protein domain common to all intermediate filament classes, most probably the alpha-helical middle portion (STEINERT *et al.* 1981 b). Presumably, this protein is responsible for the formation of the tight, wavy filament bundles characteristic of cytokeratin-containing cells.

With respect to intermediate filament-associated proteins, neurons are a special case. Although mammalian neurofilaments are composed of three polypeptides of 68, 150, and 200 kd, only the 68 kd protein can polymerize into *bona fide* 10 nm filaments by itself; the other two, alone or in combination, can not (GEISLER and WEBER 1981, LIEM and HUTCHISON 1982). They will, however, co-polymerize with the 68 kd protein; in fact, large amounts of the two high molecular weight proteins can be incorporated into neurofilaments. The 150 and 200 kd proteins may be viewed as associated proteins, but they also have structural features in common with the 68 kd protein. They may be bifunctional molecules involved in both the construction of the filament backbone *and* the formation of surface features such as projections (MINAMI *et al.* 1984, GEISLER *et al.* 1984, 1985). Indeed, wispy filamentous structures interconnecting neurofilaments into a three-dimensional network are well-known to electron microscopists. These side-arms could well represent a class of associated proteins, analogous to the high molecular weight MAPs that decorate microtubules with similar projections. That the bridges interconnecting neurofilaments may indeed be constructed, at least in part, from the 200 kd subunit is suggested by immuno-electron microscopic studies which show the projections, but not the neurofilament proper, heavily labeled with a 200 kd-specific antibody (HIROKAWA *et al.* 1984) (Figs. 31 B and C). Clearly, however, more information on the relationship between the three neurofilament subunits is needed before their specific roles can be assessed.

Another peculiartiy of neuronal tissue concerns the properties of high molecular weight MAPs, particularly MAP 1. MAPs were first described as proteins that copurify with microtubules and stimulate microtubule polymerization, but they also bind to, and cross-link, neurofilaments *in vitro* (RUNGE *et al.* 1981, LETERRIER *et al.* 1982, BLOOM and VALLEE 1983, LETERRIER *et al.* 1982). Thus, they may have two binding sites, one for microtubules and one for neurofilaments, and may serve to cross-link these two prominent neuronal cytoplasmic components. Such a function would help explain fine-structural observations of highly ordered, seemingly specific associations between the two components in some neurons (BERTOLINI *et al.* 1970). However, it remains to be unequivocally demonstrated that the interaction between the two is biologically significant, and not just a fortuitous or nonspecific epiphenomenon due to, *e.g.*, charge effects.

A number of other putative intermediate filament-associated proteins have been tentatively identified on the basis of the following three lines of evidence: 1. they are immunologically different from the proteins of the 5 intermediate filament classes; 2. they colocalize with intermediate filament-

containing structures in immunofluorescence and immunoelectron microscopy; 3. desmin- and vimentin-associated proteins redistribute coordinately
with the intermediate filaments upon colcemid treatment. Some of these
potentially very interesting proteins shall briefly be mentioned here, even
though they have not yet been characterized satisfactorily. PYTELA and
WICHE (1980) describe a set of high molecular weight components
(approximately 300,000 d) that can be isolated from cultured glioma cells.
This group of proteins, collectively termed "plectin", is distinct from high
molecular weight MAPs. They are present in intermediate filament-
enriched cytoskeletons of cultured cells and a wide variety of tissues (WICHE
et al. 1983), and are proposed to function as cross-linkers of cytoplasmic
filaments. A protein of similar molecular weight (300,000 d), but different
from plectin, was identified in baby Hamster kidney (BHK) cells by a
monoclonal antibody. The protein colocalizes with the desmin and vimentin
filaments of this cell type (YANG et al. 1985). This intermediate filament-
associated protein is also present in in vitro preparations of BHK-
intermediate filaments, where it cofractionates with the filaments during
multiple cycles of assembly/disassembly. Two other proteins of 90,000 d
and 210,000 d, respectively, are detected by monoclonal antibodies that
stain cultured cells with a characteristic intermediate filament pattern (LIN
1981). These antibodies do not cross-react with any of the major known
intermediate filament proteins. Microinjection of the anti-90 kd antibody
into living cultured fibroblasts causes the intermediate filaments to
aggregate into perinuclear bundles, while the anti-210 kd antibody does not
have this effect (LIN and FERAMISCO 1981). Two other putative associated
proteins recognized by monoclonal antibodies include a 44,500 d protein
termed epinemin present in a variety of cells (LAWSON 1983, 1984), and a
50,000 d protein which appears to be associated with desmin and vimentin
filaments in a discontinuous, dotted fashion (WANG et al. 1983). Neither of
these two components seems to be present in large amounts, and their
functions are unknown.

In summary, several potentially important intermediate filament-
associated proteins have already been identified and characterized in some
detail, while several others are just beginning to be studied. Because of the
eminent importance of proteins associated with actin and microtubules for
the function of these two cytoskeletal components, the identification of
similar proteins in intermediate filament preparations is of more than
ephemeral interest. Monoclonal antibodies have become increasingly useful
and efficient tools for the recognition of such components. However,
immunological criteria alone are insufficient to characterize a protein as
"intermediate filament-associated", and several other kinds of experimental evidence are required for determining the properties and functions
of putative associated proteins.

III.6. Phosphorylation of Intermediate Filament Proteins

An unsolved and challenging question concerning intermediate filaments is how their organization and function is regulated. LAZARIDES (1980) suggested that the disposition and function of intermediate filaments might be regulated by phosphorylation, one of the most powerful means of regulation used by cells. Indeed, soon after knowledge on the various subclasses of intermediate filament proteins began to accumulate, it was discovered that virtually all of them are phosphorylated *in vivo* and can be phorylated *in vitro*. Desmin and vimentin each appear to have 5–7 different phosphorylation sites, as shown by[32] P-labeled tryptic peptide maps (GARD and LAZARIDES 1982 a, EVAN 1984). Similarly, the 68 kd subunit of neurofilaments is phosphorylated on only a few sites, while the 150 kd and the 200 kd subunits are both highly phosphorylatable (JULIEN and MUSHYNSKI 1982, JONES and WILLIAMS 1982). Some of the keratin polypeptides are also multiply phosphorylated (SUN and GREEN 1978 b, STEINERT et al. 1982, GILMARTIN et al. 1984). The kinases that mediate the phosphorylation of most of the sites in desmin and vimentin have been demonstrated by both *in vivo* and *in vitro* studies to be cyclic AMP-dependent and intermediate-filament-specific (GARD and LAZARIDES 1982 b, O'CONNER et al. 1981, SPRUILL et al. 1983 a). Some cell types also express cyclic AMP-indenpendent phosphorylation pathways (GARD and LAZARIDES 1982 a, HUANG et al. 1984), possibly via calcium/calmodulin (SPRUILL et al. 1983 b). Both modes of intermediate filament protein phosphorylation may operate within the same cell. The relationship and differential function of the cyclic AMP-dependent and the calcium/calmodulin-dependent pathways of intermediate filament phosphorylation need to be elucidated in future studies.

Certain studies have allowed the establishment of a correlation between external stimuli and biochemical changes in intermediate filament proteins. Thus, one of the changes in cultured rat Sertoli cells in response to follicle stimulating-hormone is an increase in the phosphorylation of vimentin (SPRUILL et al. 1983 a, b). In rabbit neutrophils, treatment with chemotactic factors for only 5–10 seconds results in enhanced incorporation of phosphate into vimentin (HUANG et al. 1984). Other studies suggest that phosphorylation is developmentally regulated and may modulate intermediate filament function during cell and tissue specialization. For example, in myogenic cells, desmin and vimentin are more highly phosphorylated in mature myotubes than in immature myoblasts (GARD and LAZARIDES 1982 a). Altered levels of phosphorylation have also been linked to changes in intermediate filament distribution. A particularly intriguing observation is that monoclonal antibodies that distinguish phosphorylated an nonphosphorylated forms of neurofilaments reveal different staining patterns of neurons *in situ*. Cell bodies, dendrites, and the proximal portions of axons contain non-phosphorylated neurofilaments, whereas the neuro-filament triplet proteins in long fibers with compact assemblies of neurofila-ments are more highly phosphorylated (STERNBERGER and STERNBERGER 1983). This finding suggests that the level of phosphorylation may determine the degree of neurofilament packing which, in turn, may regulate their function. In another study, a monoclonal antibody was shown to stain vimentin filaments in a more motile subpopulation of cultured fibroblastic cells (DULBECCO *et al.* 1983). While it is not known in this case whether the antigenic site recognized by this antibody is a phosphorylation site, this remains an distinct possibility. Last but not least, increased phosphate incorporation into several cytokeratin polypeptides (BRAVO *et al.* 1982, CELIS *et al.* 1983) and vimentin (CELIS *et al.* 1983, EVANS 1984) was found in mitotic cells. These changes in phophorylation coincide temporally with changes in intermediate filament organization, such as the formation of a "cage" of intermediate filaments surrounding the spindle (*e.g.,* BLOSE *et al.* 1979, AUBIN *et al.* 1980, ZIEVE *et al.* 1980), or the transient disintegration of cytokeratin filaments in some cell types (HORVITZ *et al.* 1981, FRANKE *et al.* 1982, LANE *et al.* 1982). Collectively, these studies suggest a link between the increased incorporation of phosphate into intermediate filament proteins and several important cellular activities. The biological meaning of this correlation needs to be established in future studies. One way by which phosphorylation-dependent functional changes of intermediate filament subunits could be brought about is by an alteration of their ability to polymerize and/or to associate with each other and with other cellular structures. For example, the phosphorylation of the 150 kd neurofilament polypeptide increases its association with intact neurofilaments, whereas removal of phosphate groups induces dissociation (WONG *et al.* 1984). Thus, changes in intermediate filament protein phosphorylation might affect intermediate filament assembly and therefore might explain many of the observed structural changes.

III.7. Regulation of Intermediate Filament Assembly in the Cell

A better understanding of intermediate filament protein phosphoryla-tion may prove instrumental in understanding another elusive aspect of these filaments, namely, how their assembly and disassembly is regulated in the living cell. Phosphorylation may influence the state of assembly of some of the intermediate filament classes in some cell types, but other mechanisms are also feasible. The question of how intermediate filaments are re-distributed is particularly challenging in view of their well-documented stability. Several lines of evidence allow the conclusion that the mode of regulation of their distribution is unlike that of tubulin or actin. First, intermediate filament proteins require neither cofactors (*e.g.,* metal ions) nor high energy phosphates (*e.g.,* ATP, GTP) in order to assemble into filaments. Second, the existence of an intermediate filament-distribution center (ECKERT *et al.* 1984) with a function analogous that of a microtubule-organizing center is not generally established. Third, once assembled, intermediate filaments are insoluble in physiological buffers and solutions containing nonionic detergents, and can only be solubilized under conditions that would disassemble or denature tubulin or actin. Fourth, in contrast to tubulin and actin, cells do not seem to maintain a large pool of monomeric or nonfilamentous intermediate filament proteins; virtually all of these proteins remain with the insoluble residue (the "cytoskeleton") upon detergent extraction. This set of properties suggest that cells deal with intermediate filaments very differently than the other two major cytoplasmic filaments.

Soon after intermediate filament proteins are synthesized, they are incorporated into the cytoskeletal (polymer) fraction. Whether this transfer occurs co-translationally as the polypeptide chains are synthesized on cytoskeleton-associated polysomes (FULTON *et al.* 1980), or post-translationally from a transiently "soluble" precursor pool (BLIKSTAD and LAZARIDES 1983, MOON and LAZARIDES 1983), can not be decided with certainty for most intermediate filament proteins. It seems clear, however, that, if a precursor pool exists, intermediate filament subunits do not spend much time in it but are rapidly assembled. Once polymerized, reorganization and/or depolymerization may be accomplished by one of two possible mechanisms: a change in the degree of phosporylation (see previous section), or degradation by calcium-activated, intermediate filament-specific proteases. The most extensively investigated of these proteases are those specific for neurofilaments and vimentin/desmin. Both vertebrate (SCHLAEPFER and HASLER 1979, SCHLAEPFER and MICKO 1979, ISHIZAKI *et al.* 1983) and invertebrate neurofilaments (PANT *et al.* 1979) are susceptible to calcium-dependent proteolysis, but little is known about the subcellular distribution, cytoskeletal association, or pattern of activation of the enzymes that degrade neurofilaments. It has been speculated, however, that they are involved in the selective degradation at the tip of neurites of neurofilaments following translocation down the axon in the slow component of axoplasmic transport (*cf.,* LASEK and BLACK 1977).

Vimentin and desmin-specific proteases appear to be widely distributed in many different cell types and species ranging from fish to man (NELSON and TRAUB 1981, 1982, TRAUB and NELSON 1981). Just like the intermediate filament proteins they degrade, these enzymes appear to be evolutionarily conserved (NELSON and TRAUB 1982). Again, neither the subcellular distribution of these enzymes nor their specific function in cellular activities believed to involve intermediate filaments are known at this point.

Elucidation of these questions will constitute an important step towards understanding several fundamental properties of intermediate filaments.

III.8. Aspects of the Cellular Distribution and Function of Intermediate Filaments

It may seem paradoxical that despite rather detailed knowledge of some aspects of intermediate filament biochemistry and structure, our understanding of their function is essentially uncharted territory. While this may also be true for certain areas of the biology of microtubules and actin filaments, nowhere is this discrepancy as evident as in the field of intermediate filament research. However, even though the specific function(s) of these filaments are still unknown, intermediate filaments are believed to take part in several important cellular activities. They include transcytoplasmic and transcellular integration, interaction with other cell organelles, determination of cell shape, and involvement in cell movement and intracellular transport.

Transcytoplasmic and transcellular integration. Filaments 7–11 nm in diameter were proposed to play primarily a structural role by providing support for highly asymmetric cell extensions, as in neurons and many glial cells, or by helping maintain a flattened morphology, as in many epithelial, mesenchymal, or myogenic cells. Both their overall distribution (as visualized by specific antibodies) and their unusual stability are compatible with such a skeletal function, but direct evidence supporting this proposal is surprisingly scarce. Probably one of the best examples supporting a structural role for intermediate filaments can be found in the studies on desmin and vimentin distribution in the course of myogenesis. Early electron microscopic studies suggested that muscle cells possess a framework of structural components that maintains sarcomeric units (Z-discs, M-lines) in lateral register, and that also links these elements to the plasma membrane (see ISHIKAWA 1983 for a summary). The striking re-organization of desmin from an even distribution in myoblasts to a striated pattern coinciding with the Z-discs in myotubes, and the observation that the three-dimensional structure is maintained in potassium iodide-extracted, myosin and actin-free muscle fibers suggest that the framework is provided by intermediate filaments located specifically at the Z-discs. These and additional observations on the disposition of vimentin filaments in erythrocytes led LAZARIDES (1980) to propose that intermediate filaments form part of a transcytoplasmic matrix and serve as "mechanical integrators" of cellular space. While this descriptive term incorporates many aspects of intermediate filament distribution, it is also much too general since the same can be (and has been) said of the actomyosin system or the microtubule network. Some approaches aimed at elucidating the specific function(s) of intermediate filaments have involved attempts to disrupt intermediate filaments in living cells. This feat turned out to be more challenging than in the case of microtubules and actin filaments because of the extraordinary stability of intermediate filaments and the lack of specific, reversible inhibitors. While rapidly reversible disruption of intermediate

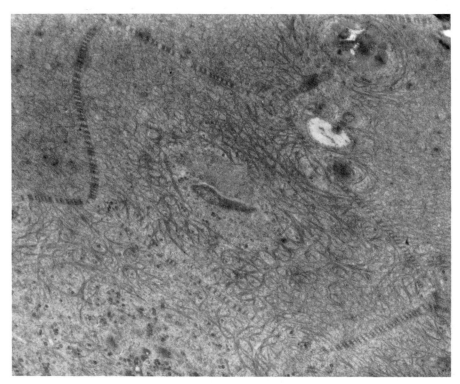

Fig. 34. Horizontal section of the epidermis of the guppy. The cell's pentagonal outline is marked by chains of desmosomes that link the tonofilament bundles of neighboring cells into a transcellular network. × 6,100

filaments in living cells has not yet been achieved, two experimental approaches were found that induce major rearrangements of cellular intermediate filaments: 1. treatment with microtubule inhibitors which causes vimentin or desmin filaments to collapse towards the cell center, forming a perinuclear "cap" (*e.g.*, GOLDMAN 1971, HYNES and DESTREE 1978, GORDON *et al.* 1978); 2. microinjection of intermediate filament-specific antibodies, which either induces filament collapse (KLYMKOWSKY 1981, LIN and FERAMISCO 1981, ECKERT *et al.* 1982, GAWLITTA *et al.* 1981), or substantial filament disruption (KLYMKOWSKY *et al.* 1983). Surprisingly, neither collapse nor disruption of intermediate filaments seems to affect cell shape. This result challenges the notion that an intact intermediate filament network is essential for cell shape generation and maintenance. One possible criticism of these microinjection experiments is that they were performed on cells in culture, where strong contacts to the substratum, or junctions with neighboring cells, might provide sufficient support to overcome the potentially shape-deforming disruption of intermediate filaments. Indeed, many epithelial sheets *in situ* (*e.g.*, epidermis) and certain epithelial models *in vitro* (*e.g.*, MDCK cells), are characterized by a fibrillar superstructure, a pan-cellular network of filament bundles that crosses intercellular borders at desmosomal junctions (Fig. 34). The contacts at these junctions might help support cell shape even after disruption of a major cytoskeletal component. These features of a transcellular network were noted in early investigations of the epidermis (*e.g.*, KROMAYER 1892), and were more recently substantiated by electron microscopic studies (*e.g.*, SCHLIWA 1975, HULL and STAEHELIN 1979, JONES *et al.* 1982, FEY *et al.* 1984). Cell contacts might help maintain the spatial disposition of individual cells, and might lend mechanical strength to the entire cell sheet or tissue.

Interaction with other organelles. Intermediate filaments have repeatedly been proposed to interact specifically with other cell organelles, including cytoskeletal components. Because many of the interactions will be discussed in more detail below (sections VIII.2., VIII.3., XI.6., XI.7.), they shall only briefly be mentioned here. An association with microtubules was suggested early on by the observation of intermediate filament collapse towards the nucleus upon treatment with antimicrotubular agents (*e.g.*, GORDON *et al.* 1978, WANG and GOLDMAN 1978) or other treatments that disrupt microtubules (*et al.* SHARPE *et al.* 1980, MARO and BORNENS 1982). Further support came from observations of a close structural relationship between these two cytoskeletal elements (GOLDMAN and FOLLETT 1969, GEIGER and SINGER 1980, BALL and SINGER 1981). Other studies demonstrate that the distribution of mitochondria is under the influence of the microtubule-intermediate filament network (*e.g.*, HEGGENESS *et al.* 1978). At least in some cases, a subpopulation of intermediate filaments rather than the microtubules may be responsible for this association (MOSE-LARSEN *et al.* 1982, SUMMERHAYES *et al.* 1983). Intermediate filaments also seem to closely appose to the nucleus (FRANKE 1971, WOODCOCK 1980, JONES *et al.* 1982). This association is strong enough to survive the isolation of a nuclear fraction from cells (STAUFENBIEL and DEPPERT 1982), supporting the suggestion that intermediate filaments anchor the nucleus and suspend it in a central position within the cell (LEHTO *et al.* 1978, FEY *et al.* 1984). However, the proposed association of mitochondria and nuclei with intermediate filaments is based on structural criteria (close apposition). Little is known about the molecular basis of this association, nor is there general consensus about its specificity.

Determination of cell shape. A role for intermediate filaments in the development and maintenance of cell shape is suggested by the congruence of cell shape and intermediate filament pattern, but direct evidence for a causal relationship is lacking. On the contrary: as already mentioned above, experiments that perturb the distribution of intermediate filaments in cultured cells (through the microinjection of specific antibodies) do not seem to perturb normal cell shape and acitivity (GAWLITTA *et al.* 1981, LIN and FERAMISCO 1981, KLYMKOWSKY *et al.* 1983). Injected cells retain their characteristic shape and are able to ruffle and translocate across the substratum just as uninjected cells.

Involvement in intracellular movement. The same microinjection studies just mentioned also seem to rule out the suggestion, made earlier (*e.g.,* WANG and GOLDMAN 1978), that intermediate filaments are involved in the transport of intracellular organelles such as lysosomes, mitochondria, and other vesicles. These functions appear to continue unperturbed in cells with a collapsed intermediate filament system.

In summary, none of the studies undertaken so far provide unequivocal evidence for a role of intermediate filaments in determining cell shape, organelle distribution, or organelle movement. As for a role in organizing cellular space, the arguments are based as much on intuition as on experimental evidence. The redistribution of intermediate filaments in myogenesis certainly leaves the strong impression that these filaments have something to do with the development of sarcomeric organization, and it should be revealing to follow this process in an experimentally induced intermediate filament-free cell system. So far, the case for a role in cytoplasmic and transcellular organization is strongest for cytokeratin filaments in epithelia, though again it is based on indirect arguments. Thus, intermediate filaments still appear as a major cellular component in search of a clearly defined function.

IV. Dynamic Aspects of Filament Assembly

Actin and tubulin differ from intermediate filament proteins in their mechanisms for assembly and in the properties of the polymers they form. Built from asymmetric subunits, actin filaments and microtubules have an intrinsic structural and functional polarity. The two ends of actin and tubulin polymers differ in both the assembly rate and the critical concentration for assembly. Unlike intermediate filament polymerization, their assembly is coupled to nucleotide triphosphate hydrolysis. However, hydrolysis is not absolutely required as assembly can also proceed in the presence of ADP (COOKE 1975, KONDO and ISHIWATA 1976, POLLARD 1974, LAL *et al.* 1984), or in the presence of nonhydrolyzable analogues of ATP or GTP (COOKE and MURDOCH 1973, WEISENBERG *et al.* 1976, PENNINGROTH and KIRSCHNER 1977). Therefore, the energy of hydrolysis does not provide energy for assembly and is not used for useful work (except, perhaps, under certain conditions: HILL 1981). What, then, might be the role of nucleotide hydrolysis during assembly? Could it be used by the cell to organize the networks of actin filaments and microtubules?

In a kinetic model, nucleotide triphosphate hydrolysis by polymerizing subunits results in a process known as head-to-tail polymerization or treadmilling. According to this model, in the steady state phase of assembly microtubules and actin filaments are believed to add subunits preferentially at one end (termed the "preferred end", "net polymerization end", or "plus end"), while simultaneously losing subunits at the same rate from the opposite end (the "net depolymerization end" or "minus end"). This process was first evaluated for actin filaments by WEGNER (1976) and was later extended to microtubules (MARGOLIS and WILSON 1978). Treadmilling implies that under conditions of steady state, subunits will flow through the polymer while the length of the polymer is held constant; the polymer functions as a "nucleotide triphosphatase".

To understand these steady state events, let us first consider a polymer which does not require hydrolysis, say, a case in which a nonhydrolyzable analog is bound to the monomer. The two reactions that occur at the two ends of such a polymer under these conditions can be written as

$$\frac{dl^+}{dt} = k_2^+ C - k_1^+ \tag{2}$$

for the plus end and

$$\frac{dl^-}{dt} = k_2^- C - k_1^- \tag{3}$$

for the minus end (KIRSCHNER 1980), where $\frac{dl^+}{dt}$ and $\frac{dl^-}{dt}$ are the net rates of elongation at the plus and minus ends, respectively, k_2^+ and k_2^- are second-order rate constants for the association reactions, k_1^+ and k_1^- are first-order rate constants for the dissociation reactions, and C is the concentration of monomers. Because the association and dissociation reactions are simple equilibria in which the reactants are the same at both ends, there is only one equilibrium constant for the two ends, which can be written as

$$K = \frac{k_2^+}{k_1^+} = \frac{k_2^-}{k_1^-} \tag{4}$$

Thus, at equilibrium the rate of addition of monomers is equal to the rate of loss at each polymer end. Both ends will have the same critical concentration C_c (defined as $1/K$). Consequently, both ends will elongate when the concentration of free monomer is above the critical concentration, and both ends will lose subunits below this concentration. The relation between the rate of elongation and monomer concentration is diagrammatically il-

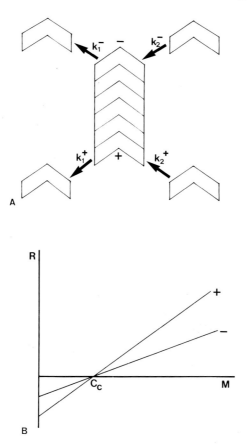

Fig. 35. *A* Diagram of bidirectional growth from a polar polymer under conditions that do not require hydrolysis. Subunits are represented by chevrons. Fast and slow-growing end are denoted by plus and minus, respectively. *B* Plot of the rate of elongation (*R*) vs. monomer concentration (*M*) for the polar polymer shown in *A*. The critical concentration (C_c) is the same for both the plus and the minus end, but the rates of elongation may be different

lustrated in Fig. 35. Interestingly, even though the ratios of the association and dissociation rate constants are the same for the two ends, the individual rates of elongation and shrinking at the two ends can still be different because of the inherent physical asymmetry of the polymer lattice: the links between the subunits at either end of the polymer, through structurally and chemically the same, are formed (or broken) via slightly different routes (KORN 1982).

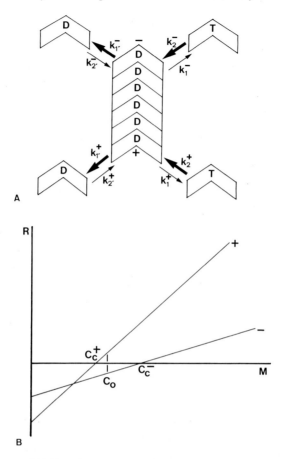

Fig. 36. *A* Diagram of bidirectional growth from a polar polymer under conditions where hydrolysis of a nucleotide triphosphate is coupled to subunit addition. *T* and *D* designate monomers with bound triphosphate or diphosphate, respectively. The kinetically important reactions are emphasized with fat arrows. *B* Plot of the rate of elongation (R) vs. monomer concentration (M) for a polymer that requires nucleotide triphosphate hydrolysis. Not only the rates of elongation, but also the critical concentrations are different for the plus and the minus end. There is an overall critical concentration (C_c) which corresponds to the steady state monomer concentration. At this point the overall rate of elongation is zero, but subunits are lost at the minus end at the same rate that subunits are added to the plus end, thus allowing for treadmilling

The introduction of hydrolysis complicates the pattern of monomer addition and loss in an interesting fashion (Fig. 36). If hydrolysis is coupled to the association step, certain reactions in this scheme can be assumed to proceed at a negligible rate (thin arrows in Fig. 36). The events at the two ends of the polymer are now asymmetric. In the kinetically significant reactions (fat arrows in Fig. 36), the rates of association and dissociation at the two ends are still given by equations identical in form to those for bidirectional growth in the absence of hydrolysis [eq. (2) and (3)]. However, k_1^+ is replaced by $k_{1'}^+$ and k_1^- is replaced by $k_{1'}^-$, and the forward and reverse rate constants at the two ends are no longer the rate constants of the same reactions. Because k_2^+/k_1^+ is not necessarity the same as $k_2^-/k_{1'}^-$, there is no overall equilibrium constant for the two ends, as in eq. (4), but rather each end may have a different equilibrium constant and, therefore, different critical concentrations (Fig. 36). Since the overall growth rate of a polymer is given by the sum of the rates at each end, there will be an overall critical concentration, C_o where the polymer does not grow or shrink, but where the association events at the plus end are counterbalanced by dissociation events at the minus end. The polymer is said to "treadmill" (for more complete discussions, see MARGOLIS and WILSON 1981, HILL and KIRSCHNER 1982, PURICH and KRISTOFFERSON 1984).

It needs to be emphasized that the phenomenon of treadmilling observed under steady state conditions is a representation of the sum of all association and dissociation events at the two ends of the polymer. Because there presumably is extensive "subunit traffic" at the two ends of the polymer, the question arises, how many subunits come on and go off before a subunit stays on the polymer, or in other words, how efficient is the flux of subunits through the polymer? WEGNER (1976) defined the efficiency of flux, s, as the ratio of the subunit flux through the polymer and the total number of association and dissociation events. An s-value of 1 would mean total efficiency, *i.e.,* every monomer that goes on the polymer also stays on, while an s-value of, say, 0.1 means that for 10 association/dissociation events, 1 subunit flows through the polymer. Another parameter of interest is the flux rate, or the number of subunits that flow through the polymer in a unit of time. Flux rate and flux efficiency are a function of the association and dissociation rate constants for monomers at the two ends. These rate constants can be determined directly in electron micrographs from the rates of pre-steady-state elongation from polymer seeds. Three such model systems have been used in the past: microtubules from flagellar axonemes (BERGEN and BORISY 1980), and actin filament bundles isolated from microvilli (POLLARD and MOOSEKER 1981) or *Limulus* sperm (BONDER *et al.* 1983). Biochemical procedures have also been used to determine efficiencies of subunit treadmilling (*e.g.,* COTE and BORISY 1981, CAPLOW *et al.* 1982). For microtubule protein containing MAPs, the s-value is reported to be 0.07. This means that for every 14 (= 1/0.07) subunit associations and dissociations, there is net gain of one subunit at the plus end and net loss of one subunit at the minus end (BERGEN and BORISY 1980). For MAP-free tubulin, the efficiency of flux seems to be significantly higher, namely 0.26 (COTE and BORISY 1980). Other studies suggest that the flux efficiency may be grossly overestimated (CAPLOW *et al.* 1982). Reported flux rates also vary, between 0.3/second for microtubule protein (MARGOLIS and WILSON 1978), 6/second for microtubule protein in the presence of ATP (MARGOLIS and WILSON 1979), and 20/second for MAP-free tubulin (COTE and BORISY 1981). For actin, the flux rate depends on the ionic environment: in KCl alone, the critical concentration is the same for both ends, so there is no flux. In KCl/magnesium, the flux rate is estimated at 0.7/second, with an s-value of 0.2 (POLLARD and MOOSEKER 1981). This value is in good agreement with s-values (0.25) reported in other studies for similar ionic conditions (WEGNER 1976, BRENNER and KORN 1983).

The possible implications of treadmilling for the regulation of actin and microtubule assembly in the cell have been evaluated (KIRSCHNER 1980, HILL and KIRSCHNER 1982), and treadmilling was proposed to be of fundamental importance for at least one cellular activity, the transport of chromosomes to the poles during mitosis (MARGOLIS *et al.* 1978). However, our view of treadmilling may require reevaluation because some features of actin and tubulin assembly *in vitro* are not in agreement with the model in its current form. One of the assumptions of head-to-tail polymerization is that nucleotide triphosphate hydrolysis is tighly coupled to the incorporation of monomers into the polymer, yielding a polymer which contains only nucleotide diphosphate. However, there are discrepancies in the rate constants for subunit addition and loss, and the rate constants predicted from the rates of ATP or GTP hydrolysis (CARLIER and PANTALONI 1981, CARLIER 1982, CAPLOW *et al.* 1982, PARDEE and SPUDICH 1982, BRENNER and KORN 1983, 1984 b). During actin and tubulin assembly, hydrolysis actually lags behind elongation. For example, under some conditions actin polymerization is complete in 10 seconds, while hydrolysis of ATP within the polymer requires 200 seconds (CARLIER *et al.* 1984 b). Therefore, under polymerizing conditions, many new triphosphate-bearing monomers can bind at the polymer ends during the time required for hydrolysis of a (once terminal) monomer. Thus, hydrolysis is not mechanistically coupled to polymerization, but occurs within the polymer in a subsequent reaction. As a consequence of the uncoupling of polymerization and hydrolysis, the end(s) of growing polymers have "caps" of subunits with bound nucleotide triphosphates, *i.e.*, ATP-actin and GTP-tubulin, respectively. Under de-polymerizing conditions, on the other hand, polymers tend to lose their nucleotide triphosphate caps quickly, exposing monomers at the ends that bear only diphosphates. Thus the molecular composition of the ends of growing and shortening polymers is different.

Experiments have established that the rates of dissociation depend on whether triphosphate or diphosphate-bearing subunits occupy the polymer ends. In the case of microtubules, the dissociation rate constants for GTP-tubulin and GDP-tubulin may differ by more than two orders of magnitude

(CARLIER *et al.* 1984 a, MITCHISON and KIRSCHNER 1984 b), while for actin, there is only a 5—10-fold difference (POLLARD 1984, LAL *et al.* 1984). The stability of the polymer, therefore, depends on which nucleotide species occupy the ends. As long as there is a stabilizing nucleotide triphosphate cap, the polymer is "protected", but exposure of diphosphate-containing subunits will lead to rapid disassembly of the polymer. Thus, the rate of gain or loss of subunits from the end of the polymer is no longer linear, as suggested in Figs. 35 and 36. Rather, there is a pronounced discontinuity in the slope of the elongation rate near the critical concentration (Fig. 37). In

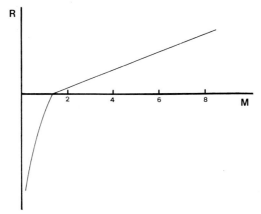

Fig. 37. Rate of tubulin polymerization (R) vs. monomer concentration (M, in micromoles) in a typical dilution experiment. Dilution of steady-state microtubules into tubulin concentrations below the critical concentration (which is about 1.7 µM) does not yield a straight line, but exhibits a marked change in slope at the critical concentration. [After CARLIER *et al.* (1984)]

the growing phase (above the x-axis), the relationship between the assembly rate and the monomer concentration still yields a straight line , in agreement with previous studies (SUMMERS and KIRSCHNER 1979, BERGEN and BORISY 1980). In the shrinking phase, however, the relationship is nonlinear. The polymer seems to undergo a phase transition at monomer concentrations close to the critical concentration. This behavior implies that the polymer can exist in one of two states, a growing phase with a triphosphate cap and a shrinking phase without such a cap (CARLIER and PANTALONI 1984, HILL and CHEN 1984, HILL 1984). Small fluctuations in monomer concentration around the critical concentration can translate into vastly different behaviors of the polymer, which is said to be "dynamically unstable" (MITCHISON and KIRSCHNER 1984 b). The effect would be further amplified if triphosphate monomers prefer "triphosphate ends" over "diphosphate ends". Then lengthening filaments would continue to grow because they can incorporate more subunits, while shortening polymers with diphosphate ends would continue to lose subunits.

Recent findings on the growth properties of microtubules either self-assembled or nucleated from microtubule seeds or organizing centers are consistent with a phase transition model for assembly (MITCHISON and KIRSCHNER 1984 a, b). For example, isolated centrosomes can be saturated with microtubules by incubation with high concentrations of tubulin. When such centrosomes are diluted into low tubulin media, the number of microtubules per centrosome decreases. However, the mean length of the remaining microtubules increases at a rate that depends on the free monomer concentration. This unexpected behavior is interpreted to mean that centrosome-associated microtubules exist as two populations, one growing and one shrinking. Thus, while the centrosome will always have an aster of microtubules that appears to be stable, each individual microtubule in this aster is potentially unstable. Similar observations were also made on free microtubules (Fig. 38). Microscopic analysis of the length distribution of microtubules at steady state indicates a time-dependent increase in average microtubule length and a decrease in the number of microtubules, while the total amount of polymer remains constant. This behavior can only be explained if the majority of microtubules is growing at the expense of a smaller number of shrinking microtubules. MITCHISON and KIRSCHNER (1984 b) estimate one shrinking microtubule for every 5 growing microtubules. It is important to stress that such unusual growth properties of both nucleated and spontaneously formed microtubules become evident only when individual microtubules rather than bulk populations are studied.

These and other data favor a model for microtubule assembly in which microtubules exist in two different states that are distinguished by the presence or absence of a GTP cap at their ends. Growing microtubules would be GTP-capped, while microtubules with GDP-subunits exposed at their ends would be unstable over a range of monomer concentrations. Transitions between the two phases are relatively rare provided the tubulin concentration is not too close to the critical concentration, as suggested by model studies (HILL 1984).

The dynamic instability model of microtubule growth is influencing the way we think about microtubule assembly, steady state kinetics, and *in vivo* microtubule dynamics. The model also applies to actin, but the effects of the phase transition on actin assembly kinetics may be more subtle. The new model seems to require a reconsideration and perhaps reinterpretation of previous experiments that propose a flux of subunits through the polymer at steady state. MITCHISON and KIRSCHNER (1984 b) suggest that previous observations on isotope uptake under steady state conditions may also be explained through microtubule length redistributions; on the other hand, treadmilling may be viewed as a special case of dynamic instability. While

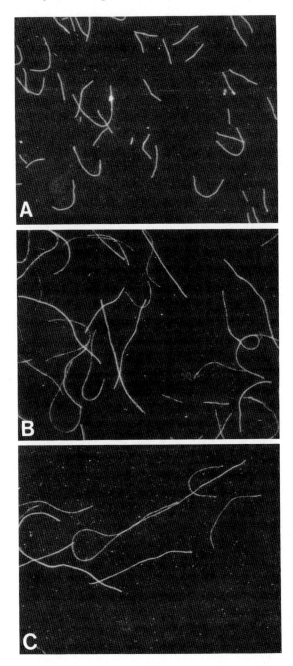

Fig. 38. Length redistribution at steady state of *in vitro*-polymerized microtubules as seen by immunofluorescence microscopy with a tubulin antibody. *A* 5 minutes, *B* 20 minutes, *C* 60 minutes ×600. (Courtesy of T. MITCHISON and M. KIRSCHNER)

this question needs to be resolved in future experiments, the implications of this model for the regulation of microtubule and actin organization in the living cell shall briefly be considered here. Built-in instability of actin and tubulin polymers based on the requirement for nucleotide hydrolysis may be advantageous to cells because such polymers could grow or shrink rapidly under nearly identical conditions. Such rapid changes are important for many cellular activities. By combining polymer growth with end-interactions (*e.g.,* capping proteins or end-to-side contacts in the case of actin, association with organizing centers or other stabilizing sites in the case of microtubules), a polymer can also be stabilized for longer time periods and will be depolymerized rapidly when these interactions are terminated. For example, a cell could utilize the phenomenon of dynamic instability to eliminate spontaneously formed polymers with both ends free, or polymers released from their anchorage at nucleating sites. This would be analogous to the role proposed for treadmilling (KIRSCHNER 1980), only that it would work much faster. Recent observations on the incorporation of microinjected, labelled tubulin into the mitotic spindle are consistent with the possibility of rapid exchange between spindle microtubules and free tubulin (SALMON *et al.* 1984). Further experiments with living cells will be required to determine whether the dynamic instability model holds true for polymer formation *in vivo*. The real role of nucleotide triphosphate hydrolysis in filament assembly in living cells may then be understood.

V. Other Filament Types

Microfilaments, microtubules, and intermediate filaments clearly are the major fibrous cytoplasmic components of eukaryotic cells, but the question remains whether any other filamentous structures have eluded identification. Such components could either be relatively minor constituents, or they might be ephemeral structures difficult to preserve and isolate. For example, the elaborate architecture of the microtrabecular lattice (see section VIII.) which is far more complex than the sum of the networks formed from the three major cytoskeletal filaments, has prompted the speculation that it contains and conceals additional cytoskeletal filaments. Is this speculation justified, and what is the evidence for the existence of other fibrous components with a cytoskeletal or motile function?

Studies over the past ten years on a variety of cell types suggest that such components do indeed exist. Based on the—admittedly sketchy— information available, these "exotic" filament systems seem to fall into three categories: 1. Contractile fibers of protozoan striated rootlet structures and nonstriated bundles of filaments known as spasmonemes and myonemes. They will be referred to as "spasmin-like" after a prominent representative. 2. Filaments forming a pervasive, elastic, cytoskeletal lattice in muscle cells. They will be referred to as "connectin/titin-like", according to a well-characterized prototype. 3. Short filaments of up to 400 nm in length interconnecting filaments of the three major cytoskeletal components. They will be referred to as "linkers". All three classes are composed of thin filaments around 3 nm in diameter (range: 2–5 nm), and do not bind heavy meromyosin and therefore are distinct from actin. Available information on the biochemical composition of some of these filaments indicates stubstantial differences in their constitutive polypeptides. For example, the spasmin-like proteins are composed of relatively small polypeptides, while the connectin-like fibers appear to be constructed from gigantic subunits in the megadalton range. The composition of most of the 3 nm connectors is unclear.

V.1. Spasmin-Like Filaments

Vorticellid ciliates possess a contractile stalk, the spasmoneme, that can contract into a bedspring-like coil within a few milliseconds. It has been calculated that the rate of contraction is at least an order of magnitude higher than that of skeletal muscle. The spasmoneme contains numerous 2–4 nm filaments aligned parallel to its long axis, as well as membraneous tubules believed to be calcium stores (AMOS 1971, AMOS et al. 1975, ROUTLEDGE et al. 1976). The major protein component of the spasmoneme is a low molecular weight protein, named spasmin, that runs as a doublet in the 20,000 d range in polyacrylamide gels. The electrophoretic mobility of the two proteins is reduced in the presence of calcium, a property spasmin shares with calmodulin. Contraction is accompanied by an apparent decrease of order within the spasmoneme, as indicated by a loss of birefringence (AMOS 1971), and apparently does not require metabolic energy. It does, however, require the presence of at least micromolar calcium. Reextension, which is much slower, occurs at submicromolar calcium levels. Contraction is elicited by the binding of calcium to spasmin, which induces a conformational change in the protein, causing the spasmoneme to shorten.

A filament system apparently homologous to the spasmoneme is found in flagellar roots of some protozoa. Rootlet structures are elongated filamentous organelles that frequently appear banded or striated. They are almost exclusively found in association with centrioles or basal bodies. Presumably, they anchor cilia or flagella to the rest of the cell, but they themselves may also serve as attachment devices for other cytoskeletal elements. Rootlet structures are contractile organelles (SALISBURY and FLOYD 1978) (Fig. 39) that seem to have an associated ATPase activity (ANDERSON 1977), but it is unclear whether this ATPase activity is related to the contractile mechanism, rootlet protein phosphorylation, or both. A major component of rootlet structures is a 20,000 d calcium-binding phosphoprotein (SALISBURY *et al.* 1984) that forms 3 nm filaments which apparently can undergo a conformational change upon calcium binding.

Motile acitivities based on 3 nm filament systems associated with striated

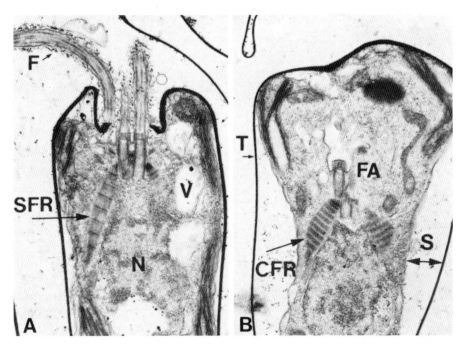

Fig. 39. Striated flagellar roots (*SFR*) in the green alga *Tetraselmis* in the absence of added calcium (*A*) and within 30 sec of a calcium shock (*B*). Both flagellar roots (*CFR*) have contracted, resulting in deflagellation, displacement of the flagellar apparatus (*FA*), and inpocketing of the plasmalemma (*S*). *F* flagella, *N* nucleus, *V* vacuole. × 15,000. (Courtesy of J. SALISBURY)

rootlet-like structures or unstriated filament bundles are apparently quite common among protozoa. For example, the ciliate *Stentor* contains so-called myoneme bundles composed of 3–4 nm filaments that undergo helical coiling upon contraction (BANNISTER and TATCHELL 1972, HUANG and PITELKA 1973). A similar helix-coil transformation may be responsible for the rapid contraction of myoneme-like structures in the *Acantharia*, a group of rhizopods whose cell body contains prominent striated bundles of 3 nm filaments (FEBVRE and FEBVRE-CHEVALIER 1982). In *Sticholonche*, a unique heliozoan which translocates by rowing movements of its microtubule-bearing axopodia, bundles of fine filaments are present at the

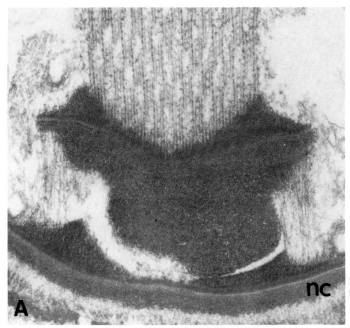

Fig. 40. Basal regions of axonemes of *Sticholonche*. *A* The dense head of the hip joint ariculates on the nuclear capsule (*nc*). Thin filaments on either side of the hip joint connect the base plate of the axoneme to the nuclear capsule. *B* Similar to *A*, but the microtubule bundle is inclined at an angle. The filaments on one side appear stretched, while on the other a series of faint bands (arrows) is visible. × 50,000.

(Courtesy of J. and M. CACHON)

base of every axopodial rod (CACHON *et al.* 1977). Axopodia are moved by alternate coiling and extension of these fine filaments (Fig. 40). Their biochemical composition is unknown. A static skeletal function, rather than a dynamic role in cell movements, may be ascribed to another protozoan organelle known as "epiplasm" (FAURE-FREMIET 1962). Epiplasm is a filamentous layer that underlies the pellicular membrane of many ciliates (*e.g.,* PITELKA 1969, ALLEN 1971, PECK 1977, HAUSSMANN and MULISH 1981). It forms a continuous sheet in isolated pellicles (VADEAUX 1976, VIGNES *et al.* 1984). Its 3 nm filaments resist extraction with 0.6 M potassium iodide, suggesting that they are not actin-like (VIGNES *et al.* 1984). Although the biochemical characterization of these filaments is in its very early stages, their major polypeptides are probably different from the spasmin-like proteins. Many other examples of non-actin-based filament bundles may be found throughout the protozoan phylum (see PITELKA 1969) and await biochemical characterization.

Rootlet structures are also quite common in many higher eukaryotic cells. Even though, with very few exceptions, nothing is known about their biochemical composition and function, it is tempting to speculate that they are composed of spasmin-like filaments. Proteins related to the 3 nm rootlet fiber protein may be present as more dispersed or unstructured assemblies in other cell types. For example, antibodies against the 20,000 d rootlet protein seem to stain diffusely the centrospheric cytoplasm of cultured cells. In as yet unknown ways, such proteins may contribute to the architecture of the cytoplasm of many cells as dispersed, less highly organized cytoskeletal elements. Even though the notion of a widely distributed class of non-actomyosin contractile proteins remains hypothetical, it seems to be more than mere coincidence that the 20,000 d striated rootlet protein shares certain features with other, more widespread calcium-modulated proteins performing a regulatory function in higher eukaryotic cells. These include calmodulin, parvalbumin, myosin light chain, and troponin. SALISBURY *et al.* (1984) speculate that the spasmin-like proteins may represent a primitive, direct motility mechanism simply dependent on calcium binding which, later in evolution, was replaced by more sophisticated contractile systems. The role of this primitve motile system may then have changed to include regulatory functions, in addition to contractile acitvities. Evidently, more work on these interesting proteins is required before their function can be defined more precisely.

Fig. 41. Titin filaments, *A* Purified, native titin after metal shadowing. × 66,000. *B, C* Isolated, negatively stained thick filaments with fibers resembling titin (arrows) running parallel to the thick filament axis. × 150,000. [From TRINICK *et al.* (1984)]

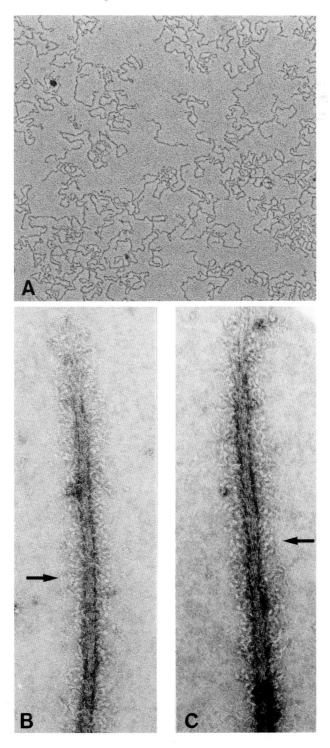

V.2. Connectin/Titin-Like Filaments

In 1976, MARUYAMA et al. described a novel filament system in muscle cells distinct from the contractile and intermediate filaments. The subunits of this apparently elastic muscle cell matrix were of unusually high molecular weight. Because of the integrative nature of this new filament network, they named the protein connectin (MARUYAMA et al. 1977). Later, WANG et al. (1979) isolated a similar, if not identical protein from skeletal muscle cells and named it titin. Connectin/titin are two closely related polypeptides (apparently a precursor-product pair) that together are the third most abundant myofibrillar protein in a wide variety of muscle tissues, representing up to 9% of the myofibrillar mass. It forms a system of long, thin filaments throughout the sarcomere that are proposed to function as an elastic cytoskeletal lattice which not only supports sarcomeric organization, but also generates passive tension upon stretching (MARUYAMA et al. 1984 b). The recent successful purification of the leading band of the connectin/titin doublet in the absence of denaturants (MARUYAMA et al. 1984 a, WANG et al. 1984, TRINICK et al. 1984) has shown the native protein to form extremely long (up to 800 nm), flexible strands of 4–5 nm diameter (Fig. 41). Individual strands can self-associate to form filamentous bundles and meshworks, supporting the notion that connectin/titin is a cell component with a skeletal function.

Using extensive extraction with guanidine-HCl in the presence of disulfide-reducing agents, LOEWY et al. (1983) have prepared similar muscle tissue "ghosts" containing only 1% of the original protein content of the tissue. Thin (3 nm) filaments are part of this matrix, but the relationship of these filaments to the connectin fibers has yet to be established.

It is unclear whether an elastic matrix similar to the connectin system is present in other cell types, or whether it is restricted to muscle cells. There is one report on the use of extensive extraction with iodide, sodium dodecyl sulfate, and urea to prepare a "ghost" of the slime mold Physarum (GASSNER et al. 1983). The slime-mold ghost also reveals a pervasive system of 2–3 nm filaments upon electron microscopic inspection. The similarities in morphological appearance and chemical stability of the muscle cell matrix and the slime-mold ghost certainly are intriguing, suggesting that the two are perhaps related. It may be worthwhile subjecting other tissues and cell types to similar extraction protocols to search for related, novel, cytoskeletal elements.

V.3. 3 nm Linkers

Extraction procedures employing mild nonionic detergents to produce characteristic "cytoskeletons" composed of the three major filament types reveal also rather short (maximum length 400 nm) filaments of approx-

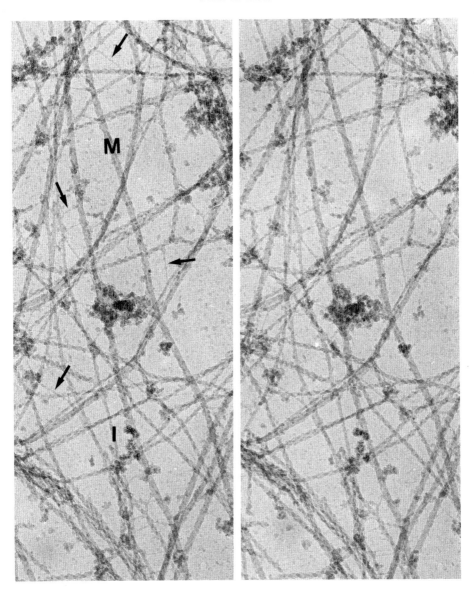

Fig. 42. Stereomicrograph form the periphery of a cultured BSC-1 cell extracted with triton X-100. The cells were labeled with myosin subfragment 1 before fixation, dehydration, and critical point-drying. 3 nm filaments (arrows) link microtubules (*M*), intermediate filaments (*I*), and arrowhead-decorated actin filaments. × 118,000

imately 3 nm diameter that do not bind heavy meromyosin (Fig. 42). In the absence of any biochemical characterization they are best described as "linkers" because their most distinguished feature seems to be that they interconnect the other three filament types (e.g., SCHLIWA and VAN BLERKOM 1981, HIROKAWA et al. 1982, HIROKAWA and TILNEY 1982, SVITKINA et al. 1984). It is not clear whether these linker elements are all composed of the same or related protein(s), or whether they constitute a biochemically heterogeneous group that shares only superficial structural similarity. For example, actin binding protein, myosin, the spectrins, high molecular weight MAPs, and the 200 kd neurofilament protein all are elongated molecules that may well appear as linker elements 2–5 nm in diameter in electron micrographs of detergent-extracted cells. In agreement with this notion, immunoelectron microscopic observations reveal the wispy links between neurofilaments in axons to be made up, at least in part, of the 200 kd neurofilament protein (HIROKAWA et al. 1984). Similarly, some of the thin filaments in the terminal web of intestinal epithelial cells are composed of myosin (HIROKAWA et al. 1982), while another set of morphologically similar filaments in the same cell region is made up of the spectrin-like protein TW 260/240 (GLENNEY et al. 1983).

Part B. Interactions

VI. General Aspects of the Three-Dimensional Organization of Cytoskeletal Components

The beginning of research on the biochemistry and supramolecular organization of the cytoskeleton is closely associated with two developments: 1. The use of antibodies to study the overall distribution of cytoskeleton-associated antigens in large populations of cells, an approach first applied at the light microscopic level and more recently at the electron microscopic level. 2. The use of nonionic detergents to remove the plasma membrane, internal membrane-bound organelles, and less stable components of the cytoplasm to generate a "cytoskeletal" preparation in which the organization of the major filament types can be studied at the electron microscopic level. Graphic images of critical point-dried or rapidly frozen preparations reveal fine details of the three-dimensional organization of filaments and their interactions, and are especially informative when viewed steroscopically. While both approaches are just ten years old, they have contributed to the establishment of a new branch of cell biological research. This section will discuss some general aspects of cytoskeletal organization as revealed by these techniques; more specialized features will then be dealt with in subsequent sections.

An overview of the nonrandom distribution of a cell's filamentous structures was made possible in the mid70ies through the application of labeled antibody procedures developed three decades earlier (see COONS 1956). The first of these studies used antibodies against actin to visualize the cellular organization of microfilament bundles (LAZARIDES and WEBER 1974). It was soon followed by similar studies with antibodies against tubulin (WEBER *et al.* 1975 a, b, BRINKLEY *et al.* 1975), and intermediate filaments (OSBORN *et al.* 1977, HYNES and DESTREE 1978). Thousands of papers followed that employed immunofluorescence microscopy or other immunocytochemical procedures (*e.g.,* immunoperoxidase; DE MEY *et al.* 1976) as a convenient assay for determining the distribution of cytoskeleton-related proteins, or for assessing the outcome of experimental treatments. These studies are summarized by WEBER and OSBORN (1979, 1982), OSBORN and WEBER (1982), and LAZARIDES (1982). While no attempt will be made to review this work, a number of important assets shall briefly be mentioned here.

Immunofluorescence microscopy was instrumental in visualizing the pattern of microtubule distribution in cultured cells. While in some respects the results of previous electron microscopic studies were simply confirmed by the new approach, three findings made immunofluorescence microscopy an important new tool that significantly advanced the field of microtubule research: 1. Single microtubules could be observed at the light microscopic level, a feat previously deemed impossible. Initial controversy was laid to rest by correlative immunofluorescence and electron microscopy of the *same* specimen (OSBORN *et al.* 1978). Visualization of single microtubules by immunofluorescence is now generally accepted. 2. The organization of the superstructure formed by the entirety of all the microtubules (the "cytoplasmic microtubule complex") was revealed. That microtubules extend from the central cell area to the cell periphery as long, unbroken, gently curving structures could only be guessed before on the basis of thin section electron microscopy, but could now be perceived at a glance. 3. Microtubules appear to grow from a limited number of organizing sites or centers located near the nucleus towards the cell periphery, upon recovery from experimentally induced microtubule depolymerization (OSBORN and WEBER 1976, FRANKEL 1976, BRINKLEY *et al.* 1981). These studies helped to advance the concept of the microtubule-organizing center as a major source for new microtubules and the generation of microtubule patterns. These and other studies strengthened the view that immunofluorescence microscopy is a useful tool that bridges the gap between light and electron microscopy, and between cell structure and biochemistry.

Much the same argument also applies to the study of microfilament and intermediate filament organization by immunofluorescence microscopy. Initially, attention was focused on the formation and fate of the prominent microfilament bundles known as stress fibers. Subsequently, the scope was broadened to include studies of the "biochemical anatomy" of these and other supramolecular filament assemblies with antibodies against myosin (WEBER and GROESCHEL-STEWART 1974), alpha-actinin (LAZARIDES and BURRIDGE 1975), filamin (WANG *et al.* 1975), and many others. In the case of

---

Fig. 43. Comparison of the three-dimensional organization of unextracted and Triton X-100 extracted whole mount preparations visualized by high voltage electron microscopy. The density of the cytoplasm of these BSC-1 cells decreases substantially upon extraction. After extraction, remnants of the cytoplasmic granules remain suspended in the network of filaments, while the mitochondria disappear. Extraction reveals details of the overall organization of the filament skeleton, such as peripheral networks and substrate-associated filament bundles.

× 5,000

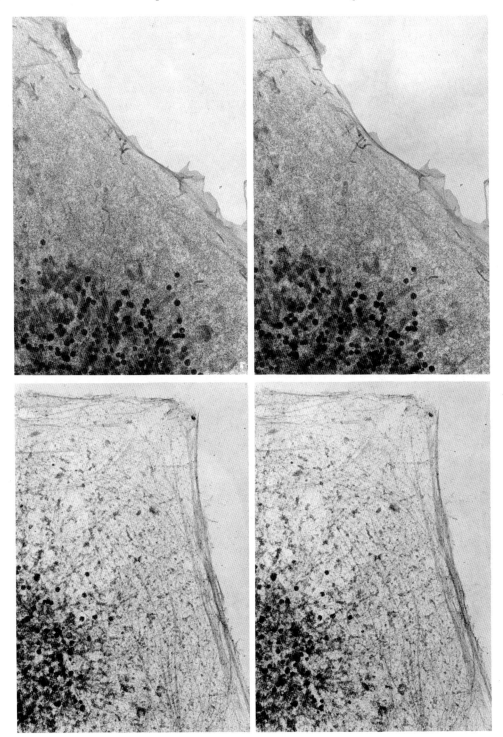

intermediate filaments, the technique of immunofluorescence microscopy was instrumental in defining the major filament subclasses and their cell and tissue specificity, and for revealing major differences in their overall patterns of organization.

Even though immunofluorescence microscopy was beginning to reveal cellular organization at the supramolecular level, it was still restricted by the limited resolution of the light microscope and two-dimensional imagery. Attempts to improve the retrieval of information from immunofluorescence micrographs by using light microscopic stereoscopy (OSBORN et al. 1978) did not receive the deserved resonance, even though the procedure produced quite remarkable results. For details of filament organization, electron microscopy still was the method of choice, but an alternative to thin sectioning of embedded specimen clearly was required.

In 1976, BROWN et al. treated cells with a nonionic detergent which removed the plasma membrane and other membraneous organelles but retained many of the more stable cell elements in their original con-figuration (see Fig. 43). Detergents had been widely used before in cell biological research, mainly for the solubilization of integral membrane proteins (see summary by HELENIUS and SIMONS 1975), but here they were "misused" in order to study the residues of this extraction process. Related studies soon followed (OSBORN and WEBER 1977, LENK et al. 1977, SMALL and CELIS 1978). The residue of detergent extraction has become known as "the cytoskeleton" *par excellance*. The result of the extraction process critically depends on the buffer system and the conditions of extraction: for example, some of the earlier studies used buffers that did not preserve microtubules, and modifications of the extraction procedure allow selective solubilization or selective stabilization of certain components. Because the complexity of the resulting cytoskeleton varies with the extraction pro-cedure employed, *the* cytoskeleton does not exist. For some, the residues of extraction are still far too complex because they include noncytoskeletal components, while for others, they are not complex enough because important, more labile components have been lost. At present, however, we will have to live with this admittedly unsatisfactory, operational definition until more rigorous criteria are established.

Another important variable that determines the character and the quality of information available form cytoskeletal preparations is the mode of specimen preparation for microscopic inspection. In negatively stained whole mount preparations the resolution is high but the three-dimensional arrangement of its components is lost (BROWN et al. 1976, OSBORN and WEBER 1978, SMALL and CELIS 1978, HOGLUND et al. 1980) (Fig. 44). Critical point-drying and scanning electron microscopy visualize the cells in three dimensions (TROTTER et al. 1978, BOYLES and BAINTON 1979, IP and FISHMAN 1979, PUDNEY and SINGER 1980), but the resolution of the

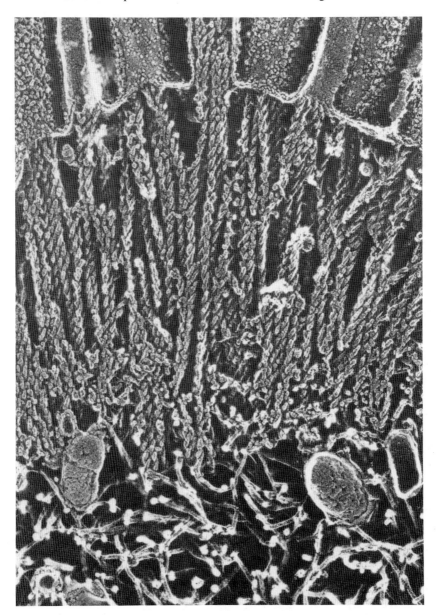

Fig. 45. Terminal web of a mouse intestinal epithelial cell decorated with sugfragment 1 before quick-freezing, deep-etching, and rotary shadowing. Decorated actin filaments assume the appearance of a two-stranded twisted rope. The twists are slightly asymmetrical, giving the filaments an arrowhead-like appearance that reveals their polarity ("barbed" ends pointing up). Intermediate filaments on the bottom of the micrograph are unlabeled. × 95,000. [From HIROKAWA *et al.* (1982)]

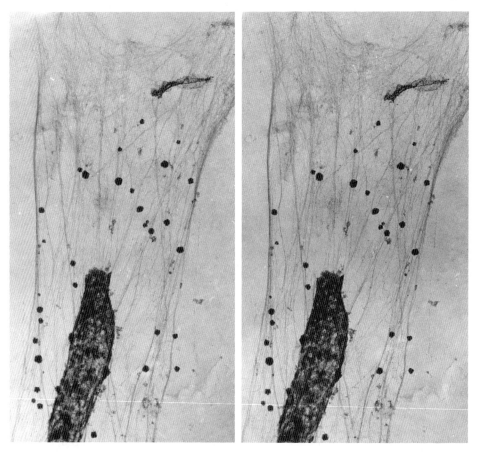

Fig. 46. Stereo micrograph of a chicken embryo fibroblast in which microtubules and intermediate filaments were solubilized with sarkosyl, an ionic detergent, revealing the major domains of the actin filament system such as networks at the leading edge, and substrate-associated cables. Some filament bundles also extend on top of the nucleus. × 6,500

of actin filaments in peripheral networks (SCHLIWA and VAN BLERKOM 1981), which results from end-to-side associations of actin filaments with each other. HARTWIG *et al.* (1980) showed that polymerization of actin in the presence of actin binding protein *in vitro* leads to the formation of a branched network with many T-shaped junctions at more or less right angles between actin filaments (see also CONDEELIS 1981). Such *in vitro* networks are not unlike those seen in extracted whole mount preparations of cells. Their structure is significantly different from polymerized pure actin, in which filaments overlap at random to form large, tangled masses (NIEDERMAN *et al.* 1983) (see Fig. 48). The fact that these *in vitro* networks

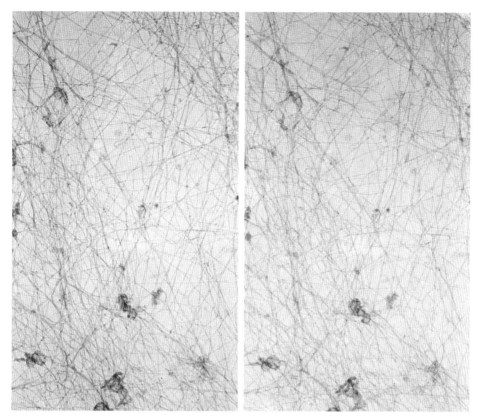

Fig. 47. Stereo micrograph of a BSC-1 cell in which microtubules and the actin filament network were solubilized with buffers of low and high ionic strength. The residual network consists of long, unbroken vimentin filaments. × 33,000

resemble those found in the periphery of motile cells encourages the belief that analysis of their properties will allow meaningful inferences on the behavior of similar networks in the living cell. The organization of cytoskeletal domain(s) occupied by intermediate filaments is best revealed after selective solubilization of the other filaments (SMALL and CELIS 1978, SCHLIWA and VAN BLERKOM 1981). The vimentin network, for example, resembles an open cage of single filaments that overlap at random at many points (Fig. 47). Remarkably few filament ends are visible in these preparations, suggesting that individual filaments are very long. Filaments of the prekeratin type form thick cables that interconnect and splay an many points and disperse into single filaments only at the very cell periphery.

Second, and perhaps more important, cytoskeletal preparations afford the impression that the three major filament types of the cytoskeleton should not be viewed as independent components that merely coexist in the

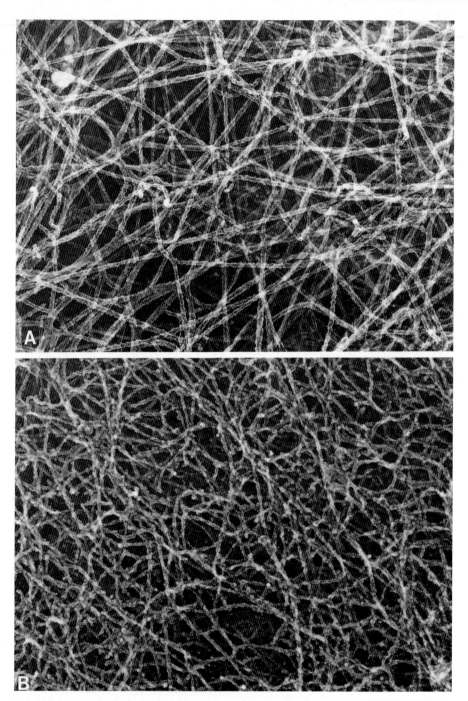

Fig. 48. *A* Purified F-actin polymerized on the surface of a grid, fixed, critical point-dried, and rotary shadowed. Actin filaments are several micrometers long and overlap or cross one another. *B* Mixture of actin and actin binding protein at a molar ratio of 25:1. The mixture forms a branching polymer network in which the actin filaments are short and straight. × 50,000. (Courtesy of R. Niederman and J. Hartwig)

cytoplasm. Rather the cytoskeleton can be thought of as a *highly interconnected structural unit* in which the functions of the different components are spatially and temporally coordinated (see Fig. 42). The structural, biochemical, and functional aspects of interactions among the cytoskeletal filament systems will be discussed in more detail in section VII. *Third*, the analysis of cytoskeletal preparations, in conjunction with studies of intact, unextracted cells, has allowed researchers to take the first steps towards a better understanding of the organization of the cytoplasmic matrix, that pervasive, gelatinous medium in which the membraneous organelles are suspended and of which the more stable filaments are but one component. Structurally, the cytoplasmic matrix is also known as the *microtrabecular lattice* (WOLOSEWICK and PORTER 1976, 1979); it is composed of slender strands of variable dimensions that form an intricate, reticular network far more complex than the networks of the more stable filaments combined. Its precise morphology, and even its existence, has aroused much discussion and some controversy, which will be dealt with in section VIII.

Immunofluorescence microscopy and electron microscopy provide static views of morphological configurations and events that in reality are in constant flux. Judging from the dynamics of the biochemical reactions of cytoskeletal proteins *in vitro* and from observations of live cells by high resolution light microscopy, changes may occur on a time scale of milliseconds. Therefore, attempts to relate static micrographs to the dynamic events of cellular motility is like reconstructing the course of a baseball game from a few snapshots of the pitcher. In light of this aspect it may seem all the more unfortunate that the term "cytoskeleton" has become so firmly established *in lieu* of a term that would also reflect the dynamics of the filament networks. However, since no one has come up with a more appropriate catchword (except for the somewhat disdainful "cytogeleton"), it simply needs to be kept in mind that the seemingly immutable "skeletal" structures pictured in micrographs are capable of rapid and dramatic reorganization.

VII. Interactions Between the Major Filamentous Components of the Cytoskeleton

Implicit in current concepts of the organization of the cytoskeleton and the cytoplasmic matrix is the idea that the networks of the major filament systems are interconnected and interacting. Cooperation among the components of the cytoskeleton is clearly involved in many aspects of cell behavior, cellular and subcellular motility, and morphogenesis. Interactions, or their purposeful absence, may be viewed as a key aspect of the organization of the cytoskeleton. Surprisingly, however, compared to the staggering amount of morphological and biochemical information on the individual filament systems and their associated proteins, the development, maintenance, and control of interfilament interactions is relatively poorly understood. Some motile activities may require only interactions between fibers of the same type. For example, the movement of the axostyle of *Saccinobaculus* is exclusively microtubule-based and crossbridge-mediated, and extension of the acrosomal process of *Thyone* or *Limulus* sperm is dependent solely on actin filaments. Other events may depend on a cooperative interaction between different filaments. Linking the networks of different fiber classes has the advantage of extending the range of their properties and helping to construct more powerful and versatile "super-structures". Thus, interactions between microtubular and actin-based networks could combine the flexural rigidity of the former with the potential for rapid changes in consistency ot the latter.

This section examines the evidence for linkages between the major fiber systems. Beginning with self-associations of microtubules, actin filaments, and intermediate filaments, some principles of filament interactions will be discussed. An analysis of associations between different fibers will follow. In each instance, the following three aspects will be considered: morphological evidence from light and electron micrographs, biochemical evidence derived from *in vitro* experiments, and cytological or physiological observations of cellular activities that can be interpreted as resulting from such interactions (*e.g.,* inhibitor experiments or other experimental treatments, micro-injection, cell behavior studies). First, however, it will be useful to consider briefly some procedures frequently employed to detect and characterize filament interactions *in vitro*.

VII.1. Some Methods for the Detection of Filament Interactions

Interactions between fibrous proteins in virtually all instances are mediated by associated proteins that specifically bind to one or more of the participating filaments. These interactions will cause changes in the "consistency" of the solution which can be detected by a variety of procedures. Associations leading to network formation, for example, may turn the filament solution into a rigid gel; associations resulting in bundle formation may also cause a change in the properties of the filament solution but will not necessarily result in gelation. The correct identification as a bundling phenomenon can only be verified by light or electron microscopy. The assays to be described here can be used to characterize the physical properties of network-forming filament solutions. A more complete description can be found in POLLARD and COOPER (1982).

Turbidity assay. Monitoring certain changes in the physical properties of a sample with time is possible by determining its turbidity. A sample is loaded into a cuvette and the absorption of the solution is determined in a spectrophotometer at 330–360 nm (e.g., POLLARD 1976). Network formation will result in a slight but distinct increase in turbidity. This assay is simple and fast, but non-quantitative, relatively insensitive, and unreliable. For example, changes in turbidity can be induced by phenomena unrelated to interactions between filaments. The turbidity assay is still useful as a quick checking procedure, but it is not widely used any more.

Test tube inversion. One of the most simple, reliable assays used early in the study of actin network formation (POLLARD 1976, HARTWIG and STOSSEL 1976), is known as test tube inversion. A sample is loaded into a cuvette or test tube and, after an appropriate incubation time, the tube is slowly inverted. This assay tests whether a solution resists flow. A liquid will flow rapidly, a viscous fluid will flow much more slowly, and a gel will behave like a solid and stay at the bottom of the tube. The assay is inexpensive, and many samples can be tested simultaneously. It is useful for determining the gel point, *i.e.,* under what conditions the transition from a viscous liquid to a solid gel occurs.

Sedimentation. When centrifuged at relatively low speed (around 10,000 g) for short time periods (5–10 minutes), noninteracting polymers will not sediment, while polymers complexed by cross-linkers will. The amount of material sedimented will be a function of the activity and amount of cross-linker in the sample, so this procedure can be used to determine and quantify its acitvity. This is another simple, inexpensive procedure. However, the assay does not distinguish between sedimentation due to interaction and sedimentation due to precipitation. Also, strong gels will not sediment at low centrifugal forces, but then such rigid gels can easily be detected by tube inversion.

Falling ball viscometry. A procedure for determining the apparent viscosity of a sample, the falling ball assay, is based on the determination of the velocity of a small steel ball falling or rolling through a sample contained in a small capillary tube. In this situation, the viscosity of the solution is given by a modified Stoke's equation,

$$\eta = 0.22 \cdot k \cdot r^2 \cdot g \cdot \sin\alpha \cdot (\delta_b - \delta_m)/v \qquad (5)$$

where η is the viscosity of the sample, k is a calibration constant required because the ball and the medium are confined to a narrow tube and inclined at an angle, r is the ball radius, g is the force of gravity, α is the inclination angle of the capillary tube, δ_b and δ_m are, respectively, the densities of the ball and the medium, and v is the velocity of the ball (MACLEAN-FLETCHER and POLLARD 1980 a). All parameters except v are known or can be determined; thus, the viscosity of the medium is inversely proportional to ball velocity. Since the viscosity of a complex fluid containing rod-shaped polymers depends on the shear rate (*i.e.,* the fluid is non-Newtonian), falling ball viscometry can only determine an *apparent viscosity*. This assay was first applied to solutions of biological macromolecules by GRIFFITH and POLLARD (1978); its theory and design was evaluated by MacLEAN-FLETCHER and POLLARD (1980 a), and it was frequently used thereafter (*e.g.,* MacLEAN-FLETCHER and POLLARD 1980 b, FOWLER and TAYLOR 1980, YIN *et al.* 1980, NUNNALLY *et al.* 1981, FOWLER *et al.* 1981, RUNGE *et al.* 1981). The falling ball assay can also be used to determine the yield strength, or the maximum force that can be applied to the system without moving the ball, if the solution is strong enough to support the ball. Falling ball viscometry is rapid, reproducible, and inexpensive (it costs about 20 c per experiment). Although there are disadvantages as well (for example, even the low shear rates applied to the sample are destructive), it is perhaps the most popular of all procedures designed to determine certain physical properties of complex mixtures of macromolecular assemblies.

Other procedures. A number of additional procedures were developed to determine other parameters that characterize interacting macromolecules. These include 1. the *gelmeter* (BROTSCHI *et al.* 1978), which provides a measure for the modulus of rigidity at the yield point of a sample, 2. the *rotational viscometer* (KUPKE and BEAMS 1977) which measures the viscosity and the yield stress, and 3. the *viscoelastometer* (FUKUDA and DATE 1963, ZANER *et al.* 1981, ZANER and STOSSEL 1982), which allows to determine the elastic and viscous moduli of samples. However, although these are important characteristics of gels, the devices are sometimes expensive, and the procedures inconvenient and time-consuming. These techniques are therefore not as widely used as some of the other procedures described above.

VII.2. Self-Associations of the Major Cytoskeletal Filament Types

VII.2.1. Actin Filaments

Of the three major filament types, the greatest potential for self-association is found in the microfilament system. Many biochemical and structural aspects of interactions among actin filaments were discussed in section I.3., so only some general considerations are appropriate here. Associations of actin filaments always seem to be mediated by specific associated proteins, even though some interactions may take place directly between filaments made from highly purified actin (MacLEAN-FLETCHER and POLLARD 1980 b). Direct interactions between actin filaments may generate a gel-like consistency of the filament solution (GRIFFITH and POLLARD 1982 a), but whether these direct actin-actin associations also occur *in vivo* remains to be demonstrated. Examples of supramolecular assemblies of F-actin mediated by actin binding proteins include the dense peripheral networks in ruffles and lamellae and stress fibers or the core bundles of rigid cell extensions.

Initially called a microfilament meshwork, the network in the periphery of many animal cells is more than an entanglement of overlapping filaments. The filaments frequently intersect and cross-connect in a perpendicular fashion to form T- and X-junctions. Auch *orthogonal networks*, in which the free path length of filaments between junctions may be quite short, are also formed upon copolymerization of actin with a cross-linker *in vitro* (HARTWIG *et al.* 1980, NIEDERMANN *et al.* 1983) (Fig. 48). Branching maximizes the solid-like elastic properties at minimal solute concentrations. By immunological criteria, peripheral ruffle-like networks of motile cells are characterizied by the presence of a variety of gel-inducing proteins, including actin-binding protein, alpha actinin, fimbrin, and others (*e.g.,* BRETSCHER and WEBER 1980, GROESCHEL-STEWART 1980, FERAMISCO and BLOSE 1980, LANGANGER *et al.* 1984, STOSSEL 1984). Even cross-linking proteins normally not found in a cell will take up their expected location after microinjection. For example, villin, a protein specific for intestinal and renal epithelial cells, will associate with actin-rich regions, notably in the cell periphery, when microinjected into tissue culture cells (BRETSCHER *et al.* 1981). Despite the durability and considerable physical strength of actin networks, small changes in the solution conditions and/or the properties of the cross-linkers may alter the physical properties of the gel and even lead to its destruction. As mentioned before, changes in actin filament length or the affinity of a cross-linker for actin filaments will determine the state of gelation. Again, specific protein factors may be involved in the execution of these changes. The interplay among all the network-forming and network-breaking parameters and their cellular mechanisms of regulation are not understood and may be difficult to resolve.

Perhaps the best-known of the cellular actin filament bundles are the stress fibers of cultured cells. These fibers contain filamin, alpha-actinin, and myosin as potential bundling factors. Other bundle structures include the cores of microvilli, the cores of stereocilia, the cell extensions known as microspikes or filopodia, microfilament bundles in certain plant cells, and the acrosomal process of sperm cells. In contrast to stress fibers, all of these are composed of actin filaments with unidirectional polarity and are packed into highly organized, frequently hexagonal, arrays (DE ROSIER and TILNEY 1982). Less highly organized and/or more ephemeral structures include the contractile ring of dividing cells (SCHROEDER 1981), the microfilament "arcs" frequently observed within the lammellae of cultured cells (SORRANO and BELL 1982, HEATH 1983), and the apical rings of epithelial cells that appear to form a contractile annulus (BURGESS 1982, HIROKAWA and TILNEY 1982, OWARIBE and MASUDA 1982). Common characteristics of these structures are the comparatively low degree of order of adjacent actin filaments, the presence of filaments of opposite polarities—a necessary orientation for a contractile function—and lateral connections to the cell membrane. Specific associated proteins can be expected to be required for the construction and maintenance of many of these superstructures, but only in some cases these factors have been identified.

In bundles formed *in vitro* form F-actin and any one of the globular bundling factors, all the helices of the laterally aggregated actin filaments are in register; the same is true for certain cellular actin bundles (TILNEY 1975, DE ROSIER *et al.* 1977, SPUDICH and AMOS 1979, DE ROSIER and EDDS 1980, TILNEY *et al.* 1980). These highly ordered bundles are qualitatively similar to the "magnesium-paracrystals" formed in the presence of high (25–50 mM) concentrations of magnesium ions (MOORE *et al.* 1970, HINSSEN 1972, SPUDICH 1973, GILLIS and O'BRIEN 1975). However, the latter are bipolar rather than unipolar, and the interfilament distances are about half those of cross-linker-induced bundles, *i.e.,* 7 nm vs. 12 nm (MATSUDAIRA *et al.* 1983). The lateral spacing in bundles induced by cross-linkers such as villin or fimbrin is similar to that between noncross-linked actin filaments packed by high-speed centrifugation, suggesting that the bundling factors act to restrict the movement of actin filaments away from one another, rather than to overcome electrostatic repulsion between actin filaments (MATSUDAIRA *et al.* 1983). Magnesium ions, on the other hand, seem to function as an ionic cross-linker. Interestingly, polyamines such as spermine and spermidine induce the formation of paracrystals with a structure similar to that of magnesium-paracrystals (GRANT *et al.* 1983). Troponin I and myelin basic protein also promote bundle formation *in vitro* when bound to actin filaments (BARYLKO and DOBROWOLSKI 1984), but the physiological significance is unclear.

Unlike many of the known microtubule bundles, some actin filament bundles are endowed with a remarkable degree of flexibility. They can be bent into supercoiled structures in which there is lateral slippage of neighboring filaments, rather than compression or stretching of filaments on opposing sides of the bundle (DE ROSIER and TILNEY 1984 b) (Fig. 49). Examples of bendable actin bundles are the stereocilia of cochlear hair cells, where the bend is transient (TILNEY *et al.* 1983), and the preformed acrosome bundle of *Limulus* sperm, where bends are built into the filament bundle (DE ROSIER and TILNEY 1984 b).

Even though actin bundles may appear as lasting cytoskeletal specializations, some are highly dynamic and may be formed and broken down within a few seconds. This is particularly evident in microspikes and filopodia, which may undergo very rapid turnover (*e.g.,* ALBRECHT-BUEHLER 1976, EDDS 1977, OTTO *et al.* 1979, SMALL 1981), or the acrosomal process of certain invertebrate sperm, which grows to a size of 90 μm within 10 seconds (TILNEY and INOUE 1982) while aligning the elongating actin filaments into the paracrystalline core bundle. Fluctuations in size, shape, and protein composition also occur in the more stable stress fibers. When rhodamine-labeled actin is microinjected into living cultured cells, it associates with the cytoskeletal assemblies known to contain actin, such as stress fibers and peripheral networks (*e.g.*, KREIS *et al.* 1979, 1982, WANG

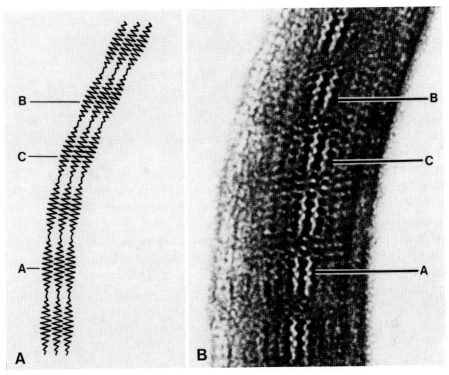

Fig. 49. Drawing of the filaments in a bend (*A*) and negatively stained image of a bend in the acrosomal bundle of a *Limulus* sperm (*B*). The two parts of the figure are aligned to show relative slippage of the filaments. *A, B* and *C* are corresponding levels in each micrograph. × 525,000. [From DE ROSIER and TILNEY (1984 b)]

et al. 1982). Labeled alpha-actinin and tropomyosin will behave in a similar fashion by associating with stress fibers in characteristic patterns (FERAMISCO 1979, WEHLAND and WEBER 1980). Labeled actin is incorporated into stress fibers much more slowly (over 20 minutes) than into peripheral cell areas, where the incorporation is essentially complete within 5 minutes (GLACY 1983). When a small area of an actin-labeled stress fiber is bleached with a focused laser microbeam, fluorescence recovers slowly but steadily with a halftime of 5–10 minutes and is completely restored in 30–40 minutes. This value is 1–2 orders of magnitude lower than the recovery time in peripheral networks bleached in a similar fashion, but it nevertheless demonstrates a considerable degree of mobility of actin within bundles (see also WANG 1984 for a discussion of stress fiber dynamics). It is difficult to tell, however, whether the exchange occurs via monomers at the ends of actin filaments within the stress fiber, or via short polymers that assemble elsewhere in the cytoplasm and then associate with the bundle. Even the core bundle of microvilli undergoes turnover of all of its components *in vivo*, albeit slower than that of the microvillar membrane proteins (STIDWILL *et al.* 1984). How the structural integrity of bundles is maintained despite the considerable mobility of their constituents is unknown.

Ordered filament bundles different from stress fibers can be induced in cultured cells by a variety of experimental treatments. Dimethyl sulfoxide (DMSO) (FUKUI 1978, FUKUI and KATSUMARU 1979, WEHLAND *et al.* 1980, SANGER *et al.* 1980) or the calcium ionophore A23187 (OSBORN and WEBER 1980) will cause the formation of *intranuclear* bundles composed of actin filaments. Forskolin and dibutryl-cyclic AMP induce the formation of morphologically similar *intracytoplasmic* bundles (OSBORN and WEBER 1984). It is not known at present whether these experimentally induced actin bundles are associated with, or require, the presence of specific actin bundling factors.

An as yet unanswered question concerns the possible role of tropomyosin in the formation of higher order structures of actin filaments. A major regulatory molecule of muscle cells, it is also present in nonmuscle cells where it associates preferentially with more stable structures, such as stress fibers, and is excluded from cell regions where actin filaments appear to undergo rapid reorganization (LAZARIDES 1975, 1976). This has led to the suggestion that different tropomyosin isoforms might regulate the interactions of actin-modulating proteins and actin (PAYNE and RUDNICK 1984). Another possibility is that is affects the polymerization and stability of actin (as it does *in vitro*). Both suggestions need to be tested in cell-physiological experiments. Interestingly, native microfilaments isolated intact from cultured cells form two subpopulations on the basis of their association with tropomyosin and alpha actinin (LIN *et al.* 1984). One has only tropomyosin associated with it, and the other contains both tropomyosin and alpha-actinin. This intriguing segregation suggests that subtle cellular regulatory mechanisms control microfilament structure and composition. The subcellular distribution of these two biochemically different classes of microfilaments has not yet been determined.

VII.2.2. Microtubules

The existence of higher order structures formed from cytoplasmic microtubules is well-known to students of microtubule biology since the early days of the post-glutaraldehyde era of electron microscopy. The best

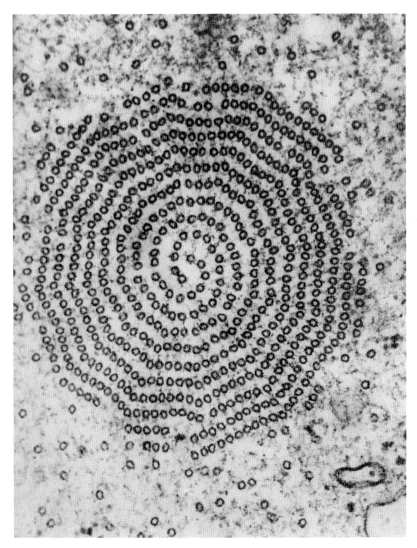

Fig. 50. Transverse section of the microtubule bundle in an axoneme of a heliozoan axopodium showing the characteristic interlocking double spiral. × 78,000

examples of microtubule-microtubule associations are found in protozoans, where microtubules are used to build a diversity of superstructures such as cell extensions, feeding organelles, tentacles, girdles, and motile machineries of astounding complexity. Some of these assemblies are so large they are visible to the naked eye. Three structures shall be discussed here because they have served as model systems, if not paradigms, for the study of microtubule biogenesis, pattern formation, interaction, and assembly: the axonemal rod of heliozoan axopodia, the cytopharyngeal basket of certain free-living ciliates, and the axostyle of certain parasitic protozoa.

The heliozoan axopodium is supported by a rod of parallel microtubules with a cross-sectional pattern of an interlocking double spiral (Fig. 50). Because of its accessiblity and lability, this organelle was used early on to study microtubule assembly/disassembly (TILNEY and PORTER 1965, 1967, TILNEY and BYERS 1968, TILNEY 1968, ROTH *et al.* 1970). Axopodial microtubules are nucleated from organizing centers composed of amor-

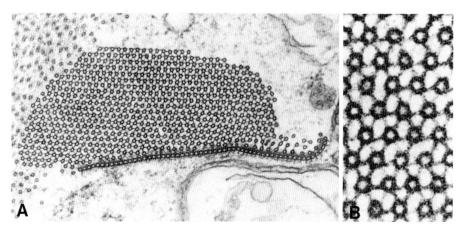

Fig. 51. Transverse section of a cytopharyngeal microtubule bundle in the ciliate *Nassula* (*A*). Microtubules are linked into a hexagonal pattern. × 43,000. *B* Higher magnification of part of a bundle. × 180,000. [From TUCKER (1979)]

phous, electron-dense material that in some species is associated with the nuclear envelope (OCKLEFORD and TUCKER 1973), while in others it forms separate, complex structures (BARDELE 1977). However, the organizing centers seem to have little to do with the formation of the bundle and the generation of pattern within it, which seems, rather, to be a postnucleation *self-linkage* process (TUCKER 1977, 1984) involving intermicrotubule links of two kinds, short intrarow bridges and long interrow links. Owing to a large spectrum of linkage possibilities, microtubule bundles show a greater polymorphism than actin bundles, and are more rigid than the latter. Microtubule connectors are also involved in the construction of regular packing patterns within the microtubule bundles forming the feeding apparatus of the ciliate *Nassula,* the cytopharyngeal basket (Fig. 51). In this case, microtubule positioning and orientation is specified by a microtubule-organizing structure acting as a *template,* in which the organizing elements are already prepositioned prior to microtubule initiation (TUCKER 1968,

1970). As in axopodial axonemes, bundle coherence and rigidity as maintained by intermicrotubule links. Unfortunately, nothing is known about their biochemical composition and properties.

A slightly different situation exists in the microtubular axostyle of *Saccinobaculus* and related species. The axostyle has both structural and motile functions. The intrarow (or intrasheet) linkages of this organelle most likely are firm, structural connectors, while a different set of projections extends from the microtubules within one sheet to those of the next, innermost sheet, but only occasionally are seen to bridge the entire distance. These links are assumed to be dynein-like in nature and are held responsible for the generation of torsional motions within the axostyle (BLOODGOOD 1975, WOODRUM and LINCK 1980).

Despite the strong structural evidence for the existence of specific microtubular linkage structures, little is known about their biochemistry. *In vitro*-studies on the protein-chemical basis of such interactions are in the early beginning. Neither microtubules formed from phosphocellulose column-purified tubulin, nor from microtubule protein containing high molecular weight MAPs, show significant interactions in tests using the falling ball assay (RUNGE *et al.* 1981). Therefore, the "classical" MAPs present in brain extracts do not seem to function as cross-linkers between microtubules. They could act as cross-linkers only if they have two binding sites for microtubules, or one binding site for microtubules and one for another MAP molecule. Neither possibility has yet been demonstrated. However, several other proteins isolated from a variety of sources have recently been shown to be able to function as cross-linkers between microtubules *in vitro*. The first is axonemal dynein (HAIMO *et al.* 1979), which bundles cytoplasmic microtubules is an ATP-dependent manner. A dynein-like protein in sea urchin extracts seems to have similar properties (HOLLENBECK *et al.* 1984). Whether these molecules might also perform motile functions in addition to structural linkage remains to be shown, for addition of ATP causes the microtubule bundles to fall apart without generating motion. Three other proteins were recently identified as potential bundling factors because they bind to and cosediment with microtubules, increase the turbidity of microtubule solutions, and induce a close association of microtubules visible by light or electron microscopy. These proteins include a 260,000 d glycoprotein from squid giant axons (MUROFUSHI *et al.* 1983), a 35,000 d protein, probably glyceraldehyde-3-phosphate dehydrogenase (KUMAGAI and SAKAI 1983), and the spectrin-like protein from brain known as fodrin (ISHIKAWA *et al.* 1983). Certain cytoplasmic extracts also seem to contain other, unidentified factor(s) with a similar property (MACRAE 1984). These observations are exciting because they make it possible to test microtubule bundling activities *in vitro*. However, it needs to be shown in each case whether a putative bundling protein performs a similar function *in vivo*. Binding to microtubules and bundling in the test tube might be nonspecific and fortuitous.

Most metazoan cells do not seem to form complex microtubular superstructures in their cytoplasm. Loosely organized bundles are present in some specialized cells, however, such as nucleated erythrocytes (NEMHAUSER *et al.* 1980, BARRETT and DAWSON 1974) and blood platelets (BEHNKE 1970), and during morphogenetic processes in certain cell types, including spermatogenesis (FAWCETT 1975), egg cell elongation in insects (TUCKER 1981), and cell wall deposition and preprophase band formation in plant cells (reviewed in LLOYD 1982). Bridges have been demonstrated in many of these assemblies, but neither their biochemical composition nor their specific role are well understood in any of these systems. Certain experimental treatments are also known to cause extensive microtubule aggregation and bundling. Thus, taxol, a low molecular weight plant product will induce the formation of parallel arrays of microtubules (SCHIFF and HORWITZ 1980, DE BRABANDER *et al.* 1981, MASUROVSKY *et al.* 1981, GREEN and GOLDMAN 1983). Taxol-induced bundling both *in vivo* and *in vitro* may require the presence of high or medium molecular weight MAPs (ALBERTINI *et al.* 1984, TURNER and MARGOLIS 1984), but the nature of the association between microtubules, MAPs, and taxol is unknown. Another intriguing, unexplained, phenomenon is the recently discovered ability of three-times cycled microtubule proteins to form an ATP-dependent microtubule "gel" (WEISENBERG and CIANCI 1984) that may undergo contraction, or syneresis. Gel "contraction" is accompanied by the formation of bundles and aster-like structures, and is typically observes 15–30 minutes after ATP addition. While this behavior of microtubule protein might well be an interesting test tube phenomenon related to the artificial conditions required for *in vitro* assembly, it might also reveal previously unknown, intrinsic properties of microtubules.

VII.2.3. Intermediate Filaments

Several classes of intermediate filaments have a strong tendency to associate with each other to form extensive, bundled arrays. These include neurofilaments, glial filaments, and the prekeratin filaments of epithelial cells. With few exceptions (BLOSE and CHACKO 1976, BLOSE 1979), highly ordered bundles of intermediate filaments are not as prominent in other cell types. As in the case of actin filaments and microtubules, interactions of intermediate filaments with themselves and with other filaments are mediated by associated proteins. Wispy connections between neurofilaments have long been known to electron microscopists (see WUERKER and KIRKPATRICK 1972). These bridges have recently been visualized by quick-freezing and deep-etching of detergent-extracted neuronal cytoskeletons as regularly spaced rods of about 2 nm diameter (HIROKAWA *et al.* 1984). They are composed, at least in part, of the 200,000 d neurofilament subunit, as shown by antibody labeling of these deep-etched preparations (HIROKAWA *et al.* 1984). This finding is in agreement with earlier studies suggesting that in self-assembling mixtures of the neurofilament triplet proteins, the 200,000 d subunit will cause the filaments to be decorated with sidearms or projections (MOON *et al.* 1981, WILLARD and SIMON 1981, GEISLER and WEBER 1981 a). Because the large neurofilament subunit may not just be a peripherally attached side-arm (as, for example, the high molecular weight MAPs of microtubules), but may also be part of the neurofilament core (OSBORN and WEBER 1983, MINAMI *et al.* 1984), neurofilaments seem to have a built-in-cross-bridge system made up of the non-alpha-helical domain of the 200,000 d polypeptide.

The keratin filament bundles of epithelial cells are characterized by the presence of a protein or proteins termed filaggrin involved in bundling. Filaggrins appear to form noncovalent linkages with the terminal domains of the keratin subunits, suggesting that the sequences of the alpha-helical keratin domains and similar domains in the cross-linker have evolved coordinately to maximize interactions (STEINERT *et al.* 1981).

While vimentin or desmin filaments generally do not form prominent bundle structures, they may interact to maintain a network-like organization. In support of this possibility, the high molecular weight protein synemin has been shown to bind to vimentin filaments at regular intervals, and has been suggested to act as an interfilament cross-linker (GRANGER and LAZARIDES 1982). While the images presented in support of this idea are highly suggestive, this possibility needs to be tested biochemically. A final question worthy of further analysis concerns the interconnection of two different kinds of intermediate filament networks in the same cell. As shown in section III.4.5., some cell types, particularly cultured cells, have more than one intermediate filament class. Thus, PtK 2 cells display both the radial vimentin filament pattern and the reticular prekeratin bundle pattern (FRANKE *et al.* 1979 a, OSBORN *et al.* 1980). Microinjection of antibodies against one or the other filament type specifically disrupts the organization of both, suggesting that the two networks are associated in some as yet unknown way (KLYMKOWSKY 1982).

VII.3. Interactions Among Different Filament Components

VII.3.1. Actin Filaments and Microtubules

The possibility of a cooperative interaction between microtubules and actin filaments (and, therefore, the actomyosin-based force-generating machinery) has been considered by investigators of directional cell locomotion, chromosome transport in mitosis, polarized intracellular organelle movements, and cell shape changes. Considerable effort was invested in attempts to demonstrate meaningful structural associations between these two major cytoskeletal systems to explain these cellular activities. However, neither for the mitotic spindle (*e.g.,* HERMAN and POLLARD 1979, FORER and JACKSON 1979, FORER *et al.* 1979, AUBIN *et al.* 1979, BARAK *et al.* 1981) nor for other microtubule-associated motile phenomena (reviewed in SCHLIWA 1984) could a convincing case be made for a functional association. The difficulty in detecting such associations resides in part in the difficulty of preserving actin for conventional electron microscopy (MAUPIN-SZAMIER and POLLARD 1978, LEHRER *et al.* 1981, SMALL 1981), and/or the problem of detecting short segments of actin filaments in the cytoplasmic matrix even if precautions for better preservation are taken (MAUPIN and POLLARD 1983, MCDONALD 1984, POLLARD *et al.* 1984). More recently, however, several pieces of evidence have renewed the idea of a functional *in vivo*-association between the two systems: 1. Detergent-extracted whole mount preparations of cultured cells reveal frequent connections between microtubules and actin filaments (SCHLIWA and VAN BLERKOM 1981, SCHLIWA *et al.* 1982 a, BRIDGMAN and REESE 1984) via lateral contacts, 3 nm linker filaments, or end-to-side associations. 2. An actin-associated protein, gelsolin, whose only known function is to sever actin filaments in the presence of micromolar calcium, inhibits fast axoplasmic transport of organelles, a process thought to be microtubule-dependent (BRADY *et al.* 1984). 3. Finally, an impressive example of a structural association between actin filaments and microtubules is found in the extensive anastomosing network of reticulopodial strands in a giant freshwater amoeba. Within these strands, microtubules and actin filaments maintain a close association throughout the network, frequently forming composite bundles where actin filaments interlace with the microtubules (Fig. 52). Whether this morphological relationship also implies a functional association remains to be demonstrated.

At the biochemical level, there can be little doubt that actin filaments and microtubules are able to interact specifically. While interactions may even take place between the monomers (VERKHOVSKY *et al.* 1981) the more significant associations are at the level of the polymer. Microtubule protein and rabbit skeletal muscle actin filaments will interact to form a gel at concentrations lower than those present in the cell (GRIFFITH and POLLARD 1978). This interaction is dependent on the presence of either high molecular weight MAPs or tau. In the absence of MAPs, mixtures of the two have a viscosity (determined by the falling-ball assay) close to the sum of the viscosities of microtubules and actin filaments alone (GRIFFITH and POLLARD 1978, 1982 b). Low concentrations of MAPs—either an unfractionated mixture, or separate fractions of high molecular weight MAPs and tau—will cross-

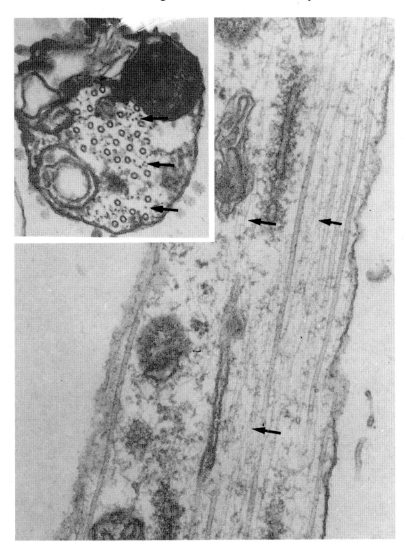

Fig. 52. Longitudinal and transverse section (insert) of a filopodial strand of the giant amoeba *Reticulomyxa*, showing parallel alignment and close association of microtubules and actin filaments (arrows). × 72,000. (Courtesy of K. McDonald)

link actin filaments into a three-dimensional gel (Sattilaro *et al.* 1981, Griffith and Pollard 1982b), or into packed bundles at higher MAP/actin ratios (Sattilaro *et al.* 1981). One millimolar ATP, AMP-PCP (a nonhydrolyzable analog), or orthophosphate seem to inhibit bundle formation and will dissociate formed bundles. The ATP-effect may be related to the recent demonstration that MAP phosphorylation influences the ability of MAPs to cross-link actin filaments (Nishida *et al.* 1981,

SELDEN and POLLARD 1983). Preparations of microtubule proteins are usually contaminated with a—MAP-associated—protein kinase. The incorporation of phosphate into either MAP 2 (7–8 phosphorylation sites) or tau (1–2 phosphorylation sites) inhibits the cross-linking ability (SELDEN and POLLARD 1983); lower levels of phosphorylation of MAP 2 (occupation of 3–4 sites) will promote it. Thus, phosphorylation of MAPs may be a physiological means by which the cell regulates microtubule-actin interactions. An important aspect of these interactions is that the association of MAPs with actin filaments is much weaker than with microtubules. While many weak bonds may be sufficient for significant interactions, slight changes in the conditions (such as MAP phosphorylation) can rapidly alter the properties of the system. Another relevant feature is that MAP 2 retains a substantial degree of molecular flexibility even when bound to microtubules (WOODY *et al.* 1983). Such flexibility may be required to accomodate various geometric configurations of microtubule-actin filament associations, *e.g.,* variations in the angle of overlap and the distance between the two. As mentioned in the discussion of actin filament networks, flexibility is an important requirement for a cross-linking protein. As might be expected, cross-linking of actin filaments by MAPs has a profound effect on the mobility of actin filaments. During fluorescence recovery of photobleached mixtures of rhodamine-labeled actin and fluorescein-labeled MAPs, ARAKAWA and FRIEDEN (1984) observed a sharp decrease in the mobility of actin filaments after addition of MAPs, presumably due to the formation of a network of nondiffusible filaments. Surprisingly, however, the MAPs themselves seem to become only partially immobilized, suggesting different classes of binding affinities for actin. Further studies using this procedure might prove useful because it is a powerful, noninvasive approach to determine supramolecular associations.

The participation of MAPs in microtubule-actin interactions in cells has yet to be demonstrated. Current lack of information can be attributed to the difficulties of identifying, isolating, and characterizing MAPs from sources other than neurons. Notwithstanding our present ignorance, the gel-forming interaction between microtubules and actin filaments might explain the "wavyness" of cytoplasmic microtubules in many cells, best demonstrated by immunofluorescence microscopy or electron microscopy of whole-mount preparations. Because microtubules are embedded in, and at many points connected to, the three-dimensional actin network, it is easy to envision that rearrangements of the actin network could influence the course of microtubules by pulling them into tortuous shapes. Three lines of indirect evidence are in agreement with this speculation: 1. endogenous cellular microtubules of detergent-extracted cultured cells are curvilinear and wavy, whereas microtubule segments polymerized onto their ends are straight (SCHLIWA and VAN BLERKOM 1981); 2. detergent-opened cytoskeletons of cells depleted of their microtubules will form a new aster of microtubules radiating from the centrosome(s) to the cell periphery when supplied with exogenous tubulin. In this aster, which did not form in the environment of an intact cell, the overwhelming majority of microtubules is perfectly straight (BRINKLEY *et al.* 1981); 3. in cells treated with cytochalasin D, the continuity of the actin network is severely disrupted and the bends in the microtubules are "straightened out" (SCHLIWA 1982). The potentially far-reaching implications of these interactions for the maintenance of cell structure and behavior clearly indicate the need for further work.

A specific association between microtubules and myosin has yet to be demonstrated. However, there are reports on the partial decoration and depolymerization of microtubules with skeletal muscle myosin (HAYASHI 1979), and the formation of interdigitating arrays of microtubules and myosin filaments in taxol-treated myoblasts (ANTIN *et al.* 1981, TOYAMA 1982). It is unclear whether these phenomena can be taken as indications for possible myosin-microtubule interactions.

VII.3.2. Actin Filaments and Intermediate Filaments

Whether actin filaments and any of the intermediate filament types interact with each other has not even begun to be approached biochemically. A few scattered cytological or physiological observations suggest that such an undertaking might be worthwhile. For example, treatment with cytochalasin B produces a star-like arrangement of both prekeratin and vimentin filaments in one epithelial cell line (CELIS *et al.* 1984). A combination of colchicine and cytochalasin induces a transition from a uniform network of keratin filament bundles to a lattice-like arrangement with membrane-associated focal centers in mouse epidermal cells (KNAPP *et al.* 1983). Either compound alone did not produce such an effect. However, such drug experiments need to be interpreted with some caution, particularly if several compounds are used in combination. So far, none of the known intermediate filament-associated proteins have been shown to have an affinity for actin filaments. Conversely, none of the actin-associated proteins has been demonstrated to bind to intermediate filaments, with the potentially important exception of spectrin (LANGLEY and COHEN 1984). The relationship between spectrin and intermediate filaments needs to be studied in more detail. If a specific association of the two can be confirmed, it would open the exciting possibility of a specific link between the plasma membrane and the intermediate filament network (see section IX.3.).

VII.3.3. Microtubules and Intermediate Filaments

Associations between microtubules and intermediate filaments have been postulated on the basis of electron microscopic observations many years ago. In neurons, neurofilaments and microtubules are closely apposed and frequently appear interconnected by filamentous crossbridges (PETERS and VAUGHN 1967, BERTOLINI et al. 1970, RAINE et al. 1971, ELLISMAN and PORTER 1980, HIROKAWA 1982). Possibly as a result of this crossbridging, microtubules and neurofilaments move together as a coherent complex in the slow component of axonal transport (LASEK and HOFFMAN 1976, BLACK and LASEK 1980, TYTELL et al. 1981, BRADY and LASEK 1982). In cultured cells, there are several indications that the distribution patterns of microtubules and intermediate filaments are not only spatially correlated, but also reflect a functional association. For example, transverse sections of cell processes occasionally reveal fairly regular arrays of the two components (GOLDMAN and FOLLET 1969, WANG and GOLDMAN 1978). In horizontal sections of cultured cells, microtubules and intermediate filaments often run parallel to one another for long distances, sometimes connected by thin bridges (FRANKE et al. 1978 c, WANG and GOLDMAN 1978). A close association between the entire microtubule and intermediate filament network has been demonstrated more recently by immunofluorescence microscopy (GEIGER and SINGER 1980, BALL and SINGER 1981). In double-labeled cells, the two networks are in some places completely superimposed (SINGER et al. 1982). A functional linkage was first suggested by the observation that treatments that disrupt or redistribute microtubules also have a profound effect on intermediate filament organization (Fig. 53). Thus colchicine, colcemid, or vinblastine treatment has been reported to induce progressive collapse and coiling of intermediate filaments until a perinuclear ring or juxtanuclear cap of tightly packed filaments is formed (GOLDMAN and KNIPE 1972, GORDON et al. 1978, LAZARIDES 1978, BENNETT et al. 1978 b, HYNES and DESTREE 1978). When microtubules are reorganized in the cell rather than completely depolymerized, the intermediate filament network changes coordinately. For example, when treatment with the microtubule-stabilizing agent taxol provokes the formation of peripheral microtubule bundle arrays, the intermediate filament network reorganizes to conform to the altered microtubule pattern (GEUENS et al. 1983, MARO et al. 1983) (Fig. 54). In apparent disagreement with these observations, the organization of intermediate filaments is unaffected in cells where microtubules were depolymerized by low temperature rather than antimicrotubule agents (VIRTANEN et al. 1980). However, this discrepancy could be resolved if the reorganization is an active, possibly energy-requiring process that would not take place at low temperature. More recent observations also indicate that complete depolymerization of microtubules is not necessary to induce intermediate filament collapse. In cells microinjected with monoclonal antibodies against tubulin, microtubules may remain intact, yet the intermediate filament system aggregates (WEHLAND and WILLINGHAM 1983, BLOSE et al. 1984). In this experiment, the attachment of the antibody to the microtubule wall seems to interfere with the association between the intermediate filaments and the microtubules. The reciprocal experiment, however, microinjection of antibodies against

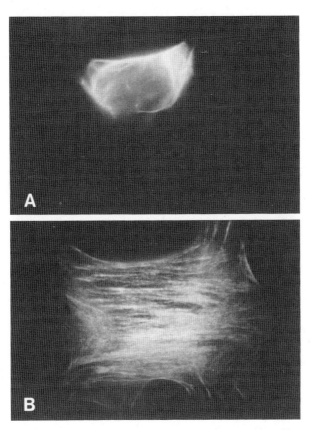

Fig. 53. Cultured PtK cell stained for vimentin intermediate filaments with an antibody against vimentin (*A*), and for actin filament with rhodamine-phalloidin (*B*). The cell was treated with 4 µg/ml nocodazole for 16 hours, which has caused the intermediate filaments to collapse towards the nucleus (*cf.,* Fig. 26). The distribution of actin filament bundles is apparently unaffected. × 400

intermediate filament proteins, does not affect the distribution of microtubules. In this case, the intermediate filaments coil up as a perinuclear bundle cross-linked by antibody molecules, but the pattern of microtubules is identical to that of uninjected cells (GAWLITTA *et al.* 1981, LIN and FERAMISCO 1981). These observations suggest that microtubule distribution is dominant over intermediate filament distribution, and that the latter require certain positional clues provided by the microtubules.

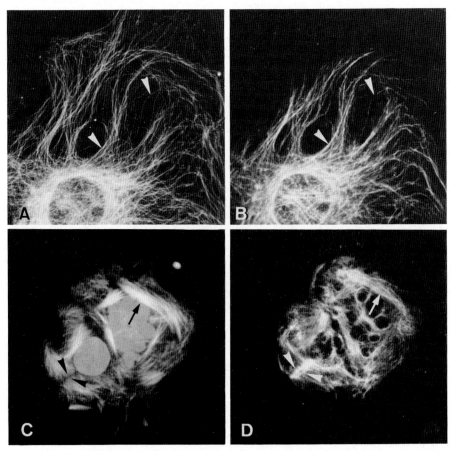

Fig. 54. Double-label immunofluorescence microscopy with antibodies against tubulin (*A, C*) and desmin (*B, D*) in epithelial MO cells. *A, B* Untreated cells. Note close correspondence of microtubule and intermediate filament patterns (arrowheads). *C, D* After treatmant with 100 µM taxol for 24 hrs. The intermediate filament pattern has changed coordinately with the microtubule pattern (arrows and arrowheads). × 500. [From GEUENS *et al*. (1983)]

It needs to be emphasized that an association of this kind only seems to exist between microtubules and the vimentin or desmin class of intermediate filaments, but not the cytokeratins. The reticular bundle organization of cytokeratin filaments is largely unaffected by antimicrotubular agents or microinjected tubulin anbibodies, and there is little, if any, overlap in the fluorescence patterns of the microtubule and keratin networks in cultured cells. The different effect of antimicrotubular agents on vimentin and cytokeratin networks is best demonstrated in cell types that express both filament classes, such as PtK cells. In this cell type, microtubule disruption induces a collapse of the vimentin network, but not the cytokeratin bundles (OSBORN et al. 1980).

These cytological observations of an apparently specific molecular association between the two filament types are supported by biochemical studies on the interaction between one of the intermediate filament types, neurofilaments, and microtubules. To the dismay of microtubule biochemists, neurofilament proteins are a tenacious contaminant of microtubule protein preparations even after repeated cycles of assembly/disassembly (BERKOWITZ et al. 1977). Neurofilaments apparently have the ability to bind and sequester high molecular weight MAPs, which may result in an inhibition or retardation of microtubule assembly (PYTELA and WICHE 1980, LETERRIER et al. 1982). Both MAP 1 and MAP 2 may be found in association with neurofilaments centrifuged out of a solution of microtubule protein. A cytological corollary of this biochemical affinity between MAPs and intermediate filaments is that in primary cultures of rat brain cells, MAP 2 will associate with perinuclear vimentin cables after experimental disassembly of microtubules (BLOOM and VALLEE 1983). Several parameters of MAP-mediated interaction between microtubules and neurofilaments were studied in vitro (RUNGE et al. 1981, AAMODT and WILLIAMS 1984 a, b). Brain neurofilaments, but not the neurofilaments prepared from spinal cord, form a viscous complex with brain microtubules, as determined by low-shear falling ball viscometry. Gel formation requires the presence of proteins resembling MAPs, which appear to be associated with brain but not spinal cord neurofilaments. However, spinal cord neurofilaments will form a complex with microtubules if brain MAPs are added to the mixture. The viscosity of microtubule-neurofilament solutions depends on the concentration of MAPs and is saturable. The conclusion is that MAP-related proteins mediate the association between the two fiber systems in vitro, presumably by acting as cross-links or bridges between them (AAMODT and WILLIAMS 1984 a). An apparent ATP requirement for the formation of the microtubule-neurofilament complex (RUNGE et al. 1981) was not confirmed in subsequent experiments and was shown to be attributable to GDP-induced disassembly of microtubules (MINAMI et al. 1982). Two enzymes which affect GDP levels are present in the in vitro preparations, a GTPase which tends to increase the level of GDP in the absence of ATP, and a nucleotide diphosphokinase, which catalyzes ATP-GTP transphosporylation and tends to decrease the level of GDP in the presence of ATP. Since GDP increases microtubule lability (see section IV.), the disruption of microtubule-neurofilament complexes in the absence of added ATP most likely was due to microtubule depolymerization, rather than labilization of MAP-mediated associations. The speculation that high molecular weight MAPs act as crossbridges is also made plausible by their molecular architecture: they are long, flexible proteins (VOTER and

ERICKSON 1982) that project from the microtubule wall (AMOS 1977, KIM *et al.* 1979). However, several additional issues remain to be addressed: an analysis of the mutual binding sites on intermediate filaments and MAPs is required; the regulation of MAP-mediated microtubule-neurofilament interaction is not understood at all; and better structural evidence (*e.g.,* by immunoelectron microscopy) that MAPs form bridges is needed. Until these questions are resolved, other modes of association also need to be considered possible. For example, one possibility supported by indirect evidence is that the 200,000 d neurofilament polypeptide is involved in network formation (MINAMI and SAKAI 1983, MINAMI *et al.* 1984). This polypeptide is capable of stimulating tubulin polymerization and network formation even when incorporated into the neurofilament. It not only participates in the construction of the neurofilament core, but also forms filamentous projections from the neurofilament that could interact with microtubules, MAPs, or both.

Complex formation between microtubules and any of the other intermediate filament classes has not yet been studied *in vitro* despite ample morphological and physiological evidence for functionally significant interactions (see discussion above). In cultured cells, a group of high molecular polypeptides related to MAPs preferentially associate with intermediate filaments (PYTELA and WICHE 1980, WICHE and BAKER 1982), but they are also present in membrane-filament attachment sites (WICHE *et al.* 1983, 1984). These proteins may serve such a linker function in a phosphorylation-dependent manner (HERRMANN and WICHE 1983). There may be other means of interactions that are tailored specifically to the association of microtubules with one or the other filament class. This clearly is a promising field for future research.

VII.4. Filament Interactions: A Perspective

One of the most striking features that emerged from the study of the three-dimensional structure of a cell's cytoskeleton is the intricacy and complexity of the interactions among its major components. The recognition of this fundamental feature was made possible through the development of specimen preparation techniques that produce graphic, three-dimensional views of the stable cytoskeletal components at the cellular (*e.g.,* immunofluorescence microscopy) and supramolecular (*e.g.,* whole mount electron microscopy) level. The concept of interaction and cooperation among the cytoskeletal elements is further supported by a spectrum of cytological observations and experiments. Compared to the remarkable progress at the structural level, there are big gaps in our knowledge of the biochemical basis for interactions. In the near future, perhaps the greatest progress can be expected in the characterization of the specific associations of the proteins that interact with either microtubules, actin filaments, or intermediate filaments alone. However, the characterization of interactions in even these "simple" systems is very tedious. To characterize the associations in more complex systems involving three or more components is more demanding still. Nevertheless, the observations on the interactions between microtubules and actin as well as microtubules and intermediate filaments encourage the belief that further efforts in this area will be rewarding. Despite the optimism for rapid progress in the near future, in should not be overlooked that the specificities of many of the interactions are not yet determined, and the binding sites and factors involved in the regulation of binding are unknown. However, it does not take much vision to predict that the field of "interactions" will continue to be an important research topic central to an understanding of the integrative function of cytoskeletal networks and the global organization of cytoplasmic space.

From the studies carried out to-date the exciting possibility seems to emerge that at least two classes of proteins are capable of interacting with all three major filament types: high molecular weight MAPs, and the proteins of the spectrin family. MAPs have a high affinity binding site for microtubules, at least one and probably two low affinity binding sites for actin filaments, and also seem to be involved in microtubule-neurofilament interactions. The proteins of the spectrin family bind specifically to actin filaments, are able to bundle microtubules, and also can interact with intermediate filaments. Coupled with their involvement in establishing and maintaining close associations between the cell membrane and the cytoskeleton (to be discussed in section IX.), the spectrin-like proteins may represent a universally important linker that integrates the entire cytoplasm and the cell membrane.

VIII. The Cytoskeleton and the Cytoplasmic Matrix

Extraction of cells with nonionic detergents not only solubilizes cellular membranes, but also liberates approximately two thirds of the cell's proteins. Many of these proteins are localized in the cytoplasmic compartment, together with the filaments and microtubules that comprise the cytoskeleton. This simple observation is a useful reminder that the cytoskeleton is but one component of a much more complex compartment known as the cytoplasmic matrix. This section will examine briefly some structural and biochemical aspects of the interrelationship of cytoskeletal and cytomatrical constituents—bearing in mind, however, that the distinction between "skeleton" and "matrix" is ill-defined and largely

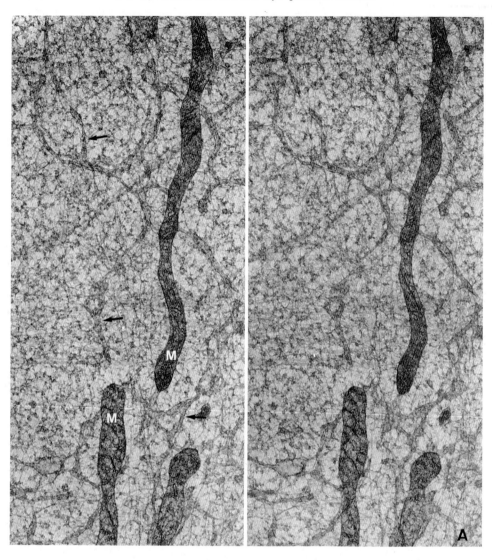

Fig. 55 A. Stereo pair of a BSC-1 cell viewed in the high voltage electron microscope after glutaraldehyde-osmium fixation and critical point-drying. This view includes elongated mitochondria (*M*) and elements of the endoplasmic reticulum (arrows). The cytoplasmic matrix shows numerous fine strands (microtrabeculae) associated with microtubules and with the upper and lower cell cortices. × 24,300. (Courtesy of K. R. Porter)

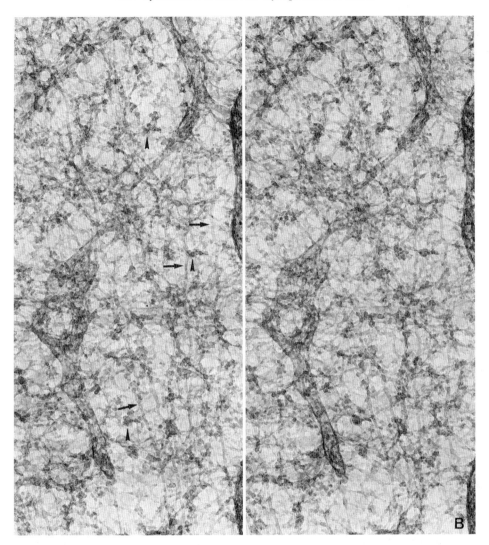

Fig. 55 B. Higher magnification of part of Fig. 55 A, allowing a better view of the intricacy of the filamentous network pervading the cytoplasm. In addition to microtubules (arrows), dense spherical particles representing ribosomes (arrowheads) are suspended in the lattice. × 60,500. (Courtesy of K. R. PORTER)

operational. Based on these considerations an attempt will be made to develop an integrative view of cytoplasmic organization.

Ever since DUJARDIN's observations, 150 years ago, the question of how the cytoplasm is organized has lingered, consciously or unconsciously, in the minds of many biologists. The history of this field (nicely summarized in PORTER 1984) has seen heated controversies between different schools of thought, interrupted by phases of relative quiescence where attention became focused on seemingly more interesting subjects. Perhaps the first climax in the discussions on cytoplasmic organization was the dispute over the granuler, alveolar, and fibrillar theories of protoplasm towards the end of the 19th century (see summaries by HEIDENHAIN 1911, WILSON 1928). These painstaking and often painful discussions on the real or artifactual nature of the structures visualized by the limited light microscopic and histologic techniques involved virtually the entire community of cytologically oriented scientists of that time. A "modern" corollary of this dispute is the discussion which has arisen over the *microtrabecular* structure of the cytoplasmic matrix proposed by KEITH PORTER and his collaborators. As an alternative to the conventional embedding/sectioning/staining scheme used for the visualization of cells in the electron microscope, cells were grown directly on electron microscopic grids, critical-point dried to preserve their three-dimensionality, and viewed as whole mount preparations. In order to penetrate the cells with the electron beam, they were examined in a high voltage electron microscope; and to appreciate the three-dimensional organization of the cells, micrographs were viewed as stereo pairs (BUCKLEY and PORTER 1975, BUCKLEY 1975, WOLOSEWICK and PORTER 1976, 1979). Visualized in this way, the cytoplasm consists of an intricate three-dimensional network of elongated, somewhat irregular strands (microtrabeculae) that vary in diameter from 2–15 nm (Fig. 55). The overall appearance of this network bears little resemblance to the network of the Triton-resistent cytoskeleton, except for the fact that both are networks. The microtrabecular lattice, as it is called now (WOLOSEWICK and PORTER 1979), is deemed to be a real cellular structure on the basis of a number of primarily morphological arguments: 1. the appearance of the microtrabecular lattice is different in cell regions that presumably perform different functions (TEMMINCK and SPIELE 1980); 2. its organization changes in a predictable fashion in different states of cellular activity, as best demonstrated by the cycles of aggregation and dispersion of pigment granules in chromatophores (BYERS and PORTER 1977); 3. treatment with metabolic and other inhibitors induces marked changes in the network's organization (LUBY and PORTER 1980, 1982, GIBBINS 1982); 4. finally, a similar network is observed in sections of cells embedded in polyethylene glycol, a water-soluble embedding medium, after the removal of the medium (WOLOSEWICK 1980, GUATELLI et al. 1982, KONDO 1984). The strands of this network are believed to represent a polymeric and polymorphic assembly that constitutes a protein-rich cytoplasmic phase separated by a water-rich solution of low molecular weight components. Through changes in its structural conformation, the network is also proposed to provide the motive force for cellular movements (BYERS and PORTER 1977, ELLISMAN and PORTER 1980), and to account for the nonrandom distribution of other cellular constituents (PORTER and MCNIVEN 1982, PORTER *et al.* 1983).

This view of cytoplasmic organization did not remain without criticism. The discussion that PORTER'S proposal aroused was soon reduced to the question of whether the images produced with the procedures used (fixation/dehydration/critical point-drying) are an accurate or artifactual representation of the organization of living cytoplasm. Initially the strands of the microtrabecular lattice were denied the status of real structures because they are not present in detergent-extracted cells (HEUSER and KIRSCHNER 1980). This argument is based on the assumption that if the microtrabeculae are filamentous polymers similar to actin filaments or microtubules, they should have similar stability. Other, more serious criticisms revolved around the morphological characteristics of micro-trabecular strands, specifically, their irregular, reticular, outline. Structures of similar morphology can be produced by osmication of actin networks (SMALL 1981), or incomplete dehydration of a variety of linear polymers such as DNA, collagen, and F-actin (RIS 1980). These observations led to the view that microtrabeculae might simply represent a distorted actin network. Another criticism is based on the inspection of rapidly frozen, deep-etched cells, or freeze-substituted and critical point-dried whole mount preparations (HEUSER and KIRSCHNER 1980, HIROKAWA and HEUSER 1981, SCHNAPP and REESE 1982, HIROKAWA 1982, BRIDGMAN and RESSE 1984). In these preparations, the cytoplasm appears to contain a network of discrete, uniform filaments embedded in, and often obscured by, a dense, granular or moderately filamentous ground substance (Figs. 5–58). Unlike fixed, critical point-dried whole mounts, images of these preparations show discrete long filaments, probably representing the major filaments of the cytoskeleton, associated with other, similarly discrete, granular material.

It is difficult to tell which technique gives a more accurate representation of cytoplasmic structure. Neither procedure is free of artifacts, but it appears that chemical fixation and critical point-drying are more likely to produce adventitious aggregation of macromolecules. The question of precisely how the cytoplasm is structured may perhaps never be solved to everyone's satisfaction. While the cytoplasmic matrix is possibly less highly organized than some of the images suggest, it is equally clear that there *is* structure and order above and beyond filaments and microtubules. Demonstrated by the various techniques, the existence of a surprisingly intricate cytoplasmic matrix is perhaps the most important lesson to be learned from the structural studies on whole, intact cells. If this conclusion is accepted, then the question arises: What is the biochemical nature and composition of these extrafilamentous components and their mode of association with the cytoskeletal fibers?

Fig. 56. Thin section of a fibroblast that was quick-frozen and freeze-substituted in acetone-osmium. The stress fibres are resolved into closely packed bundles of long filaments. Microtubules are smoothly curving through a finely floccular cytoplasmic matrix. × 85,000. Courtesy of John Heuser

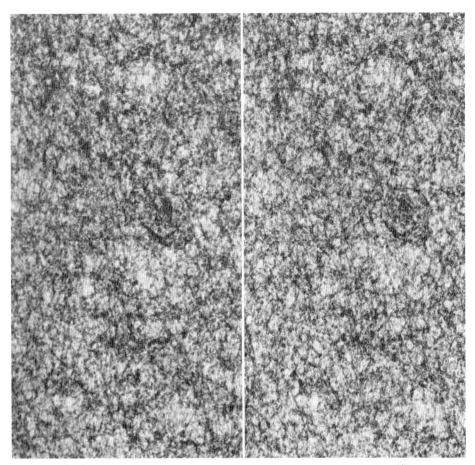

Fig. 57. Stereo micrograph of the filament meshwork in a somatic cell from *Xenopus* that was directly frozen in liquid propane/isopentane, freeze-substituted, and critical point-dried. There are many granules packed between the filaments. × 48,000. [From BRIDGMAN and REESE (1984)]

The cytoplasm of eukaryotic cells is the site of many synthetic and metabolic activities which require specific enzymes and other macromolecular factors. Further, there is at least as much nonpolymeric as polymeric actin and tubulin, all of which presumably resides in the cytoplasm. All these cytoplasmic components make up a rich macromolecular "soup" in which the concentration of protein probably exceeds 120 mg/ml. Is it possible that many of the matrix components form higher-order complexes that may take on a filamentous or granular shape, however transiently? Is there evidence for the association of cell constituents generally considered to be non-cytoskeletal with the filamentous components of the cytoskeleton? A number of observations suggest this is the case, and the evidence is strongest for the glycolytic enzymes. Originally considered to be "soluble" or "cytosolic" enzymes—an operational definition based on cell fractionation—many of these enzymes are now known to bind to structural components of the cell, as demonstrated by the following lines of experimental evidence:

1. Muscle glycolytic enzymes are nonuniformly distributed in myofibrils. By histochemical and immunological procedures, they are found to be concentrated in the I-band of striated muscle (SIGEL and PETTE 1969, DOLKEN et al. 1975). This is the region where actin filaments reside.

2. When mixed with filamentous actin, many glycolytic enzymes will bind to the filaments and cosediment when ultracentrifuged (CLARKE and MASTERS 1975, KNULL et al. 1978, WALSH et al. 1980). The affinities of different enzymes of the glycolytic pathway for actin vary (e.g., CLARKE and MASTERS 1976), but apparently all glycolytic enzymes have the capacity to bind to actin (WESTRIN and BACKMAN 1983). Some enzymes, notably aldolase, bind to actin with such avidity that they can cross-link actin filaments into rigid gels (GRIFFITH and POLLARD 1982 a), suggesting that aldolase either has two actin binding sites or can form dimers.

3. One of the most striking pieces of evidence for a structural association between enzymes of intermediary metabolism and the filamentous cytoskeleton comes from studies of axonal transport in mammalian nerves. The so-called slow component of orthograde axonal transport, in which materials move at a rate of 0.1–10 nm/day, includes two major subcomponents, designated slow components a and b. Both consist of a complex of major cytoskeletal structures: slow component a represents the microtubule-neurofilament network and little else; slow component b includes the microfilament network (with actin being the marker protein), and, among more than 100 polypeptides, the complex of metabolic enzymes. The important point is that slow component b moves along the axon as a coherent wave with little overlap with other waves of axonal transport, and little diffusional dispersion of its constituent polypeptides. The fact that enzymes of intermediary metabolism, such as enolase and creatine kinase (BRADY and LASEK 1981), move coordinately within this wave implies a structural association with the other components transported at the same rate. Slow component b appears to represent a supramolecular complex

---

Fig. 58. Quick-frozen, etched, and rotary-shadowed preparation of an untreated frog axon. In addition to a variety of membrane-bound organelles, this preparation shows microtubules (arrows) and neurofilaments (arrowheads). The cytoplasmic matrix is characterized by a dense meshwork of filamentous and granular material. × 240,000. (Courtesy of N. HIROKAWA)

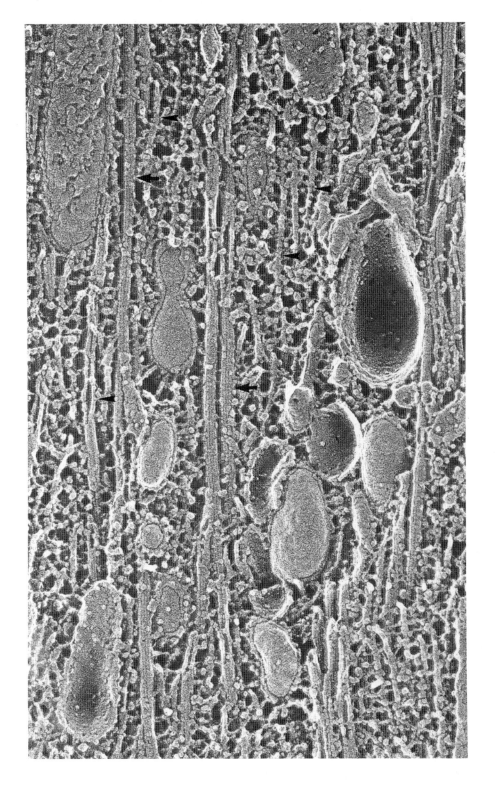

composed of a great number of diverse molecules. The "structural hypothesis" of axonal transport, which has been developed on the basis of this and other observations (see summaries by BRADY and LASEK 1982, LASEK 1982, 1984) provides a compelling argument for structural associations among diverse cytoplasmic proteins.

4. A significant aspect of enzyme binding to filamentous actin is that it may alter the functional characteristics of the enzymes. For example, both aldolase and phosphofructokinase increase their kinetic parameters upon binding to actin (KARADSHEH and UYEDA 1977, WALSH et al. 1977, HAND and SOMERO 1984). It is important to note that the state of association of an enzyme with actin may not be fixed, but rather may oscillate between the soluble and the bound state according to the local microenvironment and the metabolic status of the cell. Such behavior has been termed "ambiquitous" (WILSON 1978).

The evidence for binding of metabolic enzymes to structural components, notably actin, is compelling. One possible consequence of the association of glycolytic enzymes with actin filaments is that the pathway of glycolysis is sub-compartmentalized into a higher-order glycolytic complex. While this is not much more than attractive speculation at present, it appears plausible because actin filaments form a structural substrate onto which the complex can adsorb (CLARKE and MASTERS 1974). The formation of an adsorbed complex would be enhanced by high-affinity binding of key enzymes (i.e., aldolase, phosphofructokinase) to the actin matrix, forming an anchored complex which, in turn, might increase cooperative binding of the other components (MASTERS 1981). Such a "piggy-back" mechanism could conceivably lead to the formation of "metastable" aggregates of substantial size that could be dismantled and formed rapidly in accordance with the ambiquitous nature of the association between the enzymes and the filamentous matrix. The glycolytic enzyme complex might therefore be considered a structual, yet protean, component of the cytoplasm and might give rise to a granular or filamentous appearance in the electron microscope in carefully prepared specimens. The complex could be dispersed easily even under the mildest of conditions of cell extraction or cell fractionation. Studies employing these procedures therefore generally miss the possible structure-forming properties of "soluble" enzymes and the contributions of the stable filamentous cytoskeleton to enzyme control and metabolism.

Glycolytic enzymes provide the most striking example of associations between structural and nonstructural cell components, but there are many other cases: The regulatory subunit (R II) of the cyclic AMP-dependent protein kinase is associated with MAP 2 (VALLEE et al. 1981, MILLER et al. 1982) and other cellular proteins that have not been identified (LOHMANN et al. 1984). In cultured cells, creatine phosphokinase may be found in association with intermediate filaments by immunocytochemical criteria

(ECKERT *et al.* 1980). Certain glycosphingolipids may associate with microtubules (SAKAKIBARA *et al.* 1981). A 50,000 d protein that binds to the 5'-cap of eukaryotic messenger RNA is associated with intermediate filaments in cultured cells (ZUMBE *et al.* 1982), a finding that corroborates previous observations on the stable association of polysomes (LENK *et al.* 1977) and mRNA (NIELSEN *et al.* 1983, JEFFREY 1984) with cytoskeletal elements. In fact, this association seems to be required for translation and the cotranslational assembly of cytoskeletal proteins (FULTON *et al.* 1980, CERVERA *et al.* 1981). Finally, cytoskeletal elements may even influence, in an as yet unknown fashion, the integrity and protein synthesizing capacity of membrane-bound ribosomes (WALKER and WHITFIELD 1984).

These examples of the association and interaction of nonstructural cytoplasmic components with the stable cytoskeleton hint towards subtle interdependencies affecting many cytoplasmic constituents and metabolic or regulatory activities. Elucidating the precise mechanisms of these associations may be a formidable task. If it is considered that additional cytoplasmic components might associate with those more firmly bound to the cytoskeleton (in the "piggy-back" fashion alluded to earlier), then determining the biochemical basis of order within the cytomatrix may seem next to impossible. Hopefully, the realization of the possible existence of such subtle interactions may remove the concept of a structured cytoplasmic matrix from the realm of speculation and give it the status of a testable hypothesis in the not-to-distant future.

That the cytoplasmic matrix is substructured into macromolecular complexes is also supported by two completely different lines of indirect evidence based on biophysical approaches. The translational mobility of fluorescently labeled molecules microinjected into living cells, determined by the technique of fluorescence recovery after photobleaching, is largely independent of molecule size in the range between 10,000 and 500,000 d, even through there are cell type-specific differences (*e.g.*, WOJCIESZYN *et al.* 1981, KREIS *et al.* 1982, WANG *et al.* 1982, JACOBSON and WOJCIESZYN 1984). Because the diffusion coefficients are not inversely proportional to the molecular radius of the injected probes, this observation excludes models of the cytoplasm as a viscous solution of macromolecules. For most of the test molecules, which include insulin, bovine serum albumin, $F(ab)_2$ fragments of immunoglobulin molecules, actin, and tubulin, the diffusion coefficient is almost two orders of magnitude lower than that in dilute solutions. One explanation for this observation is that transient association with elements of the cytoplasmic matrix retards the diffusion of these molecules through the cytoplasm (JACOBSON and WOJCIESZYN 1984). In contrast, the movement of small molecules or the bulk of the cellular water itself is not markedly retarded within the cell (WOJCIESZYN *et al.* 1981, MASTRO *et al.* 1984, MASTRO and KEITH 1984). This observation makes more tenable the idea that the diffusion of macromolecules is reduced by binding to cellular structures, rather than increased cytoplasmic viscosity. Withdrawal of most of the free water from the cell almost completely stops the diffusion of molecules within the cytoplasm and compresses the cytoplasmic matrix to the point where no substructure is discernible (ALBRECHT-BUEHLER and BUSHNELL 1982, PORTER *et al.* 1982). The interdependency of the distribution of cellular water and the associations within the cytoplasm support a structural rather than viscous cytomatrical organization (KEITH 1979, CLEGG 1984 a, b, HOROWITZ and MILLER 1984).

The second physical approach suggesting the existence of structural associations among many cytoplasmic proteins is intracellular fractionation by centrifugation of intact cells (ZALOKAR 1960, KEMPNER and MILLER 1968). This approach requires cells with a tough cortex, such as the flagelate *Euglena*, that can withstand the enormous pressures generated during centrifugation and prevent bursting of the cell. Centrifugation generates stratified cells in which different layers contain different organelles. The layer at the top is organelle-free and, surprisingly, also free of "soluble" proteins (KEMPNER and MILLER 1968). The conclusion from this observation was that the cytoplasmic proteins are structure-bound and/or structure-forming, allowing their cosedimentation out from a protein-free layer.

Even the non-cytoskeletal components of the cytoplasmic matrix may be ordered and flexibly structured. "Weak" interactions among the matrix components may help establish subcompartments within the cytoplasmic supercompartment. To some extent, proteins may be forced into these interactions by the fact that the cytoplasm is so highly concentrated. For example, high concentrations of *any* globular protein promote the formation of tetramers of glyceraldehyde-3-phosphate dehydrogenase (MINTON and WILF 1981), and spectrin expresses a different capacity for associations at those high concentrations found in the subplasmalemmal layer (MORROW and MARCHESI 1981). In fact, many proteins may behave differently under "crowded" conditions (MINTON 1983), and may engage in transient interactions with other proteins. McCONKEY (1982) coined the term *quinary structure* to describe the ability of proteins for inherently transient interactions—the fifth level of organization in proteins. Quinary structure entails the ability of a protein to associate weakly with other proteins in interactions unrelated to its job. In addition to optimizing the specific function a protein performs, evolution may have selected for this property as well. These and other aspects of the unique chemistry and physiscs of the cytoplasm are briefly discussed by FULTON (1982); in fact, this paper should be required reading for cell biology students for it presents a view of cytoplasmic organization complementary to that encountered in most cell biology texts.

IX. Membrane-Cytoskeleton Interactions

IX.1. Introduction

Translation of cytoskeletal interactions and assembly/disassembly processes into useful activities such as cell translocation, shape changes, adhesion, or contact formation requires physical connections between the cytoskeleton and the plasma membrane. Because of their fundamental importance, associations between membranes and the cytoskeleton have been intensely studied in the past decade. A specific impetus was the discovery in many cell types of proteins related to the cytoskeleton of the erythrocyte, a cell that has long been a useful model for the study of cell membranes. This section examines the basic principles of membrane-cytoskeleton interactions from a general perspective; details of all the relevant model systems can not be considered here.

A priori, three modes can be postulated by which cytoskeletal elements might interact with membranes. 1. Cytoskeletal proteins could insert directly into the bilayer. In this way, a protein could function not only in membrane-cytoskeleton anchorage, but also in transmembrane linkage by providing a direct means of communication between the cell exterior and the cytoplasm. Such a function has been proposed for actin but not confirmed experimentally. 2. Cytoskeletal proteins could interact with integral membrane proteins. This form of association is used in some situations. 3. Cytoskeletal proteins could associate with the membrane indirectly by binding to peripheral membrane proteins that bind integral membrane proteins. This mode of association has been found in many systems and seems to be a common from of interaction between membranes and filaments. Frequently, linkage involves several components that form higher-order complexes.

Several experimental strategies can be used to identify and characterize membrane-cytoskeleton interactions. Membrane preperations obtained by cell fractionation can be analyzed for the presence of cytoskeletal proteins. The proteins from these fractions can be purified and used in reconstitution experiments with cellular or model membranes. Affinity columns with bound membranes can be constructed to "fish" for potential peripheral linker molecules. Immunofluorescence and immunoelectron microscopy can be used to look for close structural associations between putative linkers, filaments, and membranes. Detergent-extracted cell models can be prepared to test for the association or retention of integral membrane proteins with the stable elements of the cytoskeleton. All of these procedures have their advantages and limitations, and only a combination of several approaches can demonstrate the existence of specific interactions.

This chapter will commence with a look at the erythrocyte, a comparatively simple model in which all of the important participating molecules are known. More complex situations and interactions will then be considered in the following sections.

IX.2. Paradigm Erythrocyte

The membrane of the red blood cell is probably the best-understood of all cellular membranes in terms of molecular composition and function. Because red blood cells are readily available in large quantities free of contaminating cells, and because their membranes are comparatively simple, they were considered a good model system from which principles of membrane structure and function can be derived (STECK 1974, MARCHESI et al. 1976). The cytoskeleton of the erythrocyte is a two-dimensional weave that covers the cytoplasmic surface of the membrane. No study has produced any evidence for the existence of a three-dimensional transcellular network. The membrane with its associated cytoskeletal network can be isolated intact as a hemoglobin-free *ghost*. Ghosts are approximately the same shape as intact cells (DODGE et al. 1963), supporting the idea that the subplasmalemmal complex of proteins maintains the biconcave shape. Treatment of ghosts with nonionic detergents removes the lipid bilayer and many of the integral membrane proteins, leaving a *shell* composed of a durable and flexible protein network (YU et al. 1973). By electron-microscopic criteria, the shell is a strictly two-dimensional network with irregularly shaped interstices approximately 50 nm across (TSUKITA et al. 1980, NERMUT 1981, RIMME 1981). All of the major (and many of the minor) components of this network have been identified, and their specific interactions have been characterized (for recent reviews of the red cell cytoskeleton, see GRATZER 1981, 1983, BRANTON et al. 1981, BENNETT 1982, MARCHESI 1983, GOODMAN and SHIFFER 1983, COHEN 1983). Their properties shall now be discussed individually before considering their interactions.

IX.2.1. Spectrin

Spectrin is the best-known of all the erythrocyte cytoskeletal proteins (MARCHESI and STEERS 1968). On one-dimensional polyacrylamide gels it runs as a doublet of non-identical polypeptide chains with molecular weights of 240,000 d (alpha chain, or band 1) and 220,000 d (beta chain, or band 2). There are approximately 200,000 copies of each chain per ghost. Structural analyses of the two spectrins revealed multiple homologous but nonidentical sequences of 106 amino acids long, and approximately 12,000 d molecular weight (SPEICHER et al. 1983 a, b, SPEICHER and MARCHESI 1984). Each of these segments can fold into a triple-helical structure with a short, nonhelical, protease-sensitive region that connects neighboring helical units. None of the known sequences of other cytoskeletal proteins shows a similar domain structure. Both subunits are highly elongated molecules that associate side-by-side to form heterodimers of approximately 100 nm length, which in turn associate with each other head-to-head into (alpha, beta)$_2$ tetramers (UNGEWICKELL and GRATZER 1978, LIN and PALEK 1980, YOSHINO and MARCHESI 1984) (Fig. 59). The

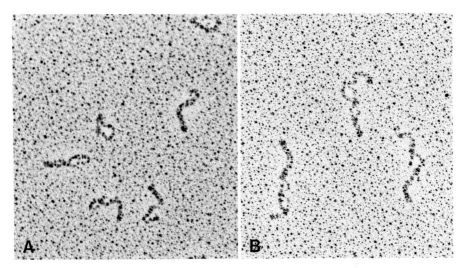

Fig. 59. Electron microscopy of rotary-shadowed erythrocyte spectrin dimers (A) and tetramers (B). × 140,000. (Courtesy of D. BRANTON, J. TYLER, and C. COHEN)

principal form of spectrin isolated from ghosts is the tetramer, which may be, therefore, the native state of the molecule in the cell (GOODMAN and WEIDNER 1980, LIN et al. 1984). Whereas few spectrin molecules appear to self-associate beyond tetramers under physiological conditions, spectrin can form higher order structures such as oligomeric aggregates or polymorphic networks under certain in vitro conditions (MORROW and MARCHESI 1981). In the tetramer, two dimers are associated head-on to form a flexible filamentous structure of approximately 200 nm contour length (SCHOTTON et al. 1979). In this respect, and in its ability to bind to actin filaments, spectrin resembles some of the high molecular weight actin-binding proteins (TYLER et al. 1980). The dimer has a binding site for ankyrin, the molecule that links it to the membrane (see next section). The alpha subunit binds calmodulin (SOBUE et al. 1981, GLENNEY et al. 1982, HUSAIN et al. 1984), but the role of calcium-dependent calmodulin binding for the function of spectrin is not clearly understood.

IX.2.2. Ankyrin

Specific, saturable, high-affinity binding sites for spectrin on the red cell membrane were demonstrated initially by rebinding spectrin to spectrin-depleted erythrocyte ghosts (BENNETT and BRANTON 1977). The protein that links spectrin to the membrane has been identified as a 200,000 d globular, slightly asymmetrical polypeptide present in approximately 100,000 copies per ghost, or one copy per spectrin tetramer. It has been given the name ankyrin[1] (BENNETT and STENBUCK 1979 a). Ankyrin binds to the beta-subunit of spectrin (CALVERT et al. 1980, MORROW et al. 1980) about 20 nm from the end involved in tetramer formation (TYLER et al. 1980) (Fig. 60). The molecule has two nonoverlapping functional domains of 55,000 and 82,000 d molecular weight that can be separated by restricted proteolysis (WEAVER and MARCHESI 1984, WEAVER et al. 1984). The smaller fragment of ankyrin retains the spectrin binding site, whereas the larger has a binding site for the integral membrane protein, band 3. Thus, ankyrin can link spectrin to the cell membrane.

IX.2.3. Band 3

Using a modification of the inside-out binding assay of BENNETT and BRANTON (1977), BENNETT and STENBUCK (1978 b) identified the purported anion transporter (termed "band 3" in the erythrocyte ghost polypeptide nomenclature; STECK 1972) as the integral membrane protein with which ankyrin associates. The most abundant integral protein of the membrane, about 1,200,000 copies of band 3 are found per ghost. The 93,000 d protein accounts for about 25% of the total membrane protein. Within the membrane, band 3 self-associates into noncovalently linked tetramers (SCHUBERT and DORST 1979, NIGG and CHERRY 1980), each of which associates with one ankyrin molecule (HARGREAVES et al. 1980, BENNETT and STENBUCK 1980). Thus there are about three times as many binding sites for ankyrin than there are ankyrin molecules. Band 3 possesses three topographically and functionally different domains, a 38,000 d outer surface segment bearing most of the associated carbohydrate (STECK et al. 1978), a 17,000 d hydrophobic fragment that traverses the membrane several times (JENNINGS and NICKNISH 1984), and a 40,000 d hydrophilic segment that extends into the cytoplasm and carries the ankyrin binding site (STECK 1978). The functional integration of the anion transport activity and ankyrin binding is not clearly understood at present.

---

[1] The same name was proposed by STEPHENS (1975 b) for two high molecular weight proteins of striated rootlets of the mussel gill epithelium.

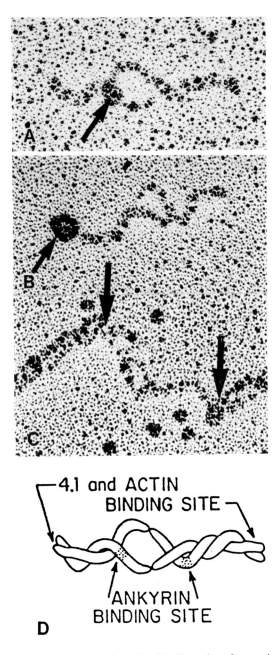

Fig. 60. Electron micrographs that localize binding sites for various molecules on spectrin. *A* Ankyrin, revealed by the position of the bound ankyrin molecule (arrow). *B* band 4.1, revealed by a ferritin-labeled band 4.1 molecule (arrow). *C* Actin, revealed by the binding of tetramers to actin filaments (arrows). × 300,000. The diagram (*D*) shows a schematic of these binding sites on the spectrin tetramer. (Courtesy of D. Branton, J. Tyler, and C. Cohen)

IX.2.4. Actin

The presence of actin in red blood cells has intrigued many investigators ever since it was first described to be a component of this cell. For many years, it was the only known cytoskeletal protein that erythrocytes shared with nonerythroid cells. The approximately 500,000 copies per cell are enough to construct 100–200 actin filaments that could traverse the red cell along its large axis. However, attempts to visualize *any* actin filament of appreciable length have been utterly unsuccessful (TILNEY and DETMERS 1975, SHEETZ and SAWYER 1978), suggesting that actin might be in a nonfilamentous state, possibly complexed with other proteins. Subsequent studies showed that an oligomeric complex of spectrin, actin, and a third protein could act as a seed for the polymerization of actin *in vitro*. This prompted the suggestion that the complex owes its seeding capability to short, oligomeric segments of actin (COHEN and BRANTON 1979, PINDER *et al.* 1979, BRENNER and KORN 1980, FOWLER and TAYLOR 1980). In contrast to earlier suggestions (PINDER and GRATZER 1983), the oligomers have now been demonstrated to be able to support the bidirectional growth of actin *in vitro* (TSUKITA *et al.* 1984, SHEN *et al.* 1984), indicating that both ends of the oligomer are free to elongate and are not complexed with end-binding proteins. The average number of monomers in these oligomeric actin segments was estimated to be between 12 and 17 (PINDER and GRATZER 1983), assuming that all the actin of the erythrocyte ghost is subdivided into segments of approximately equal length. This suggestion has been confirmed by direct visualization of erythrocyte cytoskeletal networks (see section IX.2.6.).

Myosin has recently been detected in erythrocytes (FOWLER *et al.* 1985, WONG *et al.* 1985). The approximately 5,000 copies per cell are believed to play a role in shape changes of the cells.

IX.2.5. Band 4.1

Another major component of the red cell is band 4.1, a protein composed of two chemically similar polypeptides of 80,000 and 78,000 d molecular weight (LETO and MARCHESI 1984). The two forms of this protein are present in about 100,000 copies each, matching the number of spectrin tetramers. Band 4.1 binds to those ends of spectrin dimers not involved in tetramer formation (TYLER *et al.* 1979, 1980 b, UNGEWICKELL *et al.* 1979) (Fig. 60). In addition, band 4.1 seems to anchor the spectrin-actin complex to the prominent integral membrane protein, glycophorin (COHEN and FOLEY 1982, ANDERSON and LOVRIEN 1984, SHIFFER and GOODMAN 1984). Thus band 4.1 complements the ankyrin-mediated association of the spectrin network by linking a different site of the network to another membrane protein. The relation of these two binding sites to the membrane and to one another, and their differential regulation, are presently not known.

IX.2.6. Construction of the Subplasmalemmal Network

While it is important to determine the properties of the individual proteins and to characterize their interactions one by one, only the ordered self-association of all the components will form the complex superstructure that makes up the erythrocyte shell. Whit the role of ankyrin being restricted to the attachment of this superstructure to the membrane, many research groups have focused on the association of spectrin, actin, and band 4.1, the three major components of the two-dimensional network. Two questions, in particular, have received special consideration: What are the molecular interactions in the ternary complex of spectrin, actin, and band 4.1? And, given the quantities of these three proteins in an erythrocyte ghost, precisely how are they arranged into a flexible and regular network?

Mild dissociation of the network by incubation of ghosts at low ionic strength releases a stable complex composed of these three proteins (LIN and LIN 1979, COHEN and BRANTON 1979, UNGEWICKELL et al. 1979, FOWLER and TAYLOR 1980) which can reassociate again with the erythrocyte membrane (COHEN et al. 1978, COHEN and BRANTON 1979). This finding prompted the suggestion that the ternary complex represents the unit structure of the native network. Analysis of the binary interactions between the members of this complex showed that spectrin dimers bind to actin filaments and spectrin tetramers cross-link actin filaments (BRENNER and KORN 1979, UNGEWICKELL et al. 1979, COHEN et al. 1980). In cross-linked complexes of spectrin and actin, the two ends of the spectrin tetramer are associated with the sides of actin filaments, suggesting that the binding site for actin is on the end of the spectrin molecule (COHEN et al. 1980) (Fig. 60). Both spectrin chains also interact specifically with band 4.1 in vitro (TYLER et al. 1980 b, COHEN and LANGLEY 1984), but the exact binding site has not yet been determined. An interaction of band 4.1 with actin has not yet been detected in vitro under physiological conditions and might therefore be weak if spectrin is not present. For example, band 4.1 alone does not alter the rate of actin nucleation or elongation (ELBAUM et al. 1984), whereas the spectrin/band 4.1 complex does.

A different and more exciting story unfolds if the ternary complex is considered. Perhaps its most conspicuous feature is the mutual stabilization of interactions among the three components in vitro. For example, a complex of actin filaments cross-linked by spectrin tetramers but without the participation of band 4.1 can be easily disrupted. Addition of band 4.1, either as a purified protein or complexed to spectrin, greatly promotes spectrin binding to actin and stabilizes their otherwise weak association (UNGEWICKELL *et al.* 1979, FOWLER and TAYLOR 1980, COHEN and KORSGREN 1980, COHEN and FOLEY 1982). The association constant of spectrin-actin interaction is $5 \times 10^3/M$, while the association constant of the ternary complex of actin, spectrin, and band 4.1 is about $10^{12}/M^2$ (OHANIAN *et al.* 1984). Moreover, band 4.1 confers calcium sensitivity to the complex (FOWLER and TAYLOR 1980), which does not form when the free calcium concentration is raised to micromolar levels.

The *in vitro* studies support the conjecture that the spectrin-actin band 4.1-complex is a unit structure or building block in the construction of the native network. Despite the general consensus on the involvement of this "heterotrimer" in network organization, direct visualization of the molecular architecture of intact networks has proved unexpectedly difficult. SHEN *et al.* (1984) visualized network fragments that were released from ghosts by extraction with Triton X-100, treatment with 1.5 M NaCl, and incubation with sodium phosphate. The oligomeric complexes obtained consist of short actin filaments (mean length 50 nm), linked by multiple tetramers of spectrin clustered at sites of association with band 4.1. In negatively stained preparations of hypotonically swollen and shadowed erythrocytes, BYERS and BRANTON (unpublished) observed areas of cleanly spread, intact filamentous meshworks (Fig. 61). These images provide high resolution information on the details of the structural connections within the native network. While in general confirming models of network organization inferred from the behavior of the purified components, these images reveal with remarkable clarity the long-range geometric order within the red-cell cytoskeletal superstructure.

What factors modulate and control the assembly of this multicomponent complex? Several components of the network are phosphoproteins (*e.g.,* spectrin, ankyrin, band 4.1), but whether phosphorylation plays a significant role in regulating their associations is not known. The assembly of the network must occur posttranslationally because different components are synthesized at different sites. Band 3 is cotranslationally inserted into the endoplasmic reticulum membrane and transported to the plasmalemma, while the cytoplasmic components are synthesized on free ribosomes in the cytoplasm. The stoichiometry of the network components is adjusted not during their synthesis, but posttranslationally in the course of their assembly onto the membrane. Subunits not incorporated into the network are degraded. Key elements in this scheme are the component(s) that serve(s) as the anchorage points to the membrane, *i.e.,* the anion transporter and, perhaps, glycophorin. Their availability provides an important assembly-limiting step. Further steps in the assembly scheme may exist but are not yet identified (for a review of this subject, see LAZARIDES and MOON 1984).

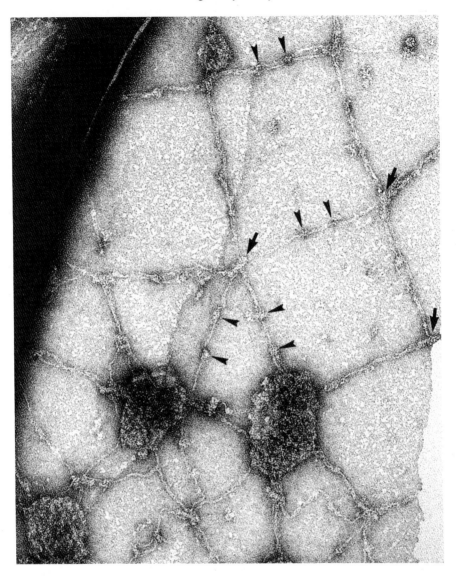

Fig. 61. Intact, spread, erythrocyte membrane skeleton visualized by negative staining. The network is formed by long spectrin molecules joined at their ends into junctions. Between 4 and 7 spectrin molecules appear to insert into each junction, which also contains a single rod of 36 nm length, presumably the actin oligomer (arrows). Up to two distinct globules are often associated with the spectrin tetramers (arrowheads). The distances between junction and globule, and between two globules on one spectrin tetramer, suggest that the globules are occupied ankyrin binding sites. × 200,000. (Courtesy of T. Byers and D. Branton)

IX.3. Proteins Related to the Erythrocyte Skeleton in Other Cell Types

For many years, the erythrocyte cytoskeleton was considered a special case which, though interesting in itself, might be irrelevant to more complex cells. This attitude was based on the assumption that the major components of the erythrocyte network are unique and not present in other cells and tissues (*e.g.,* HILLER and WEBER 1977). This view has changed with the discovery of, first, spectrin-related molecules, and then ankyrin and band 4.1-like polypeptides in non-erythroid cells. Suddenly the erythrocyte turned into one of the best-suited model systems for the study of membrane-cytoskeleton interactions.

Proteins related to alpha and beta spectrin have now been identified in many different cell types in species ranging from protozoa to man. They have not, however, been reported for plant cells. In attempting to acquire an overview of the spectrin family of proteins, the outsider is not only confronted with a spectrum of spectrins, but also with a spectrum of terminologies for the different proteins and their subunits. Unfortunately, this potential source of confusion has not yet been eliminated by unification of terms, which can only be achieved satisfactorily if the sequences of the different spectrin species are known. For sake of clarity, in this review, the term spectrin will be applied to all the subgroups of this protein family, preceded by the cell type or tissue of origin and followed by the molecular weight (in kilodaltons) of the subunit (*e.g.,* brain spectrin 240). The names used in the original papers will be given in parenthesis.

The presence of spectrin-like polypeptides in other cell types was first reported on the basis of immunological studies in 1981 (for a brief summary, see LAZARIDES and NELSON 1982). They were described as high molecular weight polypeptides of 240,000 and 230,000 d in myocytes (GOODMAN *et al.* 1981) and cultured cells (BURRIDGE *et al.* 1982), and 240,000 and 235,000 d in brain (LEVINE and WILLARD 1981). The brain spectrins (termed "fodrin") are a major neuronal constituent transported along the axon at an intermediate rate (about 50 mm/day); a small portion also comigrates with the slow components which include actin, myosin, tubulin, and neurofilament proteins (1–10 mm/day). By immunofluorescence microscopy, these proteins are localized to the periphery of axons. They have also been detected in the cortical region of many other cell types, suggesting that immunoreactive forms of brain spectrin are widely distributed. Another member of the spectrin family is a prominent component of the terminal web of the intestinal brush border (termed "TW 260/240"). Its two subunits, as suggested by their name, have molecular weights of 260,000 and 240,000 d.

Detailed protein-chemical, ultrastructural, and immunological comparison of the spectrin family of proteins has revealed the following similarities and differences:

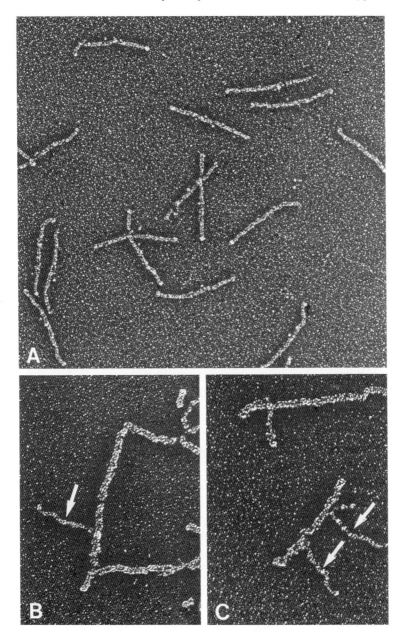

Fig. 62. *A* Pig brain spectrin (fodrin) molecules after rotary shadowing with tantalum/tungsten. × 120,000. *B,C* Interaction of brain spectrin (arrows) with actin filaments. As in erythrocyte spectrin, the binding site for actin is at the end(s) of the tetramer. × 140,000. (Courtesy of J. GLENNEY)

1. All spectrins are composed of two distinct but immunologically related subunits, one of which, the 240,000 d subunit, is very similar in both erythroid and nonerythroid cells. Muscle spectrin 230, also present in cultured cells, is closely related to erythrocyte spectrin 220 and may in fact be a beta spectrin variant (NELSON and LAZARIDES 1983). The relationship of brain spectrin 235 and brush border spectrin 260 to erythrocyte beta spectrin is not quite understood, but by immunological criteria and by peptide mapping these proteins seem to be more distantly related. All spectrins form dimers composed of one 240,000 d subunit which, in accordance with the available biochemical and immunological information, is called the "common" subunit, and one "variant" subunit. Dimers associate head-on to form tetramers approximately 200 nm long (Fig. 62).

2. The "common" spectrin subunit of 240,000 d molecular weight is a calmodulin-binding protein (DAVIS and KLEE 1981, GLENNEY et al. 1982 a, SOBUE et al. 1982, CARLIN et al. 1983, TSUKITA et al. 1983). The precise role of calmodulin binding to spectrin is not known, but the interaction of spectrin with other components of the cell cortex may be under calcium control. For example, the association of erythrocyte spectrin with actin and band 4.1 is calcium-dependent (FOWLER and TAYLOR 1980), and brain spectrin confers calcium sensitivity to the actin-activated ATPase activity of myosin (WEGNER 1984). These activities may be mediated by spectrin-associated calmodulin, but the functional significance is still obscure.

3. Analogous to erythrocyte spectrin, both brain and muscle spectrin appear to be localized close to the cell membrane, where they are assumed to be part of a submembraneous filamentous skeleton (BURRIDGE et al. 1982, BENNETT et al. 1982) (Fig. 63). A close association with components of the cell membrane is indirectly supported by the observation that brain spectrin co-caps with lectin or antibody-induced caps of cell surface receptors (LEVINE and WILLARD 1983, NELSON et al. 1983) (see Fig. 68). However, whether membrane proteins are linked, directly or indirectly, to the spectrin-like molecules can not be said with certainty. On the other hand, the localization of brush border spectrin seems to differ significantly from that of other cell types. It is present predominantly in the terminal web region, removed from the immediate vicinity of the membrane. Brush border spectrin seems to interconnect the rootlet ends of microvillar filament bundles with one another and with the cell membrane, and also associates with brush border intermediate filaments (GLENNEY 1983, HIROKAWA et al. 1983). In immunoelectron microscopy, antibodies against brush border spectrin label thin (3–5 nm) filaments of up to 200 nm length that morphologically resemble isolated spectrin tetramers. Thus, brush border spectrin primarily cross-links other filament types, rather than forming a submembranous network. Its ability to bundle actin filaments *in*

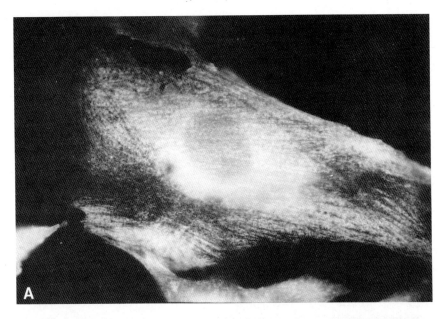

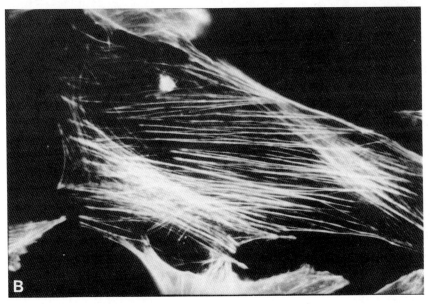

Fig. 63. NIH-3T3 cell double-labeled with antibodies against brain spectrin (*A*) and rhodamine-phalloidin to visualize actin (*B*). The cell was photographed in the same focal plane. The NIH-3T3 analog of spectrin forms a contiguous peripheral layer but is partially excluded from the region of stress fibers. × 600. (Courtesy of C. CARTER)

vitro (GLENNEY *et al.* 1982a, PEARL *et al.* 1984) is consistent with this cross-linker function.

4. Several cell types are now known to express more than one "variant" subunit of spectrin, in addition to the "common" 240,000 d subunit. While the predominant variant in neurons is brain spectrin 235, these cells also synthesize erythrocyte spectrin 220, or beta spectrin, albeit in much smaller quantities (LAZARIDES and NELSON 1983). Intestinal epithelial cells contain brain spectrin 235 als well as brush border spectrin 260. In fact, in all cases where two variant subunits were detected in one cell type, one of them is always the 235,000 d "brain-type" subunit, which therefore has the most widespread distribution of all the variant subunits (GLENNEY and GLENNEY 1983). In all cases, the variant subunits are isolated as a complex with the common subunit, and never as a complex of two variant subunits.

5. The occurrence of two different spectrin dimer complexes within the same cell raises the possibility that the two play different roles. This suggestion receives support from the finding that the two complexes found in neurons, myoblasts, and intestinal epithelial cells occupy different cytoplasmic domains and are assembled at different stages in development. In neurons, spectrin 240/220 is present in the cell body only, never in the axon, whereas spectrin 240/235 is axonally transported and associates with the axon membrane (LAZARIDES *et al.* 1984). Primary chicken myoblasts express primarily spectrin 240/235, while terminally differentiated cells gradually switch to spectrin 240/230 (NELSON and LAZARIDES 1983). Finally, spectrin 240/235 is broadly distributed along the membrane in intestinal epithelial cells, whereas spectrin 260/240 is restricted to the brush border and appears only later in development (GLENNEY and GLENNEY 1983) (Fig. 64). These findings raise interesting questions on the intracellular targeting of different spectrin isoforms and their acceptor molecules. ANDERSON (1984) discusses two factors that could be important in the segregation of the two different spectrin complexes in neurons: a) prefential assembly near the site of synthesis of the proteins by a "self-trapping" mechanism (through cross-linking) that limits the diffusion of the complex, and b) the distribution of the spectrin acceptor, perhaps an ankyrin-like molecule. One or both of these factors could be different for the two different spectrin complexes. If an ankyrin-like molecule is involved, its distribution may in turn be determined by the extracellular environment. Whether these possible mechanisms are applicable to other cell types remains to be seen.

Analogs of ankyrin, the erythrocyte membrane-network linker, are also widely distributed in cell types other than red blood cells (BENNETT 1979, DAVIS and BENNETT 1984a, NELSON and LAZARIDES 1984a, b). Polypeptides that cross-react with antibodies against erythrocyte ankyrin have an intriguing and unexpected subcellular distribution. They are found not only in association with membranes, but also within mitotic spindles and along cytoplasmic microtubules (BENNETT and DAVIS 1981). In a reciprocal experiment, a monoclonal antibody against a 200,000 d microtubule-associated polypeptide stains the periphery of erythrocytes (IZANT *et al.* 1982). Ankyrin interacts with microtubules and promotes

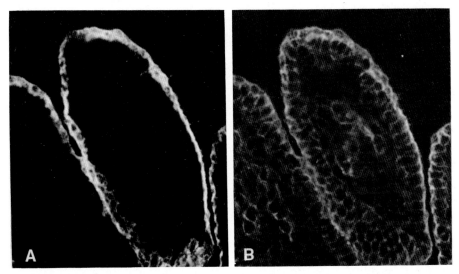

Fig. 64. Differential distribution of two variant subunits of spectrin in frozen sections of the same cells of an 18-day-old embryonic chicken intestine. *A* Brush border spectrin 260. *B* Brain spectrin 235. The 260 k antigen is located exclusively in the apical border of cells lining the lumen of the intestine. The 235 k protein is more broadly distributed. × 200. [From GLENNEY and GLENNEY (1983)]

microtubule assembly *in vitro* (BENNETT *et al.* 1982). The significance of this interaction is poorly understood, but ankyrin may link microtubules to the membrane. Protein-chemical analysis of membrane-associated forms of brain ankyrin indicates that its similarities to erythrocyte ankyrin include a molecular weight of about 200,000 d and the presence of two different functional domains, one of which can bind to spectrin (DAVIS and BENNETT 1984 a). In addition to the spectrin binding site, brain ankyrin has binding sites for tubulin and the cytoplasmic domain of band 3 (DAVIS and BENNETT 1984 b). Brain ankyrin is present in approximately the same quantities as brain spectrin tetramer, supporting the suggestion that its function in neurons might be similar to that of its erythrocyte counterpart. Because the spectrin and tubulin binding sites are different, brain ankyrin could help attach microtubules to the membrane independently of spectrin binding, as suggested above. The muscle cell equivalent of ankyrin, previously termed goblin (BEAM *et al.* 1979), is also a membrane-associated protein which, however, may serve a slightly different function than brain ankyrin. This protein codistributes with muscle spectrin at discrete foci near the Z-lines (NELSON and LAZARIDES 1984 a) and may participate in the construction of attachment sites for the contractile machinery (myofibrils) at the Z-lines.

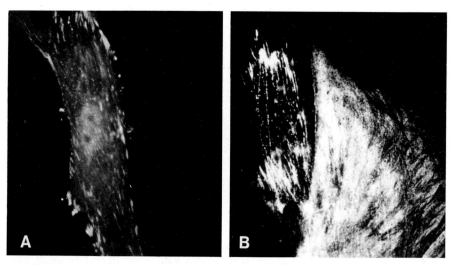

Fig. 65. Precipitation of spectrin following microinjection of a brain spectrin antibody. The cell was coinjected with rhodamine-conjugated vinculin. 3 hours later the cells were fixed, permeabilized, and stained with the antispectrin antibody. In *A*, the cells are viewed with rhodamine-specific filters to visualize the vinculin (and hence the injected cell). In *B*, the cells are visualized with fluorescein-specific filters to show the spectrin distribution. Note that the organization of vinculin appears normal, while spectrin is precipitated in the injected cell, but not the neighboring uninjected cell. × 400. [From MANGEAT and BURRIDGE (1984 a)]

To conclude the catalog of erythrocyte skeletal counterparts in other cell types, proteins related to band 4.1 and band 3 (the anion transporter), have also been detected in nonerythrocyte cells. Immunoreactive analogs of band 4.1 appear along stress fibers of cultured cells, and may be associated with actin (COHEN *et al.* 1982), while an antibody against the cytoplasmic domain of erythrocyte band 3 reacts with the membranes of a variety of cell types (DRENCKHAHN *et al.* 1984). Preliminary biochemical characterization of the band 4.1 analog of leukocytes, platelets, and lymphoid cells shows that it is a doublet with a slightly lower molecular weight than the erythrocyte counterpart, and that is does not bind spectrin (SPIEGEL *et al.* 1984). Analogs of band 4.1 are currently studied in several laboratories.

In summary, a growing body of evidence indicates the presence of proteins analogous to erythrocyte cytoskeletal proteins in a wide spectrum of animal cells. Functional studies of these molecules suggest that they play roles similar to their erythrocyte counterparts, but may have acquired additional, specific functions. Overall, these studies strengthen the view that the erythrocyte cytoskeleton is of direct relevance to the organization of the

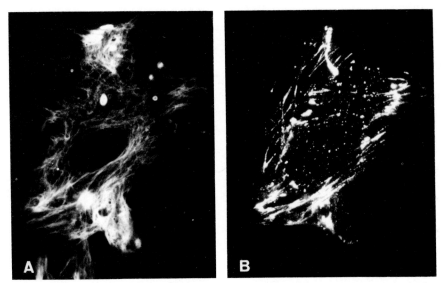

Fig. 66. Comparison of the distribution of intermediate filaments and spectrin aggregates 4 hours after the cell was microinjected with a spectrin antibody. *A* Distribution of intermediate filaments. *B* Distribution of precipitated spectrin aggregates. There is a close correspondence of the two distribution patterns. × 400. [From MANGEAT and BURRIDGE (1984a)]

cortical cytoplasm in other cell types. However, while this may be a valid generalization, like all generalizations it has to be considered *cum grano salis*. In analogy to the undisputed involvement of the spectrin network in the development and maintenance of red cell shape, one might suggest a similar function for nonerythrocyte spectrins. However, such a function is not yet supported by experimental evidence. Intracellular precipitation of spectrin in live cells by microinjected spectrin antibodies (MANGEAT and BURRIDGE 1984) does not affect overall cell shape, even though, as a result of the formation of antigen-antibody complexes, compact, irregular spectrin aggregates or clumps are formed that occasionally align with microfilament bundles (Fig. 65). At the least, these studies imply that spectrin is not the only component involved in determining cell shape in nonerythroid cells. The only cytoskeletal component whose distribution is affected by spectrin precipitation are vimentin intermediate filaments, which become distorted and condensed and frequently accumulate in regions of aggregated spectrin (Fig. 66). These experiments support the possibility, raised in studies of the association of intermediate filaments with the chicken erythrocyte membrane (GRANGER and LAZARIDES 1982), of an association between vimentin filaments and spectrin. These studies coupled with the findings that spectrin (ISHIKAWA et al. 1983, CARLIER et al. 1984 c) and ankyrin (BENNETT and DAVIS 1981, BENNETT et al. 1982) can bind to microtubules and that spectrin interacts with actin filaments (GLENNEY et al. 1982 b, PEARL et al. 1984), indicate the intriguing possibility that spectrin and its link to the membrane, ankyrin, connect *all three* major cytoskeletal filaments to the plasma membrane. Such linkages could be important in a number of membrane activities, including directed membrane traffic, control of the distribution of surface molecules, specification of the sites of endo- and exocytosis, and membrane flow during directional cell motility. On the basis of indirect evidence, cytoskeletal elements were long believed to be involved in these membrane activities. But now that the work on erythrocyte cytoskeletal proteins and analogs had identified some of the proteins involved in membrane-cytoskeletal interactions, their specific role(s) in these acitivities can be tested more directly.

IX.4. Association of Membrane Proteins with the Cytoskeleton

The fluid mosaic model views cellular membranes as a two-dimensional bilayer of oriented lipids in which proteins are freely diffusible entities floating in the lipid sea (SINGER and NICHOLSON 1972). While this model has proved correct in its basic assumptions, several lines of well-documented evidence demonstrate that many membrane proteins are highly restricted in their mobility. The rates of lateral diffusion of many proteins within the plane of the membrane are up to 10,000 times lower than those of lipids, prompting the suggestion that their movement is retarded by the interaction with submembraneous, presumably cytoskeletal, components. The analysis of these interactions has made rapid progress over the past 10 years, and several candidates of molecules that might mediate the linkage between membrane proteins and the cytoskeleton have been suggested, including spectrin and ankyrin-like proteins. In light of these advances, the identity of all the potential submembraneous linkers may be revealed within the next few years.

Evidence for an association between cell membrane proteins and the cytoskeleton comes primarily from molecular analyses of the phenomena of "patching" and "capping", and from studies on the retention of integral membrane proteins after detergent extraction (to be discussed below). The aggregation of cell surface molecules into small clusters ("patches") that eventually coalesce to form a "cap" on one pole of the cell was first described by TAYLOR *et al.* (1971) as a cellular response that ensues upon addition of antibodies directed against the cell surface-bound immunoglobulin molecules of lymphocytes. This discovery was strong experimental evidence for the fluid mosaic model because it demonstrated the ability of cell surface molecules to move laterally in the plane of the membrane. It also showed, however, that their movement can be influenced (restricted) by other, extra- and/or intracellular determinants. Since then, in addition to antibodies against cell membrane antigens, a spectrum of di- or multivalent agents ("ligands") have been shown to induce redistribution of their respective target molecules ("receptors") on the cell surface. Lectins that bind to the carbohydrate moieties of glycoproteins (YAHARA and EDELMAN 1973), and hormones or growth factors that bind to specific recptors (SCHLESSINGER *et al.* 1978) will also provoke an aggregation of their previously uniformly distributed target molecules. In other instances, receptor molecules may already be preclustered at specific sites, presumably by anchorage to some subsurface component. Examples of this include the acetylcholine receptors of neurons or myocytes, which are aggregated in postsynaptic membranes (RAFTERY *et al.* 1980), and the receptors for low density lipoprotein, which are concentrated in coated pits (GOLDSTEIN *et al.* 1979). Monovalent Fab fragments of immunoglobulin molecules, or divalent succinylated concanavalin A (instead of the tetravalent intact lectin) will not induce their respective target molecules to aggregate (SCHREINER and UNANUE 1976, EDELMAN *et al.* 1973). Patching and capping, therefore, require cross-linking of the receptors on the membrane. It became apparent early on that the integrity of cytoskeletal filaments also influences the capping process. The initial report by TAYLOR *et al.* (1971) that cytochalasin B inhibits capping was confirmed in several studies (reviewed in LOOR 1981). In some cell types, however, disruption of microfilaments by cytochalasins may have just the opposite effect (*e.g.,* SUNDQVIST and EHRNST 1976, DE GROOT *et al.* 1981). Microtubule depolymerization by colchicine or related compounds does not seem to influence capping (UNANUE *et al.* 1973) except for a promoting effect in some cell types (OLIVER *et al.* 1976) when cross-linking is induced by high doses of concanavalin A, which normally tends to prevent capping (YAHARA and EDELMAN 1973, 1975a, b). These observations were interpreted to mean that the fate of the ligand-receptor complex depends on both microfilaments and microtubules (ALBERTINI and ANDERSON 1977, BERLIN and OLIVER 1978), which together are part of a "surface-modulating assembly" (EDELMAN 1976). However, the precise nature of the association between surface receptors and cytoskeletal fibers remained unresolved.

A more direct demonstration of the involvement of cytoskeletal components in cell surface capping was achieved by two other approaches:

1. immunofluorescence and immunoelectron microscopy with labeled ligands and labeled antibodies directed against cytoskeletal proteins, and 2. isolation and biochemical charaterization of cap structures. Since the initial demonstration of actin cocapping with cross-linked surface receptors

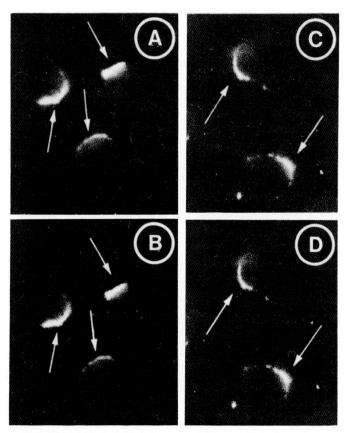

Fig. 67. Cocapping of fluorescently labeled concanavalin A (*A, C*) with actin (*B*) and myosin (*D*) in lymphocytes. × 1,500. (Courtesy of L. BOURGUIGNON)

(OLIVER *et al.* 1977, GABBIANI *et al.* 1977, BOURGUIGNON and SINGER 1977, TOH and HARD 1977), a number of major cytoskeletal and regulatory molecules that initially were more uniformly distributed at the cell cortex were found to concentrate underneath the patch or cap regions (Figs. 67 and 68). They include myosin (BOURGUIGNON and SINGER 1977, SCHREINER *et al.* 1977, BRAUN *et al.* 1978 a, b), alpha-actinin (GEIGER and SIGER 1979, HOESSLI *et al.* 1980), brain spectrin (LEVINE and WILLARD 1983, NELSON *et al.* 1983), an ankyrin-like protein (BOURGUIGNON and BOURGUIGNON 1984), and calmodulin (SALISBURY *et al.* 1981, NELSON *et al.* 1982). The

uniform   patches   cap

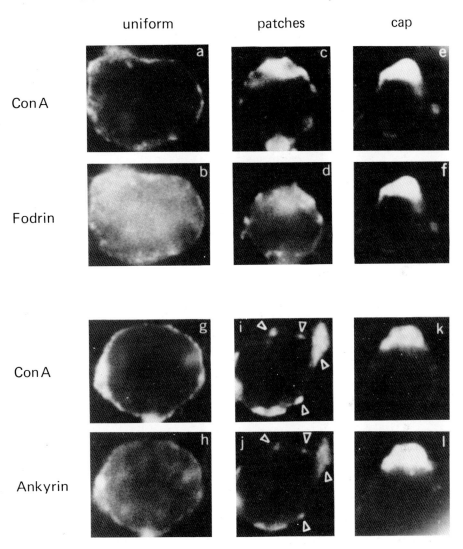

Fig. 68. Double label immunofluorescence microscopy of murine lymphocytes with fluorescent concanavalin A and antibodies against brain spectrin and ankyrin, showing copatching and cocapping of these molecules with concanavalin A receptors. × 1500. [From BOURGUIGNON and BOURGUIGNON (1984)]

association between the cross-linked receptors in caps and the underlying cytoskeleton is stable enough to survive membrane solubilization with detergents, suggesting firm transmembrane linkage. For example, ligands coupled to an electron-dense marker (*e.g.*, ferritin) are still found associated with the submembranous filament network underlying caps after detergent extraction (*e.g.*, FLANAGAN and KOCH 1978, BUTMAN *et al.* 1980,

SHETERLINE and HOPKINS 1981) (Fig. 69). Biochemical analysis of isolated patches and caps confirms the existence of a stable association between the clustered receptors and actin and myosin (CONDEELIS 1979, BOURGUIGNON and BOURGUIGNON 1981). This association is established *before* the capping stage, *i.e.,* during the energy-independent phase of receptor patching (CONDEELIS 1979, BRAUN *et al.* 1982). In light of this evidence, models for capping (and other cell surface phenomena) that propose directed flow of membrane lipids (BRETSCHER 1976) or the whole membrane (HARRIS 1976, OLIVER and BERLIN 1982) without a firm link to the underlying cytoskeleton may need to be reevaluated. Current models view capping as an energy-dependent process that seems to involve increased polymerization of actin (LAUB *et al.* 1981) at the patching site. Patches of ligand-receptor complexes linked to actin via a macromolecular assembly, possibly including spectrin, are then actively pulled into a cap by a sliding filament contractile interaction between actin and myosin, which may be regulated by kinase-mediated phosphorylation of myosin light chains (BOURGUIGNON *et al.* 1981, 1982). Calcium/calmodulin and/or cyclic AMP may be involved in this regulatory scheme in as yet unspecified ways (see LOOR 1981 and BOURGUIGNON and BOURGUIGNON 1984 for recent summaries). Whatever the details of the regulatory pathways, the transmembrane association of ligand-receptor complexes with the underlying cytoskeleton is supported by a growing body of evidence. The molecules that participate in this association, and the molecular mechanisms of interaction, need to be determined in the future. The observations of cocapping of spectrin and ankyrin-like proteins with membrane proteins seem to suggest that the linkage is mediated by these molecules, which may be additional evidence that these molecules are universal linkers in membrane-cytoskeleton interaction. If true, this idea would help unify our concept of how membranes interact with the cytoskeleton, and would strengthen the case of the erythrocyte cytoskeleton as a generally useful model system.

Additionas support for the idea of a more or less stable association between certain integral membrane proteins and underlying cytoskeletal proteins is derived from studies on the retention of membrane components after most of the lipid bilayer has been dispersed by non-ionic detergents. The residue so prepared not only retains the filamentous network, but frequently also a peripheral "surface lamina" that covers the extracted residue as if it were a continous coat (BEN-ZE'EV *et al.* 1979, FULTON *et al.*

Fig. 69. Location of ferritin-conjugated concanavalin A on capped pig polymorphonuclear leukocytes (*A*) and in detergent-extracted cytoskeletons (*B, C*). The ferrintin-labeled lectin remains associated with the cap after extraction and colocalizes with a filamentous meshwork in these thin sections of embedded cells.
    A, B × 19,000. *C* × 49,000. (Courtesy of P. SHETERLINE and C. HOPKINS)

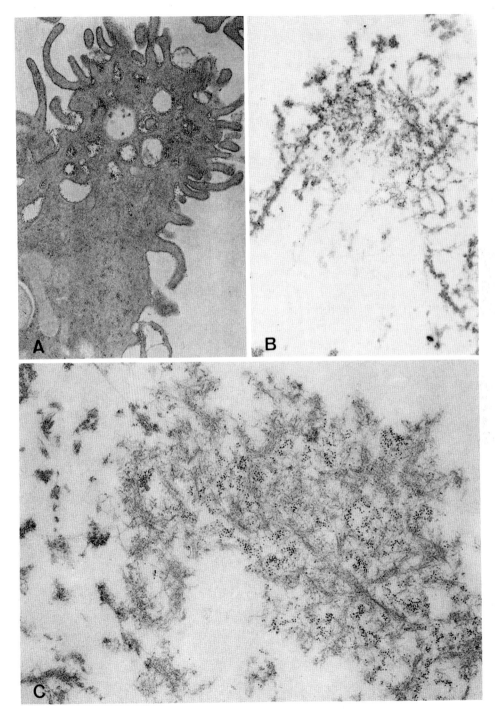

1981, MESCHER et al. 1981, LEHTO et al. 1983). The lamina contains cell surface proteins and sphingolipids and glycoplipids (STREULI et al. 1981), the latter possibly associated with the membrane proteins. Unlike the ligand-receptor complexes discussed in the previous paragraph, these membrane components are associated with the cytoskeleton in the absence of ligand-induced cross-linking. Among the proteins that are reported to remain with the filament network after extraction are a cell surface glycoprotein of 140,000 d molecular weight from fibroblast membranes (LEHTO et al. 1980, CARTER et al. 1981, LEHTO 1983), a 130,000 d neuronal glycoprotein (MOSS 1983, RANSCHT et al. 1984), and glycoproteins of about 80,000 d molecular weight from BHK cells (TARONE et al. 1984) or mammary ascites adenocarcinoma cells (CARRAWAY et al. 1983 a, b, JUNG et al. 1984). Other membrane proteins for which a similar association has been demonstrated include Na,K-ATPase (ASH et al. 1977), the cyclic AMP chemotaxis receptor of Dictyostelium (GALVON et al. 1984), and a 43,000 d protein associated with the acetylcholine receptor (WALKER et al. 1984). It is important to emphasize that these proteins are *specifically* retained upon extraction. Other membrane components are quantitatively removed, eliminating the possibility that the retention of these other proteins is due to incomplete membrane solubilization. Analysis of this still growing list of membrane proteins associated with the cytoskeleton in the absence of ligand-induced clustering suggests, at least for some of these proteins, a direct association with actin filaments. For example, the 80,000 d glyco-protein of adenocarcinoma cells forms a sedimentable complex with actin that can be dissociated only by strong denaturants (CARRAWAY et al. 1983 a). Similarly, the 43,000 d acetylcholine receptor-associated poly-peptide binds specifically to actin in nitrocellulose blots (WALKER et al. 1984). In contrast, the glycoprotein of pig intestinal microvilli seems to interact specifically with the 110,000 d protein of the microvillus core bundle (COUDRIER et al. 1983), and the 130,000 d neuronal glycoprotein is tightly associated with a complex of a number of other proteins, including actin (MOSS 1983). From these observations it appears as if several modes of membrane protein-cytoskeleton association are used in different cell types, but the scarcity of information in this young research field cautions against general statements. Only a subset of cell surface proteins are anchored to the cytoskeleton; these proteins may therefore play an important and specific role in transmembrane transfer of information to the cytoskeleton. Dissection of these specific roles is a significant goal for future studies.

In one particularly fascinating case, changes in the association of the neuronal receptor for the putative neurotransmitter L-glutamate with brain spectrin have been implicated in the regulation of the number of available receptor molecules. Selective degradation of brain spectrin by calpain, a calcium-dependent protease, leads to exposure of cryptic postsynaptic glutamate receptors and more than doubles the density of glutamate binding sites (LYNCH and BAUDRY 1984, SIMAN et al. 1984). This regulatory step may in turn affect synaptic transmission and, ultimately, the learning process. Thus a single degradation step affecting one cytoskeletal protein located at one set of specific sites within one cell type may be of consequence for the behavior (in the broadest sense) of the organism.

IX.5. Association of Cytoskeletal Fibers with Membranes

The experiments outlined in the preceeding section approach the question of interactions between membranes and the cytoskeleton "from the membrane in", by asking which membrane proteins associate with the filament network underneath, and which specific molecules do they bind. An alternative approach, "from the cytoskeleton out", determines how filaments or microtubules bind to isolated plasma membrane fragments or model membrane preparations. All three major filament types can associate with membranes in the absence of local membrane differentiations. In addition, actin and intermediate filaments participate in well-defined attachment sites, such as adhesion plaques and desmosomes (see sections IX.7.1. and IX.7.2.). Here some experiments employing the "inside-out approach" to study the biochemical basis of F-actin, intermediate filament, and microtubule associations with membranes will be considered, beginning with a look at actin filaments.

Iinitial indirect evidence for an association of *F-actin* with membranes was provided by the isolation of plasma membranes with attached cortical cytoskeleton (*e.g.,* POLLARD and KORN 1973, SPUDICH 1974; for a review of the earlier literature, see WEIHING 1979). These and related studies provided evidence for both end-on (*e.g.,* POLLARD and KORN 1973) and lateral association (*e.g.,* MUKHERJEE and STAEHELIN 1971) of actin filaments with membranes. End-on associations always seem to occur at the barbed end, or fast-growing end (*e.g.,* ISHIKAWA et al. 1969, SMALL et al. 1978, SALISBURY et al. 1980). More recently, a number of qualitative and semiquantitative assays were used to study F-actin membrane binding. In one method, the sedimentation of actin alone is compared to actin complexed to, and cross-linked by, membrane vesicle preparations (BURRIDGE and PHILLIPS 1975, TAYLOR et al. 1976, OSTLUND et al. 1977, COHEN and FOLEY 1980, WILKINS and LIN 1981, THERIEN et al. 1984). This assay demonstrates a specific interaction between actin and membranes, but it only detects qualitative differences in the ability of different membrane preparations to bind F-actin. Falling ball viscometry is more sensitive and monitors such interactions in a semiquantitative manner (FOWLER et al. 1981, LUNA et al. 1981, FOWLER and POLLARD 1982). Binding of actin to membranes significantly increases the viscosity of the mixture. Integral membrane proteins implicated to associate with actin have also been identified by two novel, related procedures: first, sedimentation binding analysis with F-actin affinity beads, and second, F-actin affinity column chromatography (Luna *et al.* 1984). Highly purified *Dictyostelium* plasma membrane fragments bind with saturation kinetics to F-actin attached to Sephacryl beads.

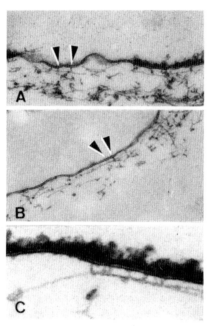

Fig. 70. Contacts between microfilaments and the cell membrane mediated by lateral bridges in isolated *Dictyostelium* cortices. The bridges in *C* exhibit a periodicity of about 36 nm, suggesting that their distribution is influenced by the helical twist of the actin filament. *A, B* × 60,000. *C* × 150,000. [From Bennett and Condeelis (1984)]

Binding is mediated by integral membrane proteins since heat-denatured, proteolyzed, or salt-extracted plasma membrane fragments, or pure lipid vesicles, do not bind actin beads. Integral membrane proteins are also invoked in the binding of actin filaments to lysosomal membranes (Mehrabian *et al.* 1984). In *Dictyostelium,* four polypeptides are particularly strong candidates for transmembrane, F-actin binding proteins (Luna *et al.* 1984). Interestingly, the plasma membrane fragments from *Distyostelium* bind preferentially to the sides of actin filaments (Fig. 70), as demonstrated by inhibition of binding by heavy meromyosin decoration of the filaments (Goodloe-Holland and Luna 1984). Disruption of the association between actin filaments and the plasma membrane by S 1-

decoration was also observed by BENNETT and CONDEELIS (1984). These studies represent first steps towards the identifiaction of integral membrane proteins that specifically interact with the sides of actin filaments. The importance of this form of interaction between actin and membranes is evident when cell membrane dynamics are considered in addition to static anchorage. Laterally bound actin filaments could interact with cytoplasmic myosin and exert a force on, and translocate, single membrane proteins, membrane domains, or entire organelles. Cellular activities for which this form of association would provide the most efficient orientation include the capping of membrane receptors (see previous section) and the constriction of the cleavage furrow. The cleavage furrow is enriched in myosin (FUJIWARA and POLLARD 1976) and contains actin filaments of opposing polartiy (SANGER and SANGER 1980) running parallel to the membrane, thus fulfilling the necessary prerequisites for a structure that generates the force for constriction. The mode of association of actin filaments with the plasma membrane in the cleavage furrow is not known, but lateral interaction with integral membrane proteins would appear to be an efficient arrangement.

Lateral and end-on associations between *microtubules* and membranes are evident in morphological studies of a wide spectrum of cell types. Examples include cilia and flagella, the cortex of plant cells, and the periphery of protozoa and nucleated red blood cells. The molecular architecture of microtubule-membrane attachment is not understood in any of these cases, but they all may be useful for biochemical dissection of the attachment sites. While a review of the voluminous and largely descriptive literature on this topic will not be attempted here, (see WEIHING 1979, DENTLER 1981, WEATHERBEE 1982 for reviews), a few general features shall briefly be considered.

The membrane enveloping cilia or flagella is attached to the axoneme by numerous bridges extending laterally from the outer doublet microtubules (DENTLER *et al.* 1980). Neither the composition nor the function of these structures is known, but they are invoked to explain processes ranging from regulation of the ordered disposition of mastigonemes (extraclelular filaments attached to the ciliary membrane) (MARKEY and BOUCK 1977, BOUCK *et al.* 1978), to the attachment and movement of particles along the outer surface of the cilium (BLOODGOOD 1977, 1982). This membrane-associated motile phenomenon is possibly mediated by the outer doublet microtubules via transmembrane glycoproteins (AGUAS and PINTO DA SILVA 1984). A different type of association is found at the distal tips of cilia where both the outer doublet and the central pair microtubules are linked to the membrane by complex structures known as distal filaments, plugs, lamellar caps, and crowns (see DENTLER 1981 for a review). These differentiations appear to plug firmly into the microtubule end, which raises interesting questions about the mechanism of microtubule elongation during ciliary growth and regeneration. The addition of subunits occurs at the distal ends of ciliary microtubules (WITMAN 1975), when these plug-like structures are already present. One possibility is that, in addition to linking the ends of microtubules and the ciliary membrane, these structures have an enzymatic activity for tubulin modification. During flagellar regeneration, new tubulin subunits are posttranslationally modified upon incorporation into the growing microtubules (L'HERNAULT and ROSENBAUM 1983, RUSSELL and GULL 1984). The putative "plugs" may not be as tight as they might appear in electron micrographs, and, while modifying tubulin subunits, may allow them to reach the growing microtubule end.

In many plant cells, microtubules are located in large numbers close to the cell membrane, often forming stunning arrays (LLOYD 1983). Bridges of unknown composition connect them to the cell membrane. These bridges in conjunction with specifically oriented microtubules, are believed to be involved in the oriented deposition of cellulose fibrils for the cell wall (for references and discussions of the hypotheses, see ROBINSON and QUADER 1982, SEAGULL and HEATH 1982), Unfortunately, biochemical analysis of the nature of the bridges extending from the microtubules to the cell membrane is technically difficult in plant cells. Recent studies on certain flagellates and nucleated red blood cells, on the other hand, suggest that these systems are potentially better suited for biochemical dissection of microtubule-membrane links. Regular arrays of microtubules closely associated with the cell membrane are a prominent feature of the cytoarchitecture of trypanosomatid and euglenoid flagellates (SOMMER 1965, HUNT and ELLAR 1974). When their cell membranes are isolated for biochemical (HUNT and ELLAR 1974) and structural analysis (MURRAY 1984 a, b), microtubules remain connected to the isolated membrane via short, compact bridges (Fig. 71). Addition of millimolar calcium to isolated membrane fragments quickly removes the microtubules, but leaves the bridge structures attached to the membrane. The membrane-microtubule complex can be reconstituted with exogenous brain tubulin, which assembles only onto those regions that were associated with microtubules *in*

Fig. 71. Microtubule-pellicle complexes in the flagellate *Distigma proteus*. *A* Transverse section showing the characteristic serrated cell boundary with underlying microtubules. × 16,500. *B* Higher magnification of isolated cell membranes with microtubules still attached in their *in situ* location. × 95,000. *C* Membrane fragment similar to (b), after washing with calcium-containing buffer. Microtubules are removed, but faint bridges in the area formerly occupied by microtubules are now seen. × 89,000. *D* Reassembly of bovine brain microtubule protein onto microtubule-depleted membrane complexes. Many microtubules form on the membrane in a pattern similar to the original arrangement. × 56,000.
[From MURRAY (1984 b)]

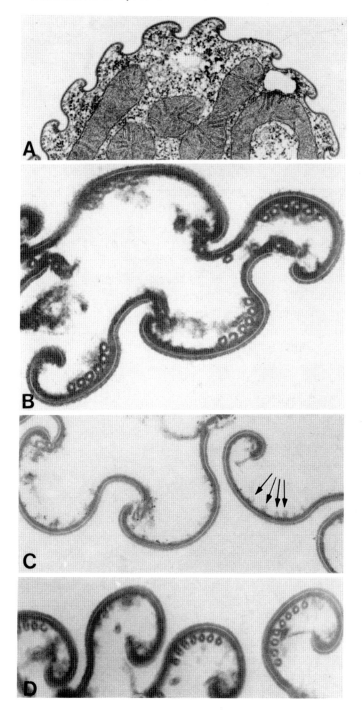

vivo (MURRAY 1984b). An analogous experiment can be done with non-mammalian erythrocytes. The marginal band of microtubules that encircles the periphery of these cells will reform in the same location upon recovery from cold-induced disassembly (COHEN *et al.* 1982, MILLER and SOLOMON 1984). More significantly, marginal bands will form in detergent-extracted cytoskeletons incubated with exogenous tubulin (SWAN and SOLOMON 1984). Thus both erythrocytes and *Euglena* cortices possess determinants of mircrotubule organization that are extrinsic to the microtubules. These determinants may be MAP-like proteins, but they are not strictly microtubule-associated. In chicken erythrocytes, they may be part of the membrane-associated "tracks" visualized on the inner face of the membrane by rotary shadowing (GRANGER and LAZARIDES 1982) (Fig. 72). Euglenoid plasma membrane fragments and chicken erythrocyte membranes therefore appear as useful systems in which to attempt a biochemical dissection of microtubule-membrane linkage.

There are only few reports of associations of *intermediate filaments* with isolated plasma membranes, aside from specific attachment sites such as desmosomes. Segments of chicken red blood cell membranes, prepared by sonication while attached to a substratum, have network fragments composed of vimentin filaments associated with them (GRANGER and LAZARIDES 1982). The mode of attachment is not known, but spectrin is possibly involved. Isolated plasma membranes of lens cells are also associated with laterally linked vimentin filaments (RAMAEKERS *et al.* 1982). The mode of linkage is not understood here either, but it could be more direct since newly synthesized vimentin has the ability to associate with the membrane in an *in vitro* translation system.

IX.6. Some Aspects of Organelle-Cytoskeleton Relationships

The preceding sections discussed associations of cytoskeletal elements with the plasmalemma. Here, evidence suggesting associations between cytoskeletal fibers and intracellular membrane-bound organelles will be examined, and implications of these associations for the nonrandom distribution of these cell components will be discussed. Research in this area is still in the "hunting and gathering" stage, though, and specific mechanisms for these associations are not yet known.

Close associations of intracellular organelles with filaments and microtubules are familiar to electron microscopists. In many cases, electron-dense bridges extend between the organelle surface and the filament(s), suggesting firm, if transient, anchorage. The linkage of synaptic vesicles and mitochondria to axonal microtubules (*e.g.*, JARLFORS and SMITH 1969, RAINE *et al.* 1971, SMITH *et al.* 1977) is a striking example, but contacts between organelles and microtubules or filaments are also common in other cell types. These associations have been recently studied in cultured cells with fluorescent antibody techniques. Double label immunofluorescence micros- copy with antibodies to cytochrome c oxidase and tubulin demonstrates a remarkable superposition of mitochondrial and microtubule distribution, at least at the level of light microscopic resolution (HEGGENESS *et al.* 1978). However, because microtubules and intermediate filaments are extensively colinear, an association between mitochondria and intermediate filaments could not be excluded. In subsequent triple-label immunofluorescence

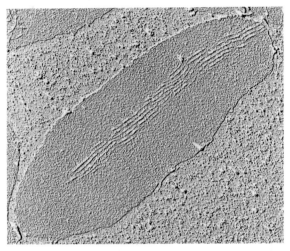

Fig. 72. Microtubule "tracks" in sonicated chicken erythrocyte plasma membrane fragments after removal of microtubules. Shadowed replica. × 9000. [From GRANGER and LAZARIDES (1982)]

studies, intermediate filaments were experimentally segregated from the microtubules; the mitochondria nonetheless remained codistributed with the microtubules (BALL and SINGER 1982), While the proteins that mediate this intertaction have not yet been identified, there is evidence for tubulin binding sites on purified mitochondria (BERNIER-VALENTIN and ROUSSET 1982). The well-known phenomenon of saltatory movement of mito- chondria through the cytoplasm seems to be microtubule-associated, but perhaps also dependent on actin (COUCHMAN and REES 1982). On the other hand, in some cells a mitochondria-specific polypeptide remains associated with intermediate filament proteins, rather than microtubules, after deter- gent extraction, suggesting that at least a subpopulation of intermediate filaments can anchor mitochondria (MOSE-LARSEN *et al.* 1982). SUMMERHAYES *et al.* (1983) suggest that both microtubules and intermediate filaments can influence the distribution of mitochondria, but neither fiber system alone is the dominant organizer.

The Golgi apparatus is intimately involved in the intracellular distribution and sorting of membrane components and vesicular organelles. It is located as a compact structure close to the nucleus, often in intimate association with the centrosome (ROBBINS and GONATAS 1964, KUPFER et al. 1982). Recent experimental studies suggest a major role for microtubules in positioning the Golgi apparatus near the nucleus and maintaining its integrity. Depolymerization of microtubules induces fragmentation of the Golgi complex and dispersion of the fragments throughout the cytoplasm (MOSKALEWSKI et al. 1975, THYBERG et al. 1980, LIN and QUEALLY 1982) (Fig. 73). Colcemid-induced dispersal of the Golgi fragments can be prevented by the microinjection of a nonhydrolyzable analog of GTP that induces the formation of twisted tubulin ribbons (WEHLAND and SANDOVAL 1983), but reconstruction of the Golgi complex requires intact microtubules. After removal of the antimicrotubular agent, when microtubules have partially reassembled from the centrosome, the dispersed Golgi elements undergo re-compaction around the centrosome (ROGALSKI and SINGER 1984 a) (Fig. 73). Somewhat unexpectedly, the microtubule stabilizing compound taxol, which induces the formation of microtubule bundles scattered in the cell periphery, will also fragment the Golgi apparatus (WEHLAND et al. 1983). The Golgi fragments, however, stay associated with the dispersed microtubule bundles (WEHLAND et al. 1983, ROGALSKI and SINGER 1984 a). Surprisingly, dispersal of the Golgi apparatus by either colcemid or taxol does not affect the transfer of plasma membrane molecules from the Golgi apparatus to the cell surface. The dispersed Golgi fragments still function correctly in the processing of membrane proteins, but insertion appears to occur all over the membrane surface, as opposed to a polarized insertion in the presence of microtubules (ROGALSKI et al. 1984). Thus the integrity of the Golgi apparatus correlates strongly with microtubule distribution, but the precise functional significance of this association still is somewhat obscure.

A third group of vesicluar cell organelles with a demonstrated affinity for microtubules *in vivo* includes lysosomes, secretory vesicles, and endosomes. These vesicular organelles frequently line up along microtubules (PHAIRE-WASHINGTON et al. 1980, ALBERTINI and HERMAN 1984), and their move-

---

Fig. 73. NRK cells double-labeled for microtubules (A, C, E) and the Golgi apparatus (B, D, F). The cells are infected with vesicular stomatitis virus (VSV), and the Golgi apparatus is revealed by an antibody against G-protein of VSV. A, B Untreated cell. C, D Microtubules disassembled with nocodazole; the Golgi apparatus is dispersed. E, F Microtubules are partially reassembled after release from nocodazole treatment; the Golgi elements are undergoing recompaction near microtubules (arrows). × 1,050. [From ROGALSKI and SINGER (1984)]

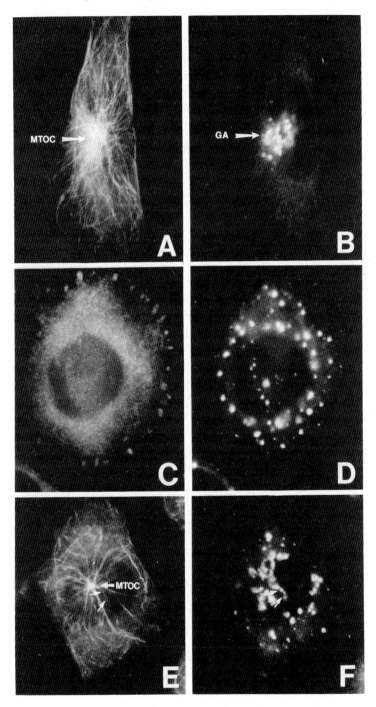

ment through the cell depends on microtubule integrity (WANG and GOLDMAN 1978). Intermediate filaments do not seem to establish or maintain this association; under conditions which segregate intermediate filaments and microtubules, these organelles remain associated with the latter (COLLOT et al. 1984). *In vitro* studies confirm that secretory vesicles can bind to microtubules (SHERLINE et al. 1977), and suggest that binding depends on the presence of MAPs (SUPRENANT and DENTLER 1982). As discussed in the previous section, secretory and lysosomal membranes can bind to actin filaments as well. Future studies will determine whether the binding of vesicle membranes to microtubules and actin filaments signals that the two fiber systems are interdependent or cooperate in directing some of the activities of these organelles (*e.g.,* movement through the cytoplasm, fusion with the cell membrane or endosomes, etc.).

The close association of membrane-bound cell organelles with elements of the cytoskeleton, though not understood biochemically, is compelling evidence for cytoskeletal control over the nonrandom distribution of intracellular inclusions. Clearly, directional movement of these organelles through the cytoplasm depends on the integrity of microtubules or actin filaments or both (reviewed in SCHLIWA 1984), but there is more to these associations than the generation of the motive force for movements. The ordered disposition of cell organelles implicates the cytoskeleton in a grander scheme of organization of cellular space. PORTER (1984) uses the term "cytoplast" to describe a unifying structure endowed with the capacity to control the distribution of all the structures it contains, and to provide the spatial and temporal information for the assembly/disassembly of its components. The cytoplast concept suggests the existence of a blueprint of cellular architecture and morphogenesis based on the principles of self-organization. While we can not yet read and fully comprehend the blueprint, we recognize, and admire, the results emerging from this construction plan of cellular architecture.

IX.7. Specialized Filamant-Membrane Assemblies

Actin and intermediate filaments form complex junctional assemblies with specialized regions in the plasmalemma distinguished from the rest of the membrane by a number of morphological and biochemical criteria. Common characterstics of such contact zones are: 1. spatial confinement to small areas of the membrane (*maculae*); 2. convergence of a large number of actin or intermediate filaments towards the contact site; 3. the presence of electron-dense material (*plaques*) where the filaments are associated with the membrane; 4. close apposition of the membranes of neighboring cells at the contact site; and 5. resistance of the attachment devices towards removal by non-ionic detergents. Two complex cytoskeleton-membrane assemblies that meet these criteria will be considered here: adherens junctions, especially adhesion plaques of cultured cells, in which actin filaments associate with specialized membrane contact sites; and desmosomes, sites where intermediate filament bundles associate with the membrane. A third site of filament-membrane attachment also to be discussed here are intestinal microvilli of the brush border, currently the best-understood microfilament organelle, where highly ordered actin filament bundles are linked to the plasmalemma end-on and laterally.

IX.7.1. Adherens Junctions and Adhesion Plaques

Adherens junctions are a group of morphologically diverse but structurally and functionally related membrane-filament contact zones. They include the *zonulae adherentes* at the periphery of epithelial cells, the *fasciae adherentes* of intercalated discs in cardiac muscle cells, and the *dense plaques* of smooth muscle. In addition, *adhesion plaques* or *focal contacts,* the specialized areas of contact of cultures cells with the surface on which they grow, belong in this category because they are probably homologous to the adherens junctions of cells in tissues. Because cultured cells are more experimentally accessible, adhesion plaques have been extensively studied as model systems for specific attachment sites of actin filament bundles with membranes. For this reason, they shall be discussed first.

Many cell types make contact with the surface of the artificial substrate (glass, plastic culture dishes, carbon films, etc.) at spatially restricted sites of close apposition between the ventral cell membrane and the substratum. While most of the cell body is separated from the substratum by more than 50 nm, the gap reduces to 10–15 nm at these sites, which were therefore considered to be areas of strong cell-substrate adhesion. Identified first by electron microscopy of transversely sectioned cells (ABERCROMBIE *et al.* 1971), these sites were shown to be associated with bundles of microfilaments, or stress fibers (BUCKLEY and PORTER 1967). These bundles contain actin filaments that appear to terminate end-on at the focal contact sites (SPOONER *et al.* 1973, GOLDMAN and KNIPE 1973, ABERCROMBIE 1980, SANGER and SANGER 1980). The distribution of these contact zones is best visualized by the technique of interference reflection microscopy, which is based on the interference of light reflected from the ventral cell surface with light reflected from the surface of the substratum (IZZARD and LOCHNER 1976, BEREITER-HAHN *et al.* 1979, GINGELL 1981). Interference reflection microscopy reveals areas of close apposition as dark gray zones because light rays reflected from the 2 surfaces interfere destructively (Fig. 74). In contrast, areas where the distance between the surfaces exceeds 20 nm appear as light grey or white (ABERCROMBIE and DUNN 1975, HEATH and DUNN 1978, WEHLAND *et al.* 1979). This technique has provided useful information about the distribution and nature of focal contacts between

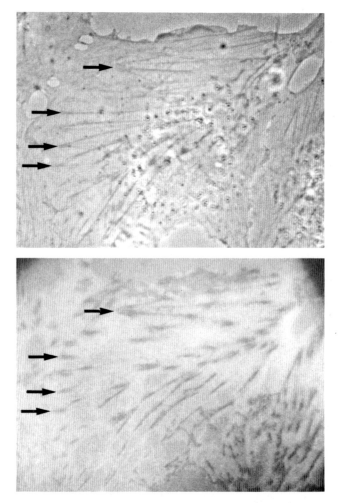

Fig. 74. Phase contrast (top) and interference reflection micrograph (bottom) of a BSC-1 cell. Note correspondence of focal adhesions and the ends of stress fibers (arrows). × 400

cells and the substratum. Because it is a noninvasive physical procedure, it can be used to study the dynamics of cell-substrate contacts in live cells. Interference reflection studies correlated with immunofluorescence microscopy confirmed that stress fibers terminate at sites of focal contact, and showed that these terminal regions of stress fibers are enriched for actin and alpha-actinin. Myosin and tropomyosin may be close to these sites, but are not necessarily part of the focal contact zone (for a recent review of stress fibers, see BYERS et al. 1984). A number of other proteins that specifically colocalize with focal contacts have attracted much attention because, unlike

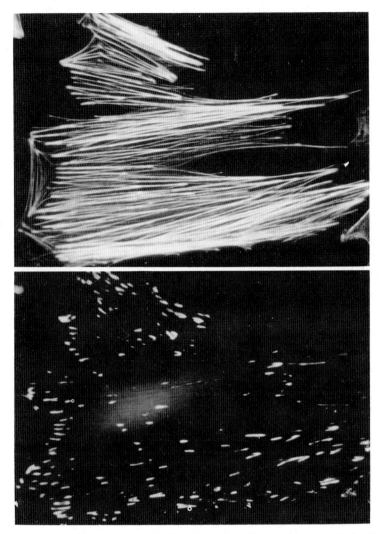

Fig. 75. A 3T3 fibroblast double-labeled for actin with rhodamine-phalloidin (top) and vinculin with an antibody (bottom). × 650

the other stress fiber components, they are found only in contact sites (see GEIGER 1983, GEIGER *et al.* 1984, MANGEAT and BURRIDGE 1984, for overviews).

The first of these adhesion plaque proteins, a 130,000 d globular polypeptide, was discovered in 1979 and named vinculin (GEIGER 1979, BURRIDGE and FERAMISCO 1980). Because of its specific localization at the ends of stress fibers (Fig. 75), it was immediately regarded as a strong candidate for a linker between actin filaments and membranes. Several

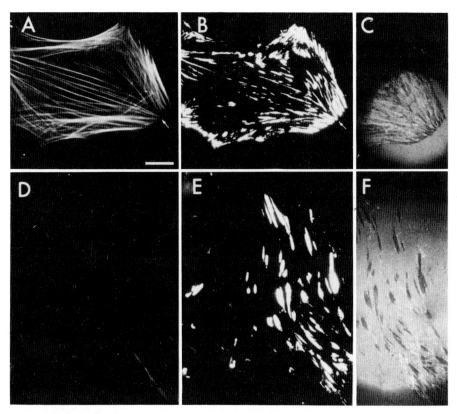

Fig. 76. Double labeling of ventral membranes of chicken fibroblasts with fluorescent phalloidin for actin (*A, D*), and with an antibody to vinculin (*B, E*). Corresponding interference reflection images are shown in (*D, F*). The ventral membrane preperation shown in (*D, E, F*) was incubated with fragmin (0.4 mg/ml) for 3 minutes to remove actin. Vinculin distribution appears unaffected. × 780.
[From Avnur *et al.* (1983)]

studies have shown that the purified protein interacts with both actin filaments and membranes. Vinculin binds with high affinity to the "barbed" end of actin filaments (Wilkins and Lin 1982), but it also seems to be able to associate with actin filaments along their length to induce the formation of bundles (Jockusch and Isenberg 1981). One or both of these effects may, however, be due to minor contaminants in conventional vinculin preparations. Diethyl-aminoethyl cellulose-purified vinculin does not bundle actin filaments (Evans *et al.* 1984). The question whether vinculin interacts with actin filaments may have to be reexamined, and the contaminants need to be characterized because they may be the more interesting components.

Evidence for an interaction of vinculin with membranes is derived from

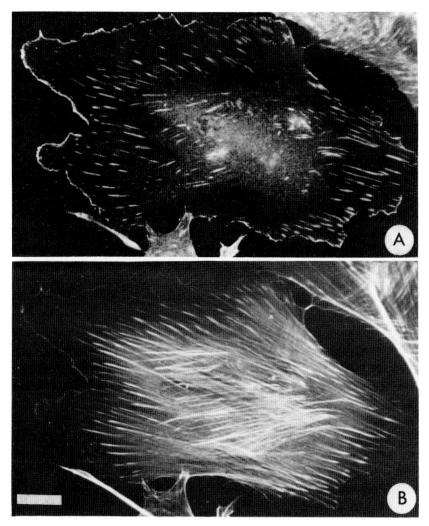

Fig. 77. Comparison of the distribution of talin (*A*) and actin (*B*) in the same cell. Talin is found at the termini of stress fibers and along much of the margin of the cell. × 600. [From BURRIDGE and CONNELL (1983)]

experiments in which the cell body of cultured cells is sheared off, leaving ventral membranes and adhesion plaques attached to the substratum in an *in situ* -configuration (AVNUR and GEIGER 1981). Actin can be removed completely from these preparations by treatment with an actin-severing protein isolated from *Physarum*, yet vinculin will remain associated with the membrane remnants (AVNUR *et al.* 1983) (Fig. 76). The firm association of vinculin with the plasma membrane demonstrated by this experiment raises the question whether it might be an integral, rather than a peripheral, membrane protein linking actin filaments to the membrane. While this suggestion has not been confirmed for vinculin itself, a closely related polypeptide of slightly higher molecular weight (150,000), termed metavinculin, was found to have solution properties of an integral membrane protein (SILICIANO and CRAIG 1982). Vinculin and metavinculin are immunologically related, but distinct proteins that are products of different genes (FERAMISCO et al. 1982); thus a precursor-product relationship can be excluded. Whether metavinculin is the sought-after membrane linker has not yet been determined experimentally; nor have the proteins to which vinculin binds in or at the membrane been identified.

Recently, another adhesion plaque-specific protein has been identified as a 215,000 d, vinculin-binding protein, termed talin (BURRIDGE and CONNELL 1983, BURRIDGE and MANGEAT 1984). Similar, if not identical polypeptides were identified as vinculin-binding proteins in a gel overlay assay (OTTO 1983, WILKINS et al. 1983). A protein of 200,000 d molcular weight has been described by MAHER and SINGER (1983) as a component of adhesion plaques, but its relationship to talin has not yet been determined. By immunofluorescence microscopy, the distribution of talin is similar to that of vinculin; it is concentrated in focal contacts, but it also weakly stains regions that coalign with underlying stress fibers, and it is slightly more concentrated at the cell margin (BURRIDGE and CONNELL 1983) (Figs. 77 and 78). Talin binds to vinculin with high affinity in vitro, as already suggested by its ability to bind iodinated vinculin in gel overlays. The biochemical properties of talin are currently being characterized. It will be of great interest to determine whether the talin-vinculin complex has actin or membrane-binding properties different from those of the separate components. Talin and vinculin may be two components of a complex of proteins, possibly including alpha-actinin and other as yet unknown polypeptides, that links the ends of actin filaments to the membrane. This possibility needs to be addressed experimentally. It will be equally important to identify the membrane "receptors" for these complexes.

Several membrane proteins are concentrated specifically at focal contacts (OESCH and BIRCHMEIER 1982, ROGALSKI and SINGER 1984). The staining patterns of these proteins, labeled with monoclonal antibodies from the outside of the cell, coincide with the patterns for adhesion plaque proteins on the inside of the membrane. These components may therefore well be transmembrane proteins. Whether these transmembrane adhesion plaque proteins interact with intracellular adhesion plaque proteins remains to be determined. Overall, the prospects for future research in this field are excellent. Given the spectrum of candidates for linker molecules already identified, the molecular anatomy of the adhesion plaques may be well understood in a few years.

As a cell changes shape, moves about the substratum, or rounds up in preparation for cell division, focal contacts wax and wane coordinately with the restructuring of the skeletal apparatus. One might ask, therefore, how stable or unstable focal contacts are, and what the diffusional or trans-

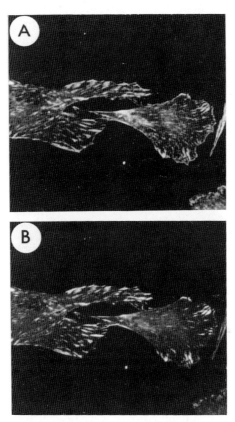

Fig. 78. Comparison of the distribution of talin (*A*) and vinculin (*B*) in the same chicken heart fibroblast. The patterns of staining are almost completely superimposable, except that talin is also present at the cell margin. × 400. [From Mangeat and Burridge (1984 b)]

lational activity of its individual components are. To begin answering these questions, Geiger *et al.* (1982) determined the lateral mobility of fluorochrome-labeled lipids and membrane proteins in focal contact areas of ventral cell membranes. Lipids were essentially free to diffuse in and out of the contact zones, whereas about 50% of the membrane proteins in these areas were essentially immobile, as determined by fluorescence recovery after photobleaching. The other 50% diffused 2–3 times more slowly than proteins outside of focal contact areas. This lower mobility may be attributed to transient weak interactions within the contact zone, or to a reduction in the area available for free diffusion, or both. The biophysical basis of the differences in the diffusibility of adhesion plaque components has not yet been determined, but their behavior can be rationalized with morphological and biochemical information. For example, GEIGER (1982) found several isoforms of vinculin that differ slightly in charge and subcellular distribution. The subcellular distribution of these isoforms may be determined by posttranslational modifications of products of the same gene. These modifications would therefore constitute molecular "address markers". Thus focal contacts have dynamic and static properties, allowing exchange of some components while at the same time maintaining associations between others.

These features are important for rapid reorganization of focal contacts. The distribution of vinculin changes rapidly upon transformation of cells with Rous sarcoma virus (DAVID-PFEUTY and SINGER 1980, SHRIVER and ROHRSCHNEIDER 1981, MARCHISIO et al. 1984), or upon treatment with tumor promoters (SCHLIWA et al. 1984). In the case of virus transformation, reorganization of vinculin appeared to correlate with increased phosphorylation of its tyrosine residues (SEFTON et al. 1981, NIGG et al. 1982, ITO et al. 1983). Phosphorylation of vinculin, therefore, was considered a candidate for the regulation of its distribution and interaction (WERTH et al. 1983). In other instances, however, reorganization of vinculin does not correlate with increased tyrosine phosphorylation (ROHRSCHNEIDER and ROSOK 1983, ROSOK and ROHRSCHNEIDER 1983, ANTLER in situ 1985). It needs to be determined whether phosphorylation of vinculin alters its binding properties for actin or the membrane. Also, vinculin is but one component of the adhesion plaque, and changes, such as phosphorylation, in other target proteins in this structure might allosterically influence associations within the complex.

Are the studies on focal adhesions in tissue culture cells relevant to cell contacts in intact tissues? Stress fiber-like filament bundles containing actin, alpha-actinin, myosin, and tropomyosin, are present in certain cells in situ, including fish scale scleroblasts (BYERS and FUJIWARA 1982) and vascular endothelial cells (WONG et al. 1983, WHITE et al. 1984). In endothelial cells, they are oriented parallel to the direction of blood flow, suggesting that they provide structural support against shear forces (BYERS et al. 1984). It has not yet been determined whether these in vivo stress fibers, which are not associated with a solid substratum, end in adhesion plaque-like structures containing vinculin and talin. Elucidation of this question may contribute to our understanding of the nature of the adhesion plaque.

Studies on adherens junctions suggest that they resemble desmosomes in structure (Fig. 79), but that they are similar in composition to focal contacts. Adherens junctions from several different tissues, including

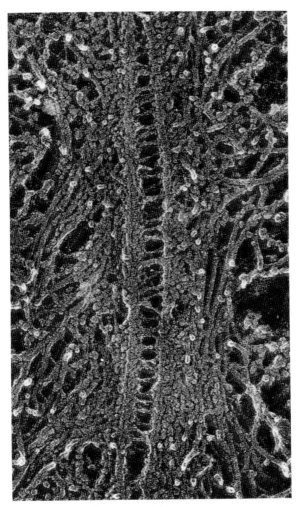

Fig. 79. Bridging structures between opposing halves of an adherens junction are revealed in this quick-frozen, deep-etched, and rotary shadowed preparation of mouse intestinal epithelial cells. The cells were extracted with saponin before freezing. × 125,000. [From HIROKAWA and HEUSER (1981)]

intestinal epithelium, smooth muscle, and cardiac muscle, contain vinculin in close association with the membrane (GEIGER *et al.* 1980, TOKUYASU *et al.* 1981) (Fig. 80), in addition to bundles of actin filaments associated with the junctional plaque. Periodically arranged, peripheral bands of vinculin-rich structures (termed "costameres") are found to encircle skeletal muscle cells perpendicular to the long axis of the muscle fiber (PARDO *et al.* 1983). The periodicity of these band corresponds to that of the adjacent sarcomeres, and the pattern of staining superimposes with that of the alpha-actinin-containing Z-lines. Thus vinculin and alpha-actinin may be part of a sarcolemma-myofibril attachment complex. As in adhesion plaques, vinculin is located closer to the membrane than alpha-actinin, as demonstrated by immunoelectron microscopy (TOKUYASU *et al.* 1981).

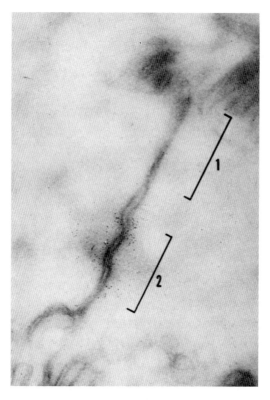

Fig. 80. Indirect immunoferritin labeling for vinculin in a frozen section of the chicken intestinal brush border. Ferritin label is located at the *zonula adherens* (bracket *2*), but not at the tight junction (bracket *1*). × 45,000. (Courtesy of B. GEIGER).

The subcellular distribution of talin in tissues has not been determined. However, given that it colocalizes with vinculin in cultured cells, that visculin is present in many different tissues, and that this protein was originally isolated from chicken gizzard smooth muscle, talin might turn out to be a component of adherens junctions in tissue cells as well. It appears likely that vinculin and talin are involved in the construction of actin-membrane adhesion sites in tissue cell associations *in situ*. Adherens junctions of tissue cells also contain specific proteins not found in focal contacts of cells in culture (VOLK and GEIGER 1984) which need to be characterized in the future. However, adherens junctions and focal contacts may well turn out to be homologous differentiations, and the study of either may prove relevant for both.

IX.7.2. Desmosomes

Of all intercellular junctions, desmosomes are probably the most conspicuous morphologically. The principal cytoplasmic components of desmosomes are intermediate filaments and a so-called dense plaque, a darkly staining lamella several nanometers thick that is closely apposed to the inner leaflet of the cell membrane. In a complete desmosomal junction between two neighboring cells, two identical half-junctions face each other across the middle of the desmosomal gap (Fig. 81). In certain epithelial tissues, desmosomal assemblies can also form along the membrane facing a basal lamina or some other extracellular material; because they lack the corresponding identical structure, they are called hemidesmosomes. In between the plasma membranes of the desmosomal halves is an amorphously stained matrix with a distinct middle lamella (*stratum centrale*) bisecting the intercellular gap. Wispy filaments extend between the outer membrane and the middle lamella. Desmosomes are typical features of epithelial cells, and they are most abundant in epidermal tissues (for review on primarily the morphological features of desmosomes, see CAMPBELL and CAMPBELL 1971, STAEHELIN 1974, WEIHING 1979). In epithelial cells, cytokeratin filaments are attached to the desmosomal plaques. Desmosome-like assemblies containing some of the same protein components as epithelial desmosomes also occur in cardiac myocytes (FRANKE *et al.* 1982c, KARTENBECK *et al.* 1983), and arachnoidal tissues (KARTENBECK *et al.* 1984). In these instances, the intermediate filaments specific for these tissues (desmin in the former, vimentin in the latter) are associated with the desmosome. Thus, contrary to a widely held belief, keratin filaments are not the only kind of intermediate filaments associated with desmosomes.

Ultrastructural studies of desmosomes demonstrate a close association of intermediate filaments specifically with the desmosomal plaque (KELLY 1966). The filaments associate with the plaque laterally rather than end-on, as in adherens junctions. They either run parallel to the cell surface and through the outer regions of the dense plaque, or they approach the desmosome perpendicular to the plane of the membrane and loop in and out of the plaque region. Thus, the predominent form of linkage to the plaque appears to be via the lateral surface of the filaments, but the molecular components involved in that linkage have not been identified.

Desmosomes have several distinctive biochemical features. Desmosome-filament complexes from tissues rich in these junctions (*e.g.,* bovine muzzle) can be prepared with relative ease (SKERROW and MATOLTSY 1974,

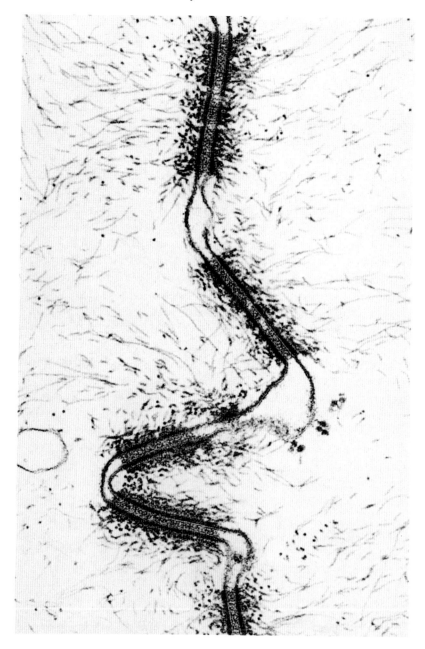

Fig. 81. Isolated membrane-desmosome-tonofilament complex from calf muzzle, showing alternating intercepts of desmosomes and interdesmosomal membrane regions. Desmosomal organization (dense plaques, bridging structures) is well preserved. × 66,000. [From Drochmans *et al.* (1979)]

DROCHMANS *et al.* 1979). Upon further subfractionation, a set of 5–6 polypeptides specifically associated with the desmosomal core are revealed (FRANKE *et al.* 1981 c, GORBSKY and STEINBERG 1981, COHEN *et al.* 1983, COWIN and GARROD 1983). Two of these, with molecular weights of 250,000 and 215,000 d, are termed desmoplakins I and II (MUELLER and FRANKE 1982), and are prominent components of the desmosomal plaque of both epithelial (FRANKE *et al.* 1981 c, GEIGER *et al.* 1983) and myocardial desmosomes (FRANKE *et al.* 1982 c, KARTENBECK *et al.* 1983). Therefore, if these proteins turn out to be involved in the linkage of the desmosome-associated filaments to the plaque, then they will do so regardless of the filament type. By immunofluorescence microscopy, antibodies against desmoplakins outline the periphery of epithelial cells with a dotted or pearlstring-like pattern, corresponding to the chains of desmosomes seen in electron micrographs of horizontally sectioned cells (*e.g.,* SCHLIWA 1975). By immunoelectron microscopy, antibodies against these polypeptides react specifically with the desmosomal plaque (FRANKE *et al.* 1981 c), but it is difficult to decide whether the components they react with are responsible for the actual linkage of the desmosome-associated filaments with the plaque proper. Several desmosomal polypeptides with molecular weights ranging from 22,000 to 150,000 d are highly glycosylated, and therefore might be membrane proteins involved in intercellular recognition and inter-membrane linkage across the desmosomal gap (GORBSKY and STEINBERG 1981, COHEN *et al.* 1983). Two of these glycoproteins are located on the surface and can be labeled with antibodies from the outside of the cell. Antibodies against these proteins applied externally to live cells will inhibit desmosome formation (COWIN *et al.* 1984).

In summary, desmosomes and adherens junctions are junctional specializations with both common and distinctive features. Both are characterized by the presence of dense plaques closely apposed to a restricted membrane area with which different types of filaments are firmly associated. In fact, the associations of the filaments with the plaque structure are so tenacious they will survive subcellular fractionation or cell death (Fig. 82). In both types of junctions certain membrane proteins are selectively clustered at the junction site and restricted from dispersion. The restriction is assumed to be based on constraints exerted by membrane-associated components such as dense plaque-proteins and cytoskeletal elements on the internal side of the membrane, and perhaps certain extracellular components. The two forms of junctions are distinguished by different sets of biochemical markers, of which vinculin (for adherens junctions) and the desmoplakins (for desmosomes) are presently the best-known. Others will undoubtedly be found and characterized in the near future. However, some of the key features of the biochemical anatomy of both junctional complexes are not understood. These are the mode of attachment of the respective intracellular filaments to the plaque, the mode of association of the plaque components with the inner membrane, and the transcellular linkages between the corresponding membrane domains of the two adjoined cells. These questions will ultimately have to be approached in reconstituted *in vitro* models using purified components. Furthermore, little

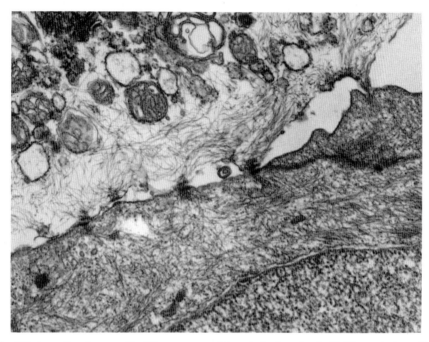

Fig. 82. A surface layer cell of the guppy epidermis is sloughed off. Though the cell is dead and lysed, the desmosomal contacts to the underlying intact cell are still maintained. × 18,000

is known about the factors that determine the site of formation of a junctional complex and how the adjacent cells communicate this positional information. For example, one could envision that the assembly of one desmosomal half in one cell will provoke the neighboring cell to assemble the corresponding structure on the opposite site. The many studies of resorption of junctional complexes after experimentally induced cell dissociation (see, for example, OVERTON 1974) have established that the modes of internalization of plaque material are different for the two kinds of junctions. Adherens junctions are released as aggregates of plaque material into the cytoplasm, whereas desmosomal material is internalized as an intact complex with attached filaments still associated with a membrane vesicle (KARTENBECK *et al.* 1982). Whether this experimentally induced pattern of resorption bears on the reorganization and dissolution of junctional complexes in living organisms remains to be demonstrated conclusively.

IX.7.3. The Brush Border of Intestinal Epithelial Cells

Several modes of actin filament-membrane interaction have been established in the brush border, and for this reason it is discussed in this section, even though the significance of brush border organization goes beyond that of a special case for membrane-cytoskeleton interactions. The brush border is perhaps the *only* highly organized cytoskeletal domain that is sufficiently stable to be isolated in intact form, in sufficient quantities, and in high purity, for detailed biochemical analyses. These attributes are reflected in its current popularity as a model system. More is known about the organization of the brush border than any other nonmuscle cytoskeletal assembly. It is likely that a detailed analysis of the brush border will provide essential information for the organization and function of other, more complex and dynamic assemblies. The growing body of information has been reviewed recently (MOOSEKER 1983, BRETSCHER 1983, several papers in PORTER and COLLINS 1983). Some of the salient features of the system will be outlined here.

IX.7.3.1. Morphology and Ultrastructure

The luminal surfaces of the intestine and the kidney tubules are lined with columnar epithelial cells whose apical surfaces are densely packed with arrays of several hundred microvilli, often arranged in a regular, hexogonal pattern. Uniform in a given cell type, the length of microvilli ranges from 0.5–2.0 µm, their width from 50–100 nm. Within the epithelium, neighboring cells are joined laterally by a belt of junctional complexes including tight junctions, adherens junctions, and desmosomes (HULL and STAEHELIN 1979) which provide a tight seal and act as rivets that distribute shearing forces throughout the epithelium. Each microvillus contains a core bundle (Fig. 83) of 10–50 actin filaments of uniform polarity attached to a dense cap structure associated with the membrane at the microvillus tip. When decorated with heavy meromyosin (BEGG et al. 1978), the "barbed" ends of the actin filaments are attached to the cap, while their "pointed" ends extend into the terminal web region. These bundles of microfilaments often form regular hexagonal arrays when viewed in cross-section. The core bundle is attached to the sides of the microvillar membrane by periodic bridges spaced 33 nm apart (e.g., MUKHERJEE and STAEHELIN 1971, MOOSEKER and TILNEY 1975, BRETSCHER and WEBER 1978). In isolated microvillar core bundles, these bridges were shown to be disposed along the

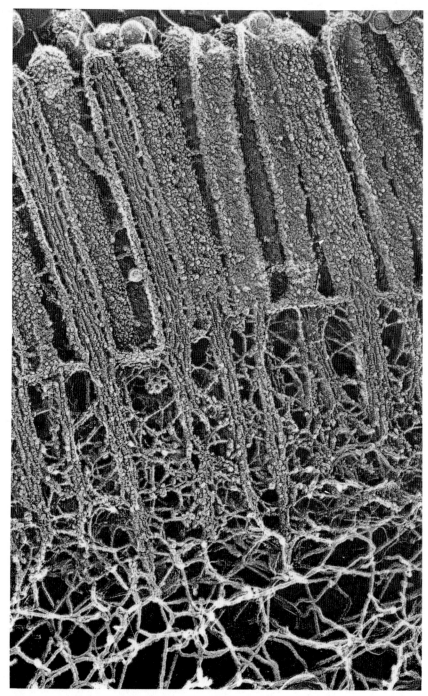

Fig. 83. Brush border of an intestinal epithelial cell of the mouse, prepared by fixation, quick-freezing, deep-etching, and rotary shadowing. The microvillar core bundles extend into the terminal web region as rootlets, where they are linked to one another by delicate filaments. The terminal web rests on a cushion of smooth intermediate filaments. × 97,000. (Courtesy of N. HIROKAWA)

bundles in a double helical array (MATSUDAIRA and BURGESS 1982 b). Within the core bundle, actin filaments are linked to one another by a different set of crossbridges, also arranged in a regular pattern.

About one quarter of the length of the microvillus core descends below the microvillus base into the terminal web region (Fig. 83). This portion of the core is termed the rootlet. The rootlets are embedded in a dense three-dimensional meshwork of filaments which includes actin, intermediate filaments, and small filaments of 3–5 nm diameter that do not decorate with heavy meromyosin and that appear to link adjacent microvillar rootlets together (BEGG et al. 1978, HIROKAWA and HEUSER 1981). The precise arrangement of actin filaments in the terminal web is not well understood. An exception is the lateral margin of the terminal web region, where a bundle of actin filaments forms a continuous belt and associates laterally with a belt-like adherens junction. Intermediate filaments of the cytokeratin type are associated with the spot desmosomes located below the adherens junction, from where they extend throughout the terminal web region, forming an open, three-dimensional network. It almost appears as if they provide a "cushion" on which the brush border rests. The transition from the apical cell region carrying the brush border to the rest of the cell body has not been studied in great detail, but there appears to be a relatively sharp demarcation since homogenization of suspensions of intestinal epithelial cells leaves only brush borders and nuclei as large intact cell fragments. This is one of the attractive features of the brush border since it essentially allows for its mass isolation in high purity. From this starting material further subfractions (e.g., microvilli, core bundles) are obtained with relative ease.

IX.7.3.2. Biochemical Anatomy

Over the past five years, all the major protein components of the brush border have been identified and more or less completely characterized. These proteins include actin, villin (95,000 d), fimbrin (68,000 d), a 110,000 d protein, calmodulin, and tropomyosin, all of which are found in the microvillous core bundle, and myosin and brush border spectrin (TW 260/240) in the terminal web. The properties of many of these proteins were discussed in section I.3. Most, if not all, of these proteins are engaged in either filament-filament or membrane-filament interactions. Linkage of the actin filaments in the microvillous core into a hexgonal bundle is mediated by villin and fimbrin. These two proteins are able to cross-link purified actin into bundles in vitro (see BRETSCHER and WEBER 1980, MOOSEKER et al. 1980, MATSUDAIRA and BURGESS 1982 a for villin; BRETSCHER 1981, GLENNEY et al. 1981 b for fimbrin). Bundles formed from fimbrin and actin resemble more closely those of microvilli in situ (MATSUDAIRA et al. 1983); fimbrin, therefore, appears to be the principal bundling factor in vivo.

Moreover, the calcium-dependent actin severing acitivity of villin suggests that villin's primary function seems to be concerned with dissolution of the microvillous core rather than maintenance. *In vitro* in the presence of micromolar calcium, microvilli break up into short segments, and the microvillar membrane vesiculates. This activity, originally (and erroneously) interpreted as "contraction" of the microvilli into the terminal web (MOOSEKER and TILNEY 1975), may have to do with microvillar length changes and assembly, but its precise function is not yet known. Rootlets are usually protected from this severing activity, possibly by tropomyosin which is found only associated with the rootlet portion of the core bundle (DRENCKHAHN and GROESCHEL-STEWART 1980). The precise disposition of villin and fimbrin in the core bundle has not yet been elucidated.

Long, thin, filaments 3–5 nm in diameter found only in the terminal web link adjacent rootlets with one another and with the other filaments. Immunoelectron microscopy implicates brush border spectrin and myosin in this form of linkage (GLENNEY *et al.* 1983, HIROKAWA *et al.* 1983), but the specific function(s) of these two proteins in maintaining brush border integrity remains to be elucidated.

The associations between filaments and the membrane within the brush border are less well understood. The core bundle interacts with the membrane at two distinct classes of sites: the microvillus tip, and the lateral surface of the microvillus membrane. The biochemical nature of the densely-staining, plaque-like region into which the actin filaments of the core bundle insert with their "barbed" ends is not known. Earlier suggestions that the tip structures contain alpha-actinin and, therefore, might represent the Z-line analog of the brush border could not be confirmed. [The biochemical candidate for brush border alpha-actinin, a 95,000 d protein, was shown to be different from alpha-actinin and later was given the name villin (BRETSCHER and WEBER 1979).] However, small amounts of alpha-actinin are present in the junctional complexes (adherens junctions) at the lateral borders of the terminal web (BRETSCHER and WEBER 1979, GEIGER *et al.* 1979). While the microvilli of intestinal epithelial cells appear to be free of alpha-actinin, this protein may be present in microvilli of other cell types. For example, on the basis of biochemical and immunological criteria, alpha-actinin is a prominent component of the branched microvilli of ascites adenocarcinoma cells (CARRAWAY *et al.* 1980), but its precise location has not been determined.

The observation that the barbed ends of actin filaments, *i.e.,* the preferred ends of filament elongation, are embedded into the dense material at the microvillus tip has puzzled investigators for years. Does it imply the filaments are "capped" or blocked at this end and inaccessible to the addition of actin monomers? This question is not purely academic because the microvillus tip behaves as a microfilament-organizing center. During recovery from pressure-induced core bundle dissolution—which leaves the dense tip structures intact—actin filaments begin to reappear in association with the tip caps (TILNEY and CARDELL 1970). MOOSEKER *et al.* (1982) have shown that isolated brush borders supplied with polymerizable actin monomers will insert new segments of F-actin between the dense tip material and the preexisting core (Fig. 84). Thus the apparently "capped" plus ends of microvillar actin filaments can add monomers while remaining associated with the membrane. The same conclusion was reached by TILNEY

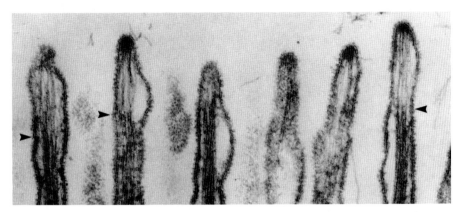

Fig. 84. Elongation of microvilli in isolated brush borders induced by poly-merization of actin from the membrane-associated ends of core filaments. Isolated brush borders were "presoaked" with 24 µM G-actin before addition of salt to induce actin assembly. The dense plaques at the tips of microvilli remain associated with the growing ends of the newly formed filaments. Arrowheads indicate junctions between presumed ends of the core filaments and the newly assembled filaments assembled fromt those ends. × 77,000. [From MOOSEKER *et al.* (1982)]

et al. (1981, 1983) on the basis of more indirect evidence in studies of the elongation of the acrosomal process in *Limulus* and *Thyone* sperm. Both types of acrosomal processes contain a bundle of actin filaments that seem to grow by addition of subunits at the membrane-associated end. Since the same may apply to growing microtubules in cilia and flagella (see section IX.5.), structures that appear as dense, ill-defined plaques in electron micrographs may turn out to be intricate molcular assemblies that provide membrane anchorage of filaments while allowing (and perhaps regulating?) subunit flow.

    Actin filaments at the periphery of the core bundle are attached laterally to the microvillar membrane by bridges whose molecular composition was revealed by combined ultrastructural and protein solubilization studies (MATSUDAIRA and BURGESS 1979). Addition of ATP to isolated core bundles which still have these side arms attached liberates the 110,000 d protein and a 17,000 d protein identified as calmodulin, and leaves core bundles that lack the lateral arms. Thus, the 110 kd protein must be a major protein of these lateral bridges, a finding later corroborated by immunoelectron microscopic localization (GLENNEY et al. 1982 b). The binding site for 110 kd protein at the microvillar membrane is presently under investigation. Suggestions that the 110 kd protein itself is a bifunctional molecule with a hydrophobic integral membrane portion and a hydrophilic cytoplasmic extension (GLENNEY and GLENNEY 1984) are not yet confirmed. On the other hand, studies by COUDRIER et al. (1983) implicate an integral membrane glycoprotein of 140,000 d molecular weight in the association of the 110,000 d protein with the membrane.

The observation of a tight association of calmodulin with the 110 kd protein is intriguing, but its functional significance is obscure. The 110 kd protein has a high affinity for calmodulin in vitro (HOWE et al. 1980, GLENNEY and WEBER 1980), but the association between the two proteins is calcium insensitive. Calmodulin is the second most abundant protein of the core bundle next to actin, and most of it is associated with the 110 kd protein. Calmodulin is not involved in the calcium-dependent solation of the core bundle, which is solely a function of the intrinsically calcium-sensitive protein, villin. Because so far there is no apparent function for calmodulin in the regulation of protein-protein interactions or other reactions in the microvillus, the suggestion was made that it simply acts as a calcium buffer. At the concentration present in the microvillus core, it is capable of buffering calcium almost to the millimolar range. Another possibility which needs to be explored is that calmodulin regulates the activity of the 110 kd protein. Recent evidence suggests that this protein is an ATPase (COLLINS and BORYSENKO 1984). The possible function of this ATPase activity is not known, but the dissociation of the 110 kd protein from the core bundle is ATP dependent (MATSUDAIRA and BURGESS 1979) and could be regulated by calmodulin. Quite clearly, the 110 kd protein-calmodulin-membrane-filament association presently is one of the most mysterious aspects of the organization of the microvillus.

Fig. 85. Quick-frozen, deep-etched, rotary shadowed brush borders of mouse intestinal epithelial cells. *A* Control preparation. Filaments 5 nm in diameter (arrowheads) cross-link adjacent rootlets. *B* Similar preparation incubated with an antibody against brain spectrin (fodrin), followed by a ferritin-conjugated secondary antibody. Clusters of ferritin molecules (arrowheads) are present in places where one would expect to see 5 nm filament linkers. × 113,000. (Courtesy of N. HIROKAWA)

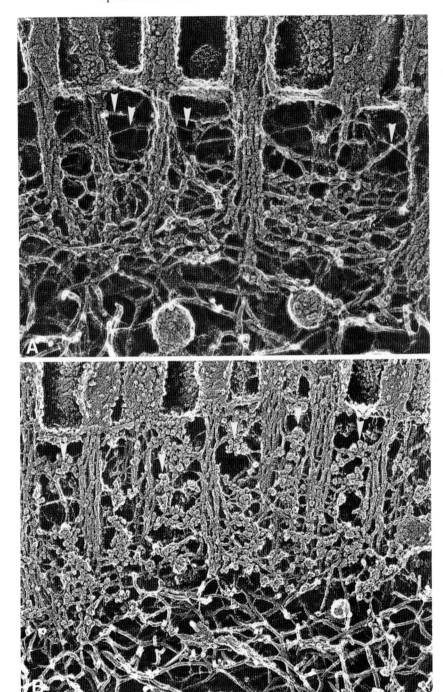

A third major site of filament-membrane interactions is in the region of the terminal web, where the 3–5 nm filaments of brush border spectrin interconnect the rootlet complexes, and link the rootlets to the plasma membrane (HIROKAWA et al. 1983) (Fig. 85). Coupled with the finding that both actin and intermediate filaments of the terminal web are associated with the junctional complexes at the lateral cell margins, the entire filament network of the terminal web is extensively cross-linked to the apical brush border membrane. This feature of the organization of the terminal web may explain the ease with which morphologically intact brush borders can be isolated.

In addition to these major brush border proteins whose functions are known, some potentially significant, but less well characterized minor components are present. For example, the microvillus contains an 80,000 d protein only about 0.7% as abundant as actin. This protein is also found in blebs and retraction fibers of other cell types (BRETSCHER 1983). The terminal web contains a 36,000 d protein which is identical with the major substrate for tyrosine phosphorylation by the src-gene product of Rous sarcoma virus (GERKE and WEBER 1984). This protein binds to spectrin and actin in a calcium-dependent manner. Its colocalization with brush border spectrin is consistent with the distribution pattern in cultured cells, where it forms a submembraneous layer similar to spectrin (see Fig. 63), and the finding that it binds to spectrin in vitro. Hopefully, some of the presently unexplained features of brush border morphology and function will be better understood if some of these minor components are characterized.

IX.7.3.3. A Note on Brush Border "Motility"

Two morphological features of microvilli-bearing epithelial cells make it difficult to think of them as "motile" systems: 1. within a given segment of the intestine or a renal tubule, all cells are morphologically remarkably similar, which argues against gross dynamic changes in the organization of intact cells; 2. the cells are part of an epithelium in which contacts between cells must be maintained at all times, precluding extensive shape changes or excursory movements of individual cells. Nevertheless, brush borders do contain myosin in two sites, indicating, at the very least, the ability to

---

Fig. 86. Contraction of the terminal web in demembranated brush borders. A Brush borders incubated in the presence of 1 µM free calcium in the absence of ATP. At this concentration of calcium, core filaments remain intact. × 33,300. B Brush border incubated in 1 µM calcium and 2 mM ATP. Presumably as a result of an active contraction of the circumferential bundle of actin filaments in the terminal web region, the brush border is constricted. × 22,500. [From KELLER and MOOSEKER (1982)]

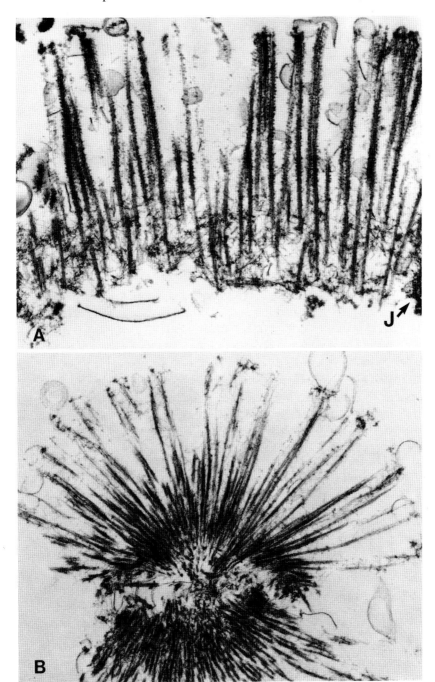

execute some form of motile activity. These two sites are the filament network of the terminal web between the microvillar rootlets, and the circumferential filament bundle associated with adherens junctions (MOOSEKER et al. 1978, DRENCKHAHN et al. 1982). While a possible role of inter-rootlet myosin in motile activities of individual microvilli has not yet been demonstrated, the myosin-containing circumferential actin bundle is most likely involved in the constriction of the terminal web elicited by the addition of ATP to either glycerinated epithelial cells or isolated brush borders (RODEWALD et al. 1976, BURGESS 1982) (Fig. 86). Contraction requires micromolar calcium and occurs over the course of 1–5 minutes both in intact and demembranated isolated brush borders. Several brush border proteins become phosphorylated under these conditions, but only the phosphorylation of myosin light chain requires calcium and correlates directly with constriction (KELLER and MOOSEKER 1982). Thus, brush border constriction is mediated by a ring of actin filaments (Fig. 87) located at the level of the adherens junctions and requires myosin light chain phosphorylation, similar to contractile activities in other nonmuscle cells. The function of brush border contractility in vivo is unknown. The force developed by constriction is unlikely to agitate the contents of the intestinal lumen, and it is difficult to see how contraction could affect the overall morphology of the epithelial sheet, or how it would induce microvillar movements. Thus, the brush border has the potential to undergo contraction or to induce calcium-mediated changes in the state of assembly of the core bundle, but whether it uses this potential, and to what purpose, is obscure.

IX.8. Transmembrane Interactions of the Cytoskeleton with the Extracellular Matrix

Fibrous components of the extracellular matrix such as collagen, fibronectin, and laminin are involved in many aspects of cell behavior, including adhesion, shape determination, shape changes, and motility. Because these same processes are based on the activities of the cytoskeleton, it seems only logical to predict the existence of direct or indirect interactions between extracellular and intracellular networks. This possibility was first suggested on the basis of electron microscopic studies that showed a close association between extracellular fibrils and the cell cortex (e.g., PORTER 1964), and later received support from a spectrum of additional observations:

1. Fibronectin, a high molecular weight glycoprotein that forms an extensive extracellular network on the cell surface (e.g., HYNES 1976, CHEN 1977, YAMADA and OLDEN 1978), is dramatically reduced in amount on the surface of cells transformed by oncogenic viruses (HYNES 1973). The loss of

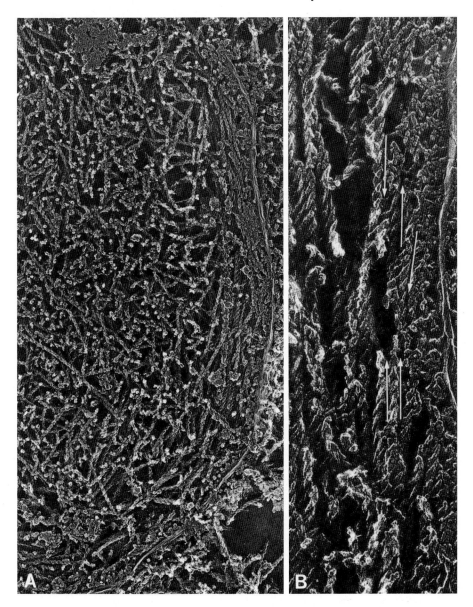

Fig. 87. Quick-frozen, deep-etched replica of the peripheral band of actin filaments at the level of the adherens junctions in a hair cell of the chick ear. The preparation was labeled with subfragment 1 before freezing. *A* Overview micrograph, showing the prominent bundle of decorated actin filaments which encircles the cell as a ring. The remainder of the apical portion of the hair cell (the "cuticular plate") contains a dense network of criss-crossing actin filaments. × 44,000. *B* Higher magnification of the actin filament bundle, which contains filaments of opposing polarity (arrows). × 123,000. (Courtesy of N. Hirokawa)

fibronectin correlates with a loss of the organization of microfilament bundles, which are much reduced or absent in transformed cells (*e.g.*, McNutt *et al.* 1973, Pollack *et al.* 1975). Conversely, addition of purified fibronectin to transformed cells which lack it causes the cells to flatten, and improves their adhesion to the substratum (Yamada *et al.* 1976 a, b, Vaheri and Mosher 1978). Flattening and increased adhesion correlate with the appearance of organized filament bundles (Ali *et al.* 1977, Willingham *et al.* 1977).

2. Mild proteolysis reduces the extracellular network of fibronectin and causes a disassembly of intracellular microfilament bundles (Pollack and Rifkin 1975).

3. Disorganization of microfilament bundles with cytochalasin B induces a corresponding disorganization of extracellular fibronectin filaments (Mautner and Hynes 1977, Kurkinen *et al.* 1978, Grinnell and Feld 1979).

4. In some cases, concentration gradients of adhesion molecules (*e.g.*, Postlewaite *et al.* 1978, Situ *et al.* 1984), or addition of purified fiibronectin to cultured cells (Ali and Hynes 1978) promotes cell motility.

5. Direct observation by immunofluorescence microscopy of cultured cells with labeled antibodies against actin and fibronectin reveals close correspondences between the two staining patterns, if only in some cell regions (Hynes and Destree 1978, Heggeness *et al.* 1978) (Fig. 88). In electron microscopic studies fibronectin bundles outside the cell appear to be continuous with microfilament bundles inside the cell at a zone of contact termed the "fibronexus" (Singer 1979).

6. Growth of cells in the presence of extracellular matrix components, *e.g.*, on the surface of, or within, collagen gels *in vitro*, greatly influences overall cell shape or the micromorphology of the cell cortex (*e.g.*, Sugrue and Hay 1981, Tomasek *et al.* 1982, Tomasek and Hay 1984). Conversely, cells cultured on or within collagen gels frequently develop enough tractional force to distort, orient, or align the fibers within the gel (Harris *et al.* 1981, Stopak and Harris 1982), demonstrating reciprocity of interactions between the cytoskeleton and the extracellular matrix.

These and other studies (discussed in more detail in Hynes 1981, Hay 1981, 1983) provide strong circumstantial evidence for an interaction between the two fiber systems. What is the molecular basis for this interaction? One possible approach is to define more clearly the specific spatial organization of the molecular components involved in these transmembrane relationships. Perhaps the best-studied of theses realtionship is that between fibronectin and actin. Following the initial demonstration by immunofluorescence microscopy of a coincidence of the staining patterns of these two proteins (Hynes and Destree 1978, Heggeness *et al.* 1978, Singer 1979), attention focused on the sites where actin filament

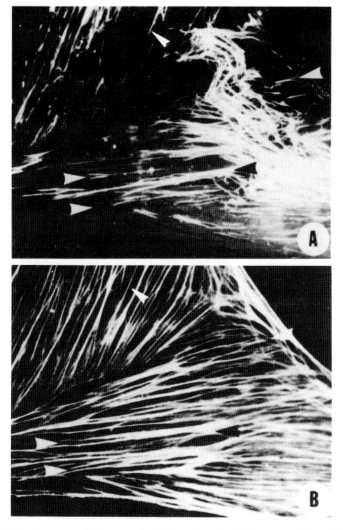

Fig. 88. Codistribution of fibronectin (*A*) and actin (*B*) in rat embryo fibroblasts. Some of the cases of colinearity are marked by arrowheads. Note, however, that many fibronectin bundles do not codistribute with actin. × 400. (Courtesy of R. HYNES)

bundles (stress fibers) associate with the membrane, the focal adhesions. The evidence for a specific localization of fibronectin at these sites of cell-substratum contact is conflicting. By interference reflection microscopy, fibronectin staining was reported to coincide with focal adhesions (HYNES *et al.* 1982, SINGER 1982) and by immunofluorescence microscopy with vinculin-containing patches (BURRIDGE and FERAMISCO 1980, SINGER and

PARADISO 1981). Based on these studies, fibronectin was hypothesized to be a constituent of, or associated with, the extracellular surface of the attachment plaque. Such an association makes intuitive sense because fibronectin is, after all, an "adhesive" protein, and focal contacts are sites of strong cell-substratum adhesion. Further, fibronectin is found in regions of cell-cell contact (FURCHT et al. 1978, SINGER 1979), and it is associated with adherens-like junctions in intact tissues (SINGER et al. 1984). However, other studies are at variance with this view. BIRCHMEIER et al. (1980) and CHEN and SINGER (1980) reported that fibronectin is absent from focal contacts of cultured cells, on the basis of immunofluorescence and immunoelectron microscopic studies. The argument that this might be due to inaccessibility of antibodies into the narrow gap between the membrane and the substratum at these sites (e.g., GRINNELL 1980; see also NEYFAKH et al. 1983) was countered by experiments in which vinculin-rich, but fibronectin-free focal contacts were labeled with antibodies to intrinsic, lectin-binding membrane proteins (CHEN and SINGER 1980, 1982). Moreover, AVNUR and GEIGER (1982 b) showed that fibronectin may be *removed* specifically from focal contact areas of cells plated onto coverslips coated with fluorescently labeled fibronectin. On the other hand, regions of close contact between the substratum and the ventral cell membrane, occasionally in the immediate vicinity of focal contacts, may be positive for fibronectin. The conclusion from all these studies is that the relationship between fibronectin distribution and focal contacts is more complex than anticipated. The controversial observations may in part be explained by the different culture conditions employed. Moreover, the work of SINGER (1982) indicates that the association of fibronectin with focal contact areas in cultured cells is both time and serum-dependent: fibronectin is absent from focal contacts in medium with high serum concentrations and at early stages in focal contact formation. In intact tissues, on the other hand, there is agreement that the sites of association between the cell surface and extracellular material are positive for fibronectin (CHEN and SINGER 1982, SINGER et al. 1984). The nature of the membrane link between the two systems is, however, obscure. *A priori*, two possibilities are feasible: the cell surface receptors that mediate adhesion of cells to fibronectin or other components of the extracellular matrix are transmembrane linkers themselves, or they are linked to a protein complex that anchors the intracellular filament network to the membrane. Affinity chromatography with the cell-attachment-promoting fragment of fibronectin has identified a 140,000 d cell surface glycoprotein as a putative receptor for fibronectin (PYTELA et al. 1985). A direct transmembrane linkage has also been postulated for the laminin receptor, an 69,000 d integral membrane protein (MALINOFF and WICHA 1983). This protein has the ability to bind to actin filaments *in vitro*, as demonstrated by binding of the receptor to immobilized actin, cosedimentation, and viscosity measurements of actin filament/receptor protein mixtures (BROWN *et al.* 1983). Because it might be responsible for the direct connection between intracellular and extracellular networks, it has been given the name "connectin"[2]. This mode of interaction undoubtedly is the most direct and perhaps the most efficient alternative for transmembrane interactions that could explain various aspects of the modulation of cell behavior by the extracellular matrix. However, other possibilities involving protein complexes that might allow for more flexibility and subtler schemes of regulation are not excluded.

---

[2] Unfortunately, this is the same name previously given to the giant cytoskeletal protein of muscle cells (see section V.2).

References

Aamodt, E. J., Williams, R. C., 1984a: Microtubule-associated proteins connect microtubules and neurofilaments *in vitro*. Biochemistry **23**, 6023–6031.

— — 1984b: Association of microtubules and neurofilmants *in vitro* is not mediated by ATP. Biochemistry **23**, 6031–6035.

Abercrombie, M., 1980: The crawling movement of metazoan cells. Proc. Roy. Soc. **B 207**, 129–147.

— Dunn, G. A., 1975: Adhesions of fibroblasts to substratum during contact inhibition observed by interference reflection microscopy. Exp. Cell Res. **92**, 57–62.

— Heaysman, J. E. M., Pegrum, S. M., 1971: The locomotion of fibroblasts in culture. IV. Electron microscopy of the leading lamella. Exp. Cell Res. **67**, 359–367.

Adelstein, R. S., Conti, M. A., 1975: Phosphorylation of platelet myosin increases actin-activated ATPase activity. Nature **256**, 597–598.

— — Hathaway, D. R., Klee, C. B., 1978: Phosophorylation of smooth muscle myosin light chain kinase by the catalytic subunit of adenosine 3′:5′-monophosphate-dependent protein kinase. J. Biol. Chem. **153**, 8347–8350.

— Eisenberg, E., 1980: Regulation and kinetics of the actin-myosin ATP interaction. Ann. Rev. Biochem. **49**, 921–956.

— Pato, M. D., Conti, M. A., 1981: The role of phosphorylation in regulating contractile proteins. Adv. Cyclic Nucleotide Res. **14**, 361–373.

Aebi, U., Fowler, W. E., Isenberg, G., Pollard, T. D., Smith, P. R., 1981: Crystalline actin sheets: their structure and polymorphism. J. Cell Biol. **91**, 340–351.

— — Rew, P., Sun, T. T., 1983: The fibrillar substructure of keratin filaments unraveled. J. Cell Biol. **97**, 1131–1143.

— Smith, R., Isenberg, G., Pollard, T. D., 1980: Structure of crystalline actin sheets. Nature **288**, 296–298.

Afzelius, B., 1959: Electron micsoscopy of the sperm tail. Results obtained with a new fixative. J. Biophys. Biochem. Cytol. **5**, 269–278.

Aguas, A. P., Pinto da Silva, P., 1984: High density of transmembrane glycoproteins on the flagellar surface of boar sperm cells. J. Cell Biol. **99**, 655–660.

Albertini, D. F., Anderson, E., 1977: Microtubule and microfilament rearrangements during capping of concanavalin A receptors on cultured ovarian granulosa cells. J. Cell Biol. **73**, 111–127.

ALBERTINI, D. F., ANDERSON, E., HERMANN, B., 1984: Cell shape and membrane receptor dynamics. Modulation by the cytoskeleton. Cell Muscle Motil. **5**, 235–253.

— — SHERLINE, P., 1984: *In vivo* and *in vitro* studies on the role of HMW-Maps in taxol-induced microtubule bundling. Eur. J. Cell Biol. **33**, 134–143.

ALBRECHT-BUEHLER, G., 1976: Filopodia of 3T3 cells: do they have a substrate-exploring function? J. Cell Biol. **69**, 275–287.

— BUSHNELL, A., 1982: Reversible compression of cytoplasm. Exp. Cell Res. **140**, 173–189.

ALI, I. U., HYNES, R. O., 1978: Effects of LETS glycoprotein on cell motility. Cell **14**, 439–446.

— MAUTNER, V. M., LANZA, R. P., HYNES, R. O., 1977: Restoration of normal morphology, adhesion, and cytoskeleton in transformed cells by addition of a transformation-sensitive surface protein. Cell **11**, 115–126.

ALLEN, C., BORISY, G. G., 1974: Structural polarity and directional growth of microtubules of *Chlamydomonas* flagella. J. Mol. Biol. **90**, 381–402.

ALLEN, R. D., 1971: Fine structure of membraneous and microfibrillar systems in the cortex of *Paramecium caudatum*. J. Cell Biol. **49**, 1–20.

AMOS, L. A., 1975: Substructure and symmetry of microtubules. In: Microtubules and Microtubule Inhibitors (BORGERS, M., DE BRABANDER, eds.), pp. 21–34. Amsterdam: North-Holland.

— 1977: Arrangement of high molecular weight associated proteins on purified mammalian brain microtubules. J. Cell Biol. **72**, 642–654.

— KLUG, A., 1974: Arrangement of subunits in flagellar microtubules. J. Cell. Sci. **14**, 523–549.

AMOS, W. B., 1971: Reversible mechanochemical cycle in the contraction of *Vorticella*. Nature **229**, 127–128.

— ROUTLEDGE, L. M., YEW, F. F., 1975: Calcium binding proteins in a vorticellid contractile organelle. J. Cell Sci. **91**, 203–213.

ANDERSON, D. J., 1984: New clues to protein localization in neurons. Trends in Neuroscience **7**, 355–357.

ANDERSON, R. A., LOVRIEN, R. E., 1984: Glycophorin is linked by band 4.1 protein to the human erythrocyte membrane skeleton. Nature **307**, 655–657.

ANDERSON, R. G. W., FLOYD, A. K., 1980: Electrophorectic analysis of basal body (centriole) proteins. Biochemistry **19**, 5625–5631.

ANDERTON, B. H., 1981: Intermediate filaments: a family of homologous structures. J. Muscle Res. Cell Motil. **2**, 141–166.

— 1983: A comprehensive catalogue of cytokeratins. Nature **302**, 211.

— THORPE, R., COHEN, J., SELVENDRAN, S., WOODHAMS, P., 1980: Specific neuronal localization by immunofluorescence of 10 nm filament polypeptides. J. Neurocytol. **9**, 835–844.

ANDREU, J. M., TIMASHEFF, S. N., 1982: Conformational states of tubulin liganded to colchicine, tropolone methyl ether, and podophyllotoxin. Biochemistry **21**, 6465–6476.

— WAGENKNECHT, T., TIMASHEFF, S. N., 1983: Polymerization of the tubulin-colchicine complex: relation to microtubule assembly. Biochemistry **22**, 1556–1566.

ANTANITUS, D. S., CHOI, B. H., LAPHAM, L. W., 1975: Immunofluorescence staining of astrocytes *in vitro* using antiserum to glial fibrillary acidic protein. Brain Res. **89**, 363–367.

ANTIN, P. B., FORRY-SHAUDIES, S., FRIEDMAN, T. M., TAPSCOTT, S. J., HOLTZER, H., 1981: Taxol induces postmitotic myoblasts to assemble interdigitating microtubule-myosin arrays that exclude actin filaments. J. Cell Biol. **90**, 300–308.

ANTLER, A. M., GREENBERG, M. E., EDELMAN, G. M., HANAFUSA, H., 1985: Increased phosphorylation of tyrosine in vinculin does not occur upon transformation by some avian sarcoma viruses. Mol. Cell. Biol. **5**, 263–267.

ARAKAWA, T., FRIEDEN, C., 1984: Interaction of microtubule-associated proteins with actin filaments. J. Biol. Chem. **259**, 11730–11734.

ARNOLD, J., 1867: Ein Beitrag zur feineren Struktur der Ganglienzellen. Virchow's Arch. **41**.

ASAI, D. J., WILSON, L., 1985: A latent activity dynein-like cytoplasmic magnesium adenosine triphosphatase. J. Biol. Chem. **260**, 699–702.

ASH, J., LOUVARD, D., SINGER, S., 1977: Antibody-induced linkages of plasma membrane proteins to intracellular actomyosin-containing filaments in cultured fibroblasts. Proc. Nat. Acad. Sci. U.S.A. **74**, 5584–5588.

AUBIN, J. E., OSBORN, M., FRANKE, W. W., WEBER, K., 1980: Intermediate filaments of the vimentin type and the cytokeratin type are distributed differently during mitosis. Exp. Cell Res. **129**, 149–165.

AVIOLO, J., LEBDUSKA, S., SATIR, P., 1984: Dynein arm substructure and the orientation of arm-microtubule attachments. J. Mol. Biol. **173**, 389–401.

AVNUR, Z., GEIGER, B., 1981 a: Substrate-attached membranes of cultured cells. Isolation and characterization of ventral cell membranes and the associated cytoskeleton. J. Mol. Biol. **153**, 361–379.

— — 1981 b: The removal of extracellular fibronectin from areas of cell substrate contact. Cell **25**, 121–132.

— SMALL, J. V., GEIGER, B., 1983: Actin-dependent association of vinculin with the cytoplasmic aspect of the plasma membrane in cell contact areas. J. Cell Biol. **96**, 1622–1630.

BACCETTI, B., BURRINI, A. G., GABBIANI, G., LEONCINI, P., RUNGGER-BRANDLE, E., 1984: Filamentous structures containing a keratin-like protein in spermatozoa of an insect, *Bacillus rossius*. J. Ultrastruct. Res. **86**, 86–92.

BAILEY, K., 1946: Tropomyosin: a new asymmetric protein component of muscle. Nature **157**, 368–369.

BALL, E. H., SINGER, S. J., 1981: Association of microtubules and intermediate filaments in normal fibroblasts and its disruption upon transformation by a temperature-sensitive mutant of Rous sarcoma virus. Proc. Nat. Acad. Sci. U.S.A. **78**, 6986–6990.

— — 1982: Mitochondria are associated with microtubules and not with intermediate filaments in cultured fibroblasts. Proc. Nat. Acad. U.S.A. **79**, 123–126.

BALLOWITZ, E., 1888: Untersuchungen über die Struktur der Spermatozoen, zugleich ein Beitrag zur Lehre vom feineren Bau der contraktilen Elemente. Arch. mikr. Anat. **32**, 401–473.

BAMBURG, J. R., HARRIS, H. E., WEEDS, A. G., 1980: Partial purification and characterization of an actin depolymerizing factor from brain. FEBS Lett. **121**, 178–182.

BANNISTER, L. H., TATCHELL, E. C., 1968: Contractility and the fiber systems of *Stentor coeruleus*. J. Cell Sci. **3**, 295–308.

BARDELE, C. F., 1977: Organization and control of microtubule pattern in centrohelidan heliozoa. J. Protozool. **24**, 9–14.

BARRETT, L. A., DAWSON, R. B., 1974: Avian erythrocyte development: microtubules and the formation of the disc shape. Dev. Biol. **36**, 72–81.

BARYLKO, B., DOBROWOLSKI, Z., 1984: Ca-calmodulin-dependent regulation of F-actin-myelin basic protein interaction. Eur. J. Cell Biol. **35**, 327–335.

BEAM, K. G., ALPER, S. L., PALADE, G. E., GREENGARD, P., 1979: Hormonally regulated phospho-protein of turkey erythrocytes. Localization to plasma membrane. J. Cell Ciol. **83**, 1–15.

BEAR, R. S., PALMER, K. J., SCHMITT, F. O., 1937: The ultrastructure of nerve axoplasm. Proc. Roy. Soc. **B 123**, 505–519.

BEGG, D. A., RODEWALD, R., REBHUN, L. I., 1978: Visualization of actin filament polarity in thin sections: Evidence for the uniform polarity of membrane-associated filaments. J. Cell Biol. **79**, 846–852.

BEHNKE, O., 1970: Microtubules in disk-shaped blood cells. Int. Rev. Exp. Pathol. **9**, 1–92.

BEN-ZE'EV, A., DUERR, A., SOLOMON, F., PENMAN, S., 1979: The outer boundary of the cytoskeleton: a lamina derived from the plasma membrane proteins. Cell **17**, 859–865.

BENNETT, G. S., FELLINI, S. A., CROOP, J. M., OTTO, J. J., BRYAN, J., HOLTZER, H., 1978 b: Differences among 100 A-filament subunits from different cell types. Proc. Nat. Acad. Sci. U.S.A. **75**, 4364–4368.

— — HOLTZER, H., 1978 a: Immunofluorescent visualization of 100 A filaments in different cultured chick embryo cell types. Differentiation **12**, 71–82.

— — TOYAMA, Y., HOLTZER, H., 1979: Redistribution of intermediate filament subunits during skeletal myogenesis and maturation *in vitro*. J. Cell Biol. **82**, 577–584.

BENNETT, H., CONDEELIS, J., 1984: Decoration of myosin subfragment-1 disrupts contacts between microfilaments and the cell membrane in isolated *Dictyostelium* cortices. J. Cell Biol. **99**, 1434–1440.

BENNETT, H. S., PORTER, K. R., 1953: An electron microscope study of sectioned breast muscle of the domestic fowl. Am. J. Anat. **93**, 61–105.

BENNETT, V., 1979: Immunoreactive forms of human erythrocyte ankyrin are present in diverse cells and tissues. Nature **281**, 597–599.

— 1982: The molecular basis for membrane-cytoskeleton association in human erythrocytes. J. Cell Biochem. **18**, 49–65.

— BRANTON, D., 1977: Selective association of spectrin with the cytoplasmic surface of human erythrocyte plasma membranes. J. Biol. Chem. **252**, 2753–2763.

— DAVIS, J., 1981: Erythrocyte ankyrin: immunoreactive analogues are associated with mitotic structures in cultured cells and with microtubules in brain. Proc. Nat. Acad. Sci. U.S.A. **78**, 7550–7554.

— — 1982: Immunoreactive forms of human erythrocyte ankyrin are localized in mitotic structures in cultured cells and are associated with microtubules in brain. Cold Spring Harbor Symp. Quant. Biol. **46**, 647–658.

— — FOWLER, W. E., 1982: Brain spectrin, a membrane-associated protein related in structure and function to erythrocyte spectrin. Nature **299**, 126–131.

— STENBUCK, P. J., 1979 a: Identification and partial purification of ankyrin, the high affinity membrane attachment site for human erythrocyte spectrin. J. Biol. Chem. **254**, 2533–2541.

— — 1979 b: The membrane attachment protein for spectrin is associated with band 3 in human erythrocyte membranes. Nature **280**, 468–473.

— — 1980: Association between ankyrin and the cytoplasmic domain of band 3 isolated from the human erythrocyte membrane. J. Biol. Chem. **255**, 6424–6432.

BEREITER-HAHN, J., FOX, C. H., THORELL, B., 1979: Quantitative reflection contrast microscopy of living cells. J. Cell Biol. **82**, 767–779.

BERGEN, L. G. BORISY, G. G., 1980: Head-to-tail polymerization of microtubules *in vitro*: electron microscope analysis of seeded assembly. J. Cell Biol. **84**, 141–150.

BERKOWITZ, S. A., KATAGIRI, J., BINDER, H. K., WILLIAMS, R. C., 1977: Separation and characterization of microtubule proteins from calf brain. Biochemistry **16**, 5610–5617.

BERLIN, R. D., OLIVER, J. M., 1978: Analogous ultrastructure and surface properties during capping and phagocytosis. J. Cell Biol. **77**, 789–804.

BERNIER-VALENTIN, F., ROUSSET, B., 1982: Interaction of tubulin with rat liver mitochondria. J. Biol. Chem. **257**, 7092–7099.

BERNS, M. W., RATTNER, J. B., BRENNER, S., MEREDITH, S., 1977: The role of the centriolar region in animal cell mitosis. J. Cell Biol. **72**, 351–372.

BERNSTEIN, B. W., BAMBURG, J. R., 1982: Tropomyosin binding to F-actin protects the F-actin from disassembly by brain actin depolymerizing factor (ADF). Cell Motil. **2**, 1–8.

BERSHADSKY, A. D., GELFAND, V. I., 1981: ATP-dependent regulation of cytoplasmic microtubule disassembly. Proc. Nat. Acad. Sci. U.S.A. **78**, 3610–3613.

— — 1983: Role of ATP in the regulation of stability of cytoskeletal structures. Cell Biol. Int. Rep. **7**, 173–187.

BERTOLINI, B., MONACO, G., ROSSI, A., 1970: Ultrastructure of a regular arrangement of microtubules and neurofilaments. J. Ultrastruct. Res. **33**, 173–186.

BETHE, A., 1897: Über die Primitivfibrillen in den Ganglienzellen vom Menschen und anderen Wirbeltieren. Morph. Arbeiten, Vol. 8.

BETTEX-GALAND, M., LUSCHER, E. F., 1959: Extraction of an actomyosin-like protein from human thrombocytes. Nature **184**, 276–278.

BHATTACHARYYA, B., WOLFF, J., 1976: Tubulin aggregation and disaggregation-mediation by 2 distinct vinblastine-binding sites. Proc. Nat. Acad. Sci. U.S.A. **73**, 2375–2378.

BIBRING, T., BAXANDALL, J., DENSLOW S., WALKER, B., 1976: Heterogeneity of the alpha subunit of tubulin and the variability of tubulin within a single organism. J. Cell. Biol. **69**, 301–312.

BIELSCHOWSKY, M., POLLAK, B., 1907: Über die fibrilläre Struktur der Ganglienzellen. J. f. Psychol. Neurol. **10**, 274–281.

— WOLFF, M., 1904: Zur Histologie der Kleinhirnrinde. J. f. Psychol. Neurol. **4**, 1–23.

BINDER, L. I., ROSENBAUM, J. L., 1978: The *in vitro* assembly of flagellar outer doublet tubulin. J. Cell Biol. **79**, 500–515.

BIRCHMEIER, C., KREIS, T. E., EPPENBERGER, H. M., WINTERTHALER, K. H., BIRCHMEIER, W., 1980: Corrugated attachment membrane in Wi-38 fibroblasts: alternating fibronectin fibers and actin-containing focal contacts. Proc. Nat. Acad. Sci. U.S.A. **77**, 4108–4112.

BISWAS, B. B., BANERJEE, A. C., BHATTACHARYYA, B., 1981: Tubulin and the microtubule system in cellular growth and development. Subcell. Biochem. **8**, 123–183.

BLACK, M. M., KURDYLA, J. T., 1983: Microtubule-associated proteins of neurons. J. Cell Biol. **97**, 1020–1028.

— LASEK, R. J., 1980: Slow components of axonal transport: two cytoskeletal networks. J. Cell Biol. **86**, 616–623.

BLAKESLEE, A., 1937: Dedoublement du nombre de chromosomes chez les plantes par traitement chimique. C. R. Acad. Sci. (Paris) **205**, 476–479.

BLIKSTAD, I., LAZARIDES, E., 1983: Vimentin filaments are assembled from a soluble precursor in avian erythroid cells. J. Cell Biol. **96**, 1803–1808.

— MARKEY, F., CARLSSON, L., PERSSON, T., LINDBERG, U., 1978: Selective assay of monomeric and filamentous actin in cell extracts, using inhibition of DNase I. Cell **15**, 935–943.

BLOODGOOD, R. A., 1976: Biochemical analysis of axostyle motility. Cytobios **14**, 101–120.

— 1977: Motility occurring in association with the surface of the *Chlamydomonas* flagellum. J. Cell Biol. **75**, 983–989.

— 1982: Flagellum as a model system for studying dynamic cell-surface events. Cold Spring Harbor Quant. Biol. **46**, 683–693.

BLOOM, G. S., LUCA, F. C., VALLEE, R. B., 1984 b: Widespread cellular distribution of MAP-1A (microtubule-associated protein 1A) in the mitotic spindle and on interphase microtubules. J. Cell Biol. **98**, 331–340.

— SCHOENFELD, T. A., VALLEE, R. B., 1984 a: Widespread distribution of the major polypeptide component of MAP 1 (microtubule-associated protein 1) in the nervous system. J. Cell Biol. **98**, 320–330.

— VALLEE, R. B., 1983: Association of MAP 2 with microtubules and intermediate filaments in cultured brain cells. J. Cell Biol. **96**, 1523–1531.

BLOSE, S., CHACKO, S., 1976: Rings of intermediate (100 A) filament bundles in the perinuclear region of vascular endothelial cells. J. Cell Biol. **70**, 459–466.

BLOSE, S. H., 1979: Ten-nanometer filaments and mitosis: maintenance of structural continuity in dividing endothelial cells. Proc. Nat. Acad. Sci. U.S.A. **76**, 3372–3376.

— MELTZER, D. I., FERAMISCO, J. R., 1984: 10-nm filaments are induced to collapse in living cells microinjected with monoclonal and polyclonal antibodies against tubulin. J. Cell Biol. **98**, 847–858.

BLUMENTHAL, D. K., STULL, J. T., 1980: Activation of skeletal muscle myosin light chain kinase by Ca and calmodulin. Biochemistry **19**, 5608–5612.

BODIAN, D., 1936: A new method for staining nerve fibers and nerve endings in mounted paraffin sections. Anat. Rec. **65**, 89–97.

BONDER, E. M., FISHKIND, D. J., MOOSEKER, M. S., 1983: Direct measurement of critical concentrations and assembly rate constants at the two ends of an actin filament. Cell **34**, 491–501.

BORDAS, J., MANDELKOW, E.-M., MANDELKOW, E., 1983: Stages of tubulin assembly and disassembly studied by time-resolved synchrotron x-ray scattering. J. Mol. Biol. **164**, 89–135.

BORGERS, M., DE BRABANDER, M., 1975: Microtubules and Microtubule Inhibitors. Amsterdam: North-Holland.

BORISY, G. G., TAYLOR, E. W., 1967 a: The mechanism of action of colchicine. Binding of colchicine-^{3}H to cellular protein. J. Cell Biol. **34**, 525–534.

— — 1967 b: The mechanism of action of colchicine. Colchicine binding to sea urchin eggs and the mitotic apparatus. J. Cell Biol. **34**, 535–548.

BOUCHARD, P., PENNINGROTH, S. M., CHEUNG, A., GAGNON, C., BARDIN, C. W., 1981: erythro-9-3-(2-hydroxynonyl)adenine is an inhibitor of sperm motility that blocks dynein ATPase and protein carboxymethylase activities. Proc. Nat. Acad. Sci. U.S.A. **78**, 1033–1036.

BOUCK, G. B., ROGALSKI, A., VALAITIS, A., 1978: Surface organization and composition of *Euglena*. II. Flagellar mastigonemes. J. Cell Biol. **77**, 805–826.

BOURGUIGNON, G. J., BOURGUIGNON, L. Y. W., 1981: Isolation and characterization of a lymphocyte cap structure. Biochim. Biophys. Acta **646**, 109–118.

BOURGUIGNON, L. Y. W., BOURGUIGNON, G. J., 1984: Capping and the cytoskeleton. Int. Rev. Cytol. **87**, 195–224.

— NAGPAL, M. L. BALAZOVICH, K., GUERRIERO, V., MEANS, A. R., 1982: Association of myosin light chain kinase with lymphocyte membrane-cytoskeleton complex. J. Cell Biol. **95**, 793–797.

— — HSING, Y. C., 1981: Phosphorylation of myosin light chain during capping of mouse T-lymphoma cells. J. Cell Biol. **91**, 889–894.

— SINGER, S. J., 1977: Transmembrane interactions and the mechanism of capping of surface receptors by their specific ligands. Proc. Nat. Acad. Sci. U.S.A. **74**, 5031–5035.

BOWERS, B., KORN, E. D., 1968: The fine structure of *Acanthamoeba castellanii*. I. The trophozoite. J. Cell Biol. **39**, 95–111.

BOYLES, J., BAINTON, D. F., 1979: Changing patterns of plasma membrane-associated filaments during the initial phases of polymorphonuclear leucocyte adherence. J. Cell Biol. **82**, 347–368.

BRADY, S. T., LASEK, R. J., 1981: Nerve-specific enolase and creatine phos-phokinase in axonal transport: soluble proteins and the axoplasmic matrix. Cell **23**, 515–523.

— — 1982: Axonal transport: a cell biological method for studying proteins that associate with the cytoskeleton. Meth. Cell Biol. **25**, 365–398.

— — ALLEN, R. D., YIN, H. L., STOSSEL, T. P., 1984: Gelsolin inhibition of fast axonal transport indicates a requirement for actin microfilaments. Nature **310**, 56–58.

BRANTON, D., COHEN, C. M., TYLER, J., 1981: Interaction of cytoskeletal proteins on the human erythrocyte membrane. Cell **24**, 24–32.

BRAUN, J., FUJIWARA, K., POLLARD, T. D., UNANUE, E., 1978 a: Two distinct mechanisms for the redistribution of lymphocyte surface macromolecules: I. Relationship to cytoplasmic myosin. J. Cell Biol. **79**, 409–418.

— — — — 1978 b: Two distinct mechanisms for the redistribution of lymphocyte surface macromolecules: II. Contrasting effects of local anesthetics and a calcium ionophore. J. Cell Biol. **79**, 419–426.

— HOCHMAN, P. S., UNANUE, E. R., 1982: Ligand-induced association of surface immunoglobulin with the detergent-insoluble cytoskeletal matrix of the B-lymphocyte. J. Immunol. **128**, 1198–1204.

BRECKLER, J. LAZARIDES, E., 1982: Isolation of a new high molecular weight protein associated with desmin and vimentin filaments from avian embryonic skeletal muscle. J. Cell Biol. **92**, 795–806.

BRENNER, S. L., BRINKLEY, B. R., 1982: Tubulin assembly sites and the organization of microtubule arrays in mammalian cells. Cold Spring Harbor Quant. Biol. **46**, 241–254.

— KORN, E. D., 1979: Spectrin-actin interaction. Phosphorylated and dephosphorylated spectrin tetramer cross-link F-actin. J. Biol. Chem. **254**, 8620–8627.

— — 1980 a: The effects of cytochalasins on actin polymerization and actin ATPase provide insights into the mechanism of polymerization. J. Biol. Chem. **255**, 841–844.

— — 1980 b: Spectrin/actin complex isolated from sheep erythrocytes accelerates actin polymerization by simple nucleation. J. Biol. Chem. **255**, 1670–1676.

— — 1981: Stimulation of actin ATPase activity by cytochalasins provides evidence for a new species of monomeric actin. J. Biol. Chem. **256**, 8663–8670.

— — 1983: On the mechanism of actin monomer-polymer subunit exchange at steady state. J. Biol. Chem. **258**, 5013–5020.

— — 1984: Evidence that F-actin can hydrolyze ATP independent of monomer-polymer end interactions. J. Biol. Chem. **259**, 1441–1446.

BRETSCHER, A., 1981: Fimbrin is a cytoskeletal protein that crosslinks F-actin *in vitro*. Proc. Nat. Acad. Sci. U.S.A. **78**, 6849–6853.

— 1983 a: Microfilament organization in the cytoskeleton of the brush border. In: Cell and Muscle Motility (SHAY, J. W., ed.), Vol. 4, pp. 239–268. New York: Plenum.

— 1983 b: Purification of an 80,000-dalton protein that is a component of the isolated microvillus cytoskeleton, and its localization in nonmuscle cells. J. Cell Biol. **97**, 425–432.

— 1984: Smooth muscle caldesmon. J. Biol. Chem. **259**, 12873–12880.

— OSBORN, M., WEHLAND, J., WEBER, K., 1981: Villin associates with specific microfilamentous structures as seen by immunofluorescence microscopy on tissue sections and cells microinjected with villin. Exp. Cell Res. **135**, 213–219.

— WEBER, K., 1978: Purification of microvilli and an analysis of the protein components of the microfilament core bundle. Exp. Cell Res. **116**, 397–412.

— — 1979: Villin: The major microfilament-associated protein of the intestinal microvillus. Proc. Nat. Acad. Sci. U.S.A. **76**, 2321–2325.

— — 1980 a: Villin is a major protein of the microvillus cytoskeleton which binds both G- and F-actin in a calcium-dependent manner. Cell **20**, 839–847.

— — 1980 b: Fimbrin, a new microfilament-associated protein present in microvilli and other cell surface structures. J. Cell Biol. **86**, 335–340.

BRETSCHER, M. S., 1976: Directed lipid flow in cell membranes. Nature **260**, 21–23.

BRIDGMAN, P. C., REESE, T. S., 1984: The structure of cytoplasm in directly frozen cultured cells. I. Filamentous meshworks and the cytoplasmic ground substance. J. Cell. Biol. **99**, 1655–1668.

BRINKLEY, B. R., COX, S. M., PEPPER, D. A., WIBLE, L., BRENNER, S. L., PARDUE, R. L., 1981: Tubulin assembly sites and the organization of cytoplasmic microtubules in cultured mammalian cells. J. Cell Biol. **90**, 554–562.

— FISTEL, S. H., MARCUM, J. M., PARDUE, R. L., 1980: Microtubules in cultured cells: indirect immunofluorescence staining with tubulin antibody. Int. Rev. Cytol. **63**, 59–95.

— FULLER, G. M., HIGHFIELD, D. P., 1975: Cytoplasmic microtubules in normal and transformed cells in culture: analysis by tubulin antibody and immunofluorescence. Proc. Nat. Acad. Sci. U.S.A. **72**, 4981–4985.

BRODY, I., 1960: The ultrastructure of the tonofibrils in the keratinization process of normal human epidermis. J. Ultrastruct. Res. **4**, 264–297.

BROKAW, C. J., 1972: Flagellar movement, a sliding filament model. Science **178**, 455–462.

BROSCHAT, K. O., STIDWILL, R. P., BURGESS, D. R., 1983: Phosphorylation controls brush border motility by regulating myosin structure and association with the cytoskeleton. Cell **35**, 561–571.

BROTSCHI, E. A., HARTWIG, J. H., STOSSEL, T. P., 1978: The gelation of actin by actin-binding protein. J. Biol. Chem. **253**, 8988–8993.

BROWN, S. S., LEWINSON, W., SPUDICH, J., 1976: Cytoskeletal elements of chick embryo fibroblasts revealed by detergent extraction. J. Supramol. Struct. **5**, 119–130.

— MALINOFF, H. L., WICHA, M. S., 1983: Connectin: Cell surface protein that binds both laminin and actin. Proc. Nat. Acad. Sci. U.S.A. **80**, 5927–5930.

— SPUDICH, J. A., 1979: Cytochalasin inhibits the rate of elongation of actin fragments. J. Cell Biol. **83**, 657–662.

— YAMAMOTO, K., SPUDICH, J. A., 1982: A 40,000 dalton protein from *Dictyostelium discoideum* affects assembly properties of actin in a Ca-dependent manner. J. Cell Biol. **93**, 205–210.

BRULET, P., BABINET, C., KEMLER, R., JACOB, F., 1980: Monoclonal antibodies against troph-ectoderm-specific markers during mouse blastocyst formation. Proc. Nat. Acad. Sci. U.S.A. **77**, 4113–4117.

BRUNKE, K. J., YOUNG, E. E., BUCHBINDER, B. U., WEEKS, D. P., 1982: Coordinate regulation fo the four tubulin genes of *Chlamydomonas reinhardi*, Nucleic Acids Res. **71**, 749–767.

BRYAN, J., 1976: A quantitative analysis of microtubule elongation. J. Cell Biol. **3**, 747–767.

— KANE, R. E., 1978: Separation and interaction of the major components of sea urchin actin gel. J. Mol. Biol. **125**, 207–224.

— KURTH, M. C., 1984: Actin-gelsolin interactions. J. Biol. Chem. **259**, 7480–7487.

BUCKLEY, I. K., 1975: Three-dimensional fine structure of cultured cells: possible implications for subcellular motility. Tissue Cell **7**, 51–72.

— PORTER, K. R., 1975: Electron microscopy of critical point dried cultured cells. J. Microsc. **104**, 107–120.

BULINKSI, J. C., BORISY, G. G., 1979: Self-assembly of microtubules in extracts of cultured HeLa cells and the identification of HeLa microtubule-associated proteins. Proc. Nat. Acad. Sci. U.S.A. **76**, 293–297.

— — 1980: Microtubule-associated proteins from cultured HeLa cells. J. Biol. Chem. **255**, 11570–11576.

BUNGE, M. B., BUNGE R. P., RIS, H., 1961: Ultrastructural study of remyelination in an experimental lesion in adult cat spinal cord. J. Biophys. Biochem. Cytol. **10**, 67–94.

BURGESS, D. R., 1982: Reactivation of intestinal epithelial cell brush border motility: ATP-dependent contraction via a terminal web contractile ring. J. Cell Biol. **95**, 853–863.

BURNS, R. G., ISLAM, K., 1982: Characterization of the chick brain high molecular weight multiple phosphorylated microtubule-associated protein. Eur. J. Biochem. **122**, 25–29.

— — 1984: Direct incorporation of microtubule oligomers at high GTP concentrations. FEBS Lett. **173**, 67–76.

— — CHAPMAN, R., 1984: The multiple phosphorylation of the microtubule-associated protein MAP2 controls the MAP2 : tubulin interaction. Eur. J. Biochem. **141**, 609–615.

— POLLARD, T. D., 1974: A dynein-like protein from brain. Fed. European Biochem. Soc. Letters **40**, 274–280.

BURRIDGE, K., CONNELL, L., 1983: A new protein of adhesion plaques and ruffling membranes. J. Cell Biol. **97**, 359–367.

— FERAMISCO, J., 1980: Microinjection and localization of a 130 K protein in living fibroblasts: a relationship to actin and fibronectin. Cell **19**, 587–595.

— FERAMISCO, J. R., 1982: α-actinin and vinculin from nonmuscle cells: Calcium-sensitive interactions with actin. In: Cold Spring Harbor Symposia on Quantitative Biology, Vol. XLVI, Part 2, pp. 587–597. Cold Spring Harbor: Cold Spring Harbor Laboratory.

— KELLY, T., MANGEAT, P., 1982: Nonerythrocyte spectrins: actin-membrane attachment proteins occurring in many cell types. J. Cell Biol. **95**, 478–486.

— MANGEAT, P., 1984: An interaction between vinculin and talin. Nature **308**, 744–745.

— PHILLIPS, J. H., 1975: Association of actin and myosin with secretory granule membranes. Nature **254**, 526–529.

BURTON, P. R., HIMES, R. H., 1978: Electron microscope studies of pH effects on assembly of tubulin free of associated proteins. J. Cell Biol. **77**, 120–133.

— HINKLEY, R. E., 1974: Further electron microscopic characterization of axoplasmic microtubules of the ventral cord of the crayfish. J. Submicrosc. Cytol. **6**, 311–326.

— — PIERSON, G. B., 1975: Tannic acid-stained microtubules with 12, 13, and 15 protofilaments. J. Cell Biol. **65**, 227–233.

— PAIGE, J. L., 1981: Polartiy of axoplasmic microtubules in the olfactory nerve of the bullfrog. Proc. Nat. Acad. Sci. U.S.A. **78**, 3269–3273.

BUTMAN, B. T., BOURGUIGNON, G. J., BOURGUIGNON, L. Y. W., 1980: Lymphocyte capping induced by polycationized ferritin. J. Cell. Physiol. **105**, 7–15.

BYERS, B., SHRIVER, K., GOETSCH, L., 1978: The role of spindle pole bodies and modified microtubule ends in the initiation of microtubule assembly in *Saccharomyces cerevisiae*. J. Cell Sci. **30**, 331–352.

BYERS, H. R., FUJIWARA, K., 1982: Stress fibers in cells *in situ*: Immunofluorescent visualization with anti-actin, anti-myosin and anti-alpha-actinin. J. Cell Biol. **93**, 804–811.

— PORTER, K. R., 1977: Transformation in the structure of the cytoplasmic ground substance in erythrophores during pigment aggregation and dispersion. I. A study using whole-cell preparations in stereo high voltage electron microscopy. J. Cell Biol. **75**, 541–558.

— WHITE, G. E., FUJIWARA, K., 1984: Organization and function of stress fibers in cells *in vitro* and *in situ*. In: Cell and Muscle Motility (SHAY, J. W., ed.), Vol. 5, pp. 83–137. New York: Plenum.

CACHON, J., CACHON, M., TILNEY, L. G., TILNEY, M., 1977: Movement generated by interaction between the dense material at the ends of microtubules and non-actin containing microfilaments in *Sticholonche zanclea*. J. Cell Biol. **72**, 314–338.

CAJAL, S. R., 1899–1904: Textura del sisterna nervioso del hombre y de los vertebrados, Vol. 2. Madrid: Nicolas Moya.

CALARCO-GILLAM, P. D., SIEBERT, M. C., HUBBLE, R., MITCHISON, T., KIRSCHNER, M., 1983: Centrosome development in early mouse embryos as defined by an autoantibody against pericentriolar material. Cell **35**, 621–629.

CAMPBELL, A. K., 1983: Intracellular Calcium—Its Universal Role as Regulator. New York: J. Wiley.

CAMPBELL, R. D., CAMPBELL, J. H., 1971: Origin and continuity of desmosomes. In: Origin and Continuity of Cell Organelles (REINERT, J., URSRPUNG, H., eds.), pp. 261–298. Berlin-Heidelberg-New York: Springer.

CAPETANAKI, Y. G., NGAI, J., LAZARIDES, E., 1984: Characterization and regulation in the expression of a gene coding for the intermediate filament protein desmin. Proc. Nat. Acad. Sci. U.S.A. **81**, 6909–6913.

CAPLOW, M., LANGFORD, G., ZEEBERG, B., 1982: Concerning the efficiency of the treadmilling phenomenon with microtubules. J. Biol. Chem. **257**, 15012–15021.

CARLIER, M.-F., 1982: Guanosine-5′-triphosphate hydrolysis and tubulin polymerization. Mol. Cell. Biochem. **47**, 97–113.

— 1983: Kinetic evidence for a conformation change of tubulin preceding microtubule assembly. J. Biol. Chem. **258**, 2415–2420.

— HILL, T. L., CHEN, Y.-D., 1984 a: Interference of GTP hydrolysis in the mechanism of microtubule assembly: an experimental study. Proc. Nat. Acad. Sci. U.S.A. **81**, 771–775.

— PANTALONI, D., 1981: Kinetic analysis of guanosine-5′-triphosphate hydrolysis associated with tubulin polymerization. Biochemistry **20**, 1918–1924.

CARLIER, M.-F., PANTALONI, D., KORN, E. D., 1984 b: Evidence for an ATP cap at the ends of actin filaments and its regulation of the F-actin steady state. J. Biol. Chem. **259**, 9983–9986.

— SIMON, C., CASSOLY, R., PRADEL, L.-A., 1984 c: Interaction between microtubule-associated protein tau and spectrin. Biochimie **66**, 305–311.

— — PANTALONI, D., 1984 d: Polymorphism of tubulin oligomers in the presence of microtubule-associated proteins. Implications in microtubule assembly. Biochemistry **23**, 1582–1590.

CARLIN, R. K., BARTELT, D. C., SIEKEVITZ, P., 1983: Identification of fodrin as a major calmodulin-binding protein in postsynaptic density preparations. J. Cell Biol. **96**, 443–448.

CARLSSON, L., NYSTROM, L.-E., SUNDKVIST, I., MARKEY, F., LINDBERG, U., 1977: Actin polymerizability is influenced by profilin, a low molecular weight protein in nonmuscle cells. J. Mol. Biol. **115**, 465–483.

CAROTHERS CARRAWAY, C. A., JUNG, G., CARRAWAY, K. L., 1983 a: Isolation of actin-containing transmembrane complexes from ascites adenocarcinoma sublines having mobile and immobile receptors. Proc. Nat. Acad. Sci. U.S.A. **80**, 430–434.

— — CRAIK, J. R., RUBIN, R. W., CARRAWAY, K. L., 1983 b: Identification of a cytoskeleton-associated glycoprotein from isolated microvilli of a mammary ascites tumor. Exp. Cell Res. **143**, 303–308.

CARRAWAY, K. L., CERRA, R. F., JUNG, G., CARRAWAY, C. A. C., 1982: Membrane associated actin from the microvillar membranes of acites tumor cells. J. Cell Biol. **94**, 624–630.

— HUGGINS, J. W., CERRA, R. F., YELTMAN, D. R., CAROTHERS CARRAWAY, C. A., 1980: α-actinin-containing branched microvilli isolated from an ascites adenocarcinoma. Nature **285**, 508–510.

CARTER, W., HAKOMORI, S., 1981: A new cell surface, detergent-insoluble glyco-protein matrix of human and hamster fibroblasts. J. Biol. Chem. **256**, 6953–6960.

CELIS, J. E., LARSEN, P. M., FEY, S. J., CELIS, A., 1983: Phosphorylation of keratin and vimentin polypeptides in normal and transformed mitotic human epithelial amnion cells: behavior of keratin and vimentin filaments during mitosis. J. Cell Biol. **97**, 1429–1434.

— SMALL, J. V., LARSEN, P. M., FEY, S. J., DE MEY, J., CELIS, A., 1984: Intermediate filaments in monkey kidney TC 7 cells: focal centers and interrelationship with other cytoskeletal systems. Proc. Nat. Acad. Sci. U.S.A. **81**, 1117–1121.

CERVERA, M., DREYFUSS, G., PENMAN, S., 1981: Messenger RNA is translated when associated with the cytoskeletal framework in normal and VSV infected HeLa cells. Cell **23**, 113–120.

CHALFIE, M., THOMSON, J. N., 1979: Organization of neuronal microtubules in the nematode *Caenorhabditis elegans*. J. Cell. Biol. **82**, 278–289.

— — 1982: Structural and functional diversity in the neuronal microtubules of *Caenorhabditis elegans*. J. Cell Biol. **93**, 15–23.

CHANG, K. S., ZIMMER, W. E., BERGSMA, D. J., DODGSON, J. B., SCHWARTZ, R. J., 1984: Isolation and characterization of six different chicken actin genes. Mol. Cell. Biol. **4**, 2498–2508.

CHEN, L. B., 1977: Alteration in cell surface LETS protein during myogenesis, Cell **10**, 393–400.

CHEN, W. T., SINGER, S. J., 1980: Fibronectin is not present in the focal adhesions formed between normal cultured fibroblasts and their substrate. Proc. Nat. Acad. Sci. U.S.A. **77**, 7318–7322.

— — 1982: Immunoelectron microscopic studies of the sites of cell-substratum and cell-cell contacts in cultured fibroblasts. J. Cell Biol. **95**, 205–222.

CHIU, F.-C., KOREY, B., NORTON, W. T., 1980: Intermediate filaments from bovine, rat and human CNS: Mapping analysis of the major proteins. J. Neurochem. **34**, 1149–1159.

CLARK, T. G., ROSENBAUM, J. L., 1982: Pigment particle translocation in detergent-permeabilized melanophores of *Fundulus heteroclitus*. Proc. Nat. Acad. Sci. U.S.A. **79**, 4655–4659.

CLARKE, F. M., MASTERS, C. J., 1974: On the association of glycolytic components in skeletal muscle extracts. Biochim. Biophys. Acta **358**, 193–207.

— — 1975: On the association of glycolytic enzymes with structural proteins of skeletal muscle. Biochim. Biophys. Acta **381**, 37–46.

— — 1976: Interactions between muscle proteins and gycolytic enzymes. Int. J. Biochem. **7**, 359–365.

CLEGG, J. S., 1984 a: Intracellular water and the cytomatrix: Some methods of study and current views. J. Cell Biol. **99**, 167s–171s.

— 1984 b: Properties and metabolism of the aqueous cytoplasm and its boundaries. Am. J. Physiol. **246**, R133–R151.

CLEVELAND, D. W., 1982: Treadmilling of tubulin and actin. Cell **28**, 689–691.

— 1983: The tubulins: from DNA to RNA to protein and back again. Cell **34**, 330–332.

— LOPATA, M. A., MacDONALD, R. J., COWAN, N. J.,RUTTER, W. J., KIRSCHNER, M., 1980: Number and evolutionary conservation of α and β tubulin and cytoplasmic β and γ actin genes using specific cloned cDNA probes. Cell **20**, 95–105.

COHEN, C. M., 1983: The molecular organization of the red-cell membrane skeleton. Sem. Hematol. **20**, 141–158.

— BRANTON, D., 1979: The role of spectrin in erythrocyte membrane stimulated actin polymerization. Nature **279**, 163–165.

— FOLEY, S. F., 1980: Spectrin-dependent and -independent association of F-actin with the erythrocyte membrane. J. Cell Biol. **86**, 694–698.

— — 1982: The role of band 4.1 in the association of actin with erythrocyte membranes. Biochim. Biophys. Acta **688**, 691–701.

— — KORSGREN, C., 1982: A protein immunologically related to erythrocyte band 4.1 is found on stress fibers of non-erythroid cells. Nature **299**, 648–650.

— JACKSON, P. L., BRANTON, D., 1978: Actin-membrane interactions: associations of G-actin with the red cell membrane. J. Supramol. Struct. **9**, 113–124.

— KORSGREN, C., 1980; Band 4.1 causes spectrin-actin gels to become thixotropic. Biochem. Biophys. Res. Commun. **97**, 1429–1435.

— LANGLEY, R. C., 1984: Functional characterization of human erythrocyte spectrin α and β chains: association with actin and erythrocyte protein 4.1. Biochemistry **23**, 4488–4495.

COHEN, C. M., TYLER, J. M., BRANTON, D., 1980: Spectrin-actin associations studied by electron microscopy of shadowed preparations. Cell 21, 875–883.

COHEN, S. M., GORBSKY, G., STEINBERG, S., 1983: Immunochemical characterization of related families of glycoproteins in desmosomes. J. Biol. Chem. 258, 2621–2627.

COHEN, W. D., BARTELT, D., JAEGER, R., LANGFORD, G., NEMHAUSER, I., 1982: The cytoskeletal system of nucleated erythrocytes. I. Comparison and function of major elements. J. Cell Biol. 93, 828–838.

COLLINS, J. H., BORYSENKO, C., CHIAO, J., LAMPERSKI, B., 1984: The 110,000 dalton actin- and calmodulin-binding protein from intestinal brush border is a myosin-like ATPase. J. Biol. Chem. 259, 14128–14135.

— COTE, G. P., KORN, E. D., 1982 b: Localization of the three phosphorylation sites on each heavy chain of Acanthamoeba myosin II to a segment at the end of the tail. J. Biol. Chem. 257, 4529–4534.

— KORN, E. D., 1980: Actin activation of Ca-sensitive Mg-ATPase activity of Acanthamoeba myosin II is enhanced by dephosphorylation of its heavy chain. J. Biol. Chem. 255, 8011–8014.

— KUZNICKI, J., BOWERS, B., KORN, E. D., 1982 a: Comparison of the actin binding and filament formation properties of phosphorylated and dephosphorylated Acanthamoeba myosin II. Biochemistry 21, 6910–6915.

COLLOT, M., LOUVARD, D., SINGER, S. J., 1984: Lysosomes are associated with microtubules and not with intermediate filaments in cultured fibroblasts. Proc. Nat. Acad. Sci. U.S.A. 81, 788–792.

COLUCCIO, L. M., TILNEY, L. G., 1984: Phalloidin enhances actin assembly by preventing monomer dissociation. J. Cell Biol. 99, 529–535.

CONDEELIS, J., VAHEY, M., 1982: A calcium- and pH-regulated protein from Dictyostelium discoideum that cross-links actin filaments. J. Cell Biol. 94, 466–471.

CONDEELIS, J. S., 1979: Isolation of Con A caps during various stages of their formation and their association with actin and myosin. J. Cell Biol. 80, 751–758.

— 1981: Microfilament-membrane interactions in cell shape and motility. In: International Cell Biology, 1980–1981 (SCHWEIGER, H. G., ed.), pp. 306–320. Berlin-Heidelberg-New York: Springer.

— TAYLOR, D. L., 1977: The contractile basis of amoeboid movement. V. The control of gelation, solation, and contraction in extracts from Dictyostelium discoideum. J. Cell Biol. 74, 901–927.

CONNOLLY, J. A., KALNINS, V. I., CLEVELAND, D. W., KIRSCHNER, M. W., 1977: Immunofluorescent staining of cytoplasmic and spindle microtubules in mouse fibroblasts with antibody to tau protein. Proc. Nat. Acad. Sci. U.S.A. 74, 2437–2440.

— — CLEVELAND, D. W. KIRSCHNER, M. W., 1978: Intracellular localization of the high molecular weight microtubule accessory protein by indirect immunofluorescence. J. Cell Biol. 76, R781–R786.

COOKE, P., 1976: A filamentous cytoskeleton in vertebrate smooth muscle fibers. J. Cell Biol. 68, 539–556.

— 1983: Organization of contractile fibers in smooth muscle. In: Cell and Muscle

Motility, Vol. 3 (Dowben, R. M., Shay, J. W. eds.), pp. 57–77. New York: Plenum.

Cooke, R., 1975: The role of bound nucleotide in the polymerization of actin. Biochemistry 14, 3250–3256.

— Murdoch, L., 1973: Interaction of actin with analogs of adenosine triphosphate. Biochemistry 12, 3927–3932.

Coons, A. H., 1956: Histochemistry with labeled antibody. Int. Rev. Cytol. 5, 1–23.

Cooper, J. A., Blum, J. D., Pollard, T. D., 1984: *Acanthamoeba castellanii* capping protein: properties, mechanism of action, immunologic cross-reactivity, and localization. J. Cell Biol. 99, 217–225.

— Buhle, E. L., Walker, S. B., Tsong, T. Y., Pollard, T. D., 1983: Kinetic evidence for a monomer activation step in actin polymerization. Biochemistry 22, 2193–2202.

Cote, G. P., Collins, J. H., Korn, E. D., 1981: Identification of three phosphorylation sites on each heavy chain of *Acanthamoeba* myosin II. J. Biol. Chem. 256, 12811–12816.

Cote, R. A., Borisy, G. G., 1981: Head-to-tail polymerization of microtubules *in vitro*. J. Mol. Biol. 150, 577–602.

Couchman, J. R., Rees, D. A., 1982: Organelle-cytoskeleton relationships in fibroblasts: mitochondria, Golgi apparatus, and endoplasmic reticulum in phases of movement and growth. Eur. J. Cell Biol. 27, 47–54.

Coudrier, E. Reggio, H. Louvard, C., 1983: Characterization of an integral membrane glycoprotein associated with the microfilaments of pig intestinal microvilli. EMBO J. 2, 469–475.

Cowan, N. J., Dudley, L., 1983: Tubulin isotypes and the multigene tubulin families. Int. Rev. Cytol. 85, 147–173.

Cowin, P., Garrod, D. R., 1983: Antibodies to epithelial desmosomes show wide tissue and species cross-reactivity. Nature 302, 148–150.

— Mattey, D., Garrod, D., 1984: Identification of desmosomal surface components (desmocollins) and inhibition of desmosome formation by specific Fab′. J. Cell Sci. 70, 41–60.

Craig, R., Smith, R., Kendrick-Jones, J., 1983: Light-chain phosphorylation controls the conformation of vertebrate non-muscle and smooth muscle myosin molecules. Nature 302, 436–439.

Craig, S. W., Pardo, J. V., 1979: Alpha-actinin localization in the junctional complex of intestinal epithelial cells. J. Cell Biol. 80, 203–210.

— Pollard, T. D., 1982: Actin-binding proteins. Trends in Biochem. Sci. 7, 88–92.

— Powell, L. D., 1980: Regulation of actin polymerization by villin, a 95,000-dalton cytoskeleton component of intestinal brush borders. Cell 22, 739–746.

Czosnek, H., Soifer, D., Wisniewski, H. M., 1980: Studies on the biosynthesis of neurofilament proteins. J. Cell Biol. 85, 725–734.

D'Angelo Siliciano, J., Craig, S. W., 1982: Meta-vinculin–a vinculin-related protein with solubility properties of a membrane protein. Nature 300, 533–535.

Dahl, D., 1976: Glial fibrillary acidic protein from bovine and rat brain. Degradation in tissues and homogenates. Biochim. Biophys. Acta 420, 142–154.

DAHL, D., BIGNAMI, A., 1973: Immunochemical and immunofluorescence studies of the glial fibrillary acidic protein in vertebrates. Brain Res. **61**, 279–293.

DALE, B. A., 1977: Purification and characterization of a basic protein from the *stratum corneum* of mammalian epidermis. Biochim. Biophys. Acta **491**, 193–202.

— HOLBROOK, K. A., STEINERT, P. M., 1978: Assembly of *stratum corneum* basic protein and keratin filaments in macrofibrils. Nature **276**, 729–731.

DAMSKY, C. H., BUCK, C. A., HORWITZ, A. F., 1983: Localization of a surface membrane substratum adhesion glycoprotein using a monoclonal antibody. J. Cell Biol. **97**, 99 a.

DANIELS, M. P., 1973: Fine structural changes in neurons and nerve fibers associated with colchicine inhibition of nerve fiber formation *in vitro*. J. Cell Biol. **58**, 463–470.

DAVID-PFEUTY, T., SINGER, S. J., 1980: Altered distributions of the cytoskeletal proteins vinculin and α-actinin in cultured fibroblasts transformed by Rous sarcoma virus. Proc. Nat. Acad. Sci. U.S.A. **77**, 6687–6691.

DAVIES, P. J. A., KLEE, C. B., 1981: Calmodulin-binding proteins: a high molecular weight calmodulin-binding protein from bovine brain. Biochem. Int. **3**, 203–212.

DAVIS, J. Q., BENNETT, V., 1984 a: Brain ankyrin. Purification of a 72,000 MW spectrin-binding domain. J. Biol. Chem. **259**, 1874–1881.

— — 1984 b: Brain ankyrin. A membrane-associated protein with binding sites for spectrin, tubulin, and the cytoplasmic domain of the erythrocyte anion channel. J. Biol. Chem. **259**, 13550–13559.

DAVISON, P. F., JONES, R., 1980: Neurofilament proteins of mammals compared by peptide mapping. Brain Res. **182**, 470–473.

— WINSLOW, B., 1974: The protein subunit of calf brain neurofilament. J. Neurobiol. **5**, 119–133.

DE BRABANDER, M., AERTS, F., VAN DE VEIRE, R., BORGERS, M., 1975: Evidence against interconversion of microtubules and filaments. Nature **253**, 119–120.

— DE MEY, J., 1980: Microtubules and Microtubule Inhibitors. Amsterdam: Elsevier.

— GEUENS, G., NUYDENS, R., WILLEBRORDS, R., DE MEY, J., 1981: Taxol induces the assembly of free microtubules in living cells and blocks the organizing capacity of the centrosomes and the kinetochores. Proc. Nat. Acad. Sci. U.S.A. **78**, 5608–5612.

DE BRABANDER, M. J., VAN DE VEIRE, R. M., AERTS, F., BURGESS, M., JANSSEN, P. A., 1976: The effects of methyl-5-(2-thienyl-carbonyl)-1H-benzimidazol-2-yl carbamate (R 17934), a new synthetic antitumoral drug interfering with microtubules, in mammalian cultured cells *in vitro*. Cancer Res. **36**, 905–916.

DE GROOT, C., WORMEESTER, J., MANGUS-SMET, C., 1981: Capping of surface immunoglobulin on rabbit and mouse lymphocytes. II. Cytoskeletal involvement in different sub-populations. Eur. J. Cell Biol. **25**, 202–211.

DE HARVEN, E., BERNHARD, W., 1956: Étude au microscope électronique de l'ultrastructure du centriole chez les vertébres. Z. Zellforsch. **45**, 378–498.

DE MEY, J., HOEBEKE, J., DE BRABANDER, M., GEUENS, G., JONIAN, M., 1976:

Immunoperoxidase visualization of microtubules and microtubular proteins. Nature **264**, 273–275.

— MOEREMANS, M., GEUENS, G., NUYDENS, R., VAN BELLE, H., DE BRABANDER, M., 1980: Immunocytochemical evidence for the association of calmodulin with microtubules of the mitotic apparatus. In: Microtubules and Microtubule Inhibitors (DE BRABANDER, M., DE MEY, J., eds.), pp. 227–241. Amsterdam: Elsevier.

DEROSIER, D. J., EDDS, K. T., 1980: Evidence for fascin cross-links between actin filaments in coelomocyte filopodia. Exp. Cell Res. **126**, 490–498.

— MANDELKOW, E., SILLIMAN, A., TILNEY, L. G., KANE, R., 1977: The structure of actin-containing filaments from two types of nonmuscle cells. J. Mol. Biol. **113**, 679–695.

— TILNEY, L. G., 1982: How actin filaments pack into bundles. Cold Spring Harbor Symp. Quant Biol. **46**, 525–540.

— — 1984 a: The form and function of actin. A product of its unique design. In: Cell and Muscle Motility, Vol. 5 (SHAY, J. W., ed.), pp. 139–169. New York: Plenum.

— — 1984 b: How to build a bend into an actin bundle. J. Mol. Biol. **175**, 57–73.

— — BONDER, E. M., FRANKL, P., 1982: A change in twist of actin provides the force for the extension of the acrosomal process in *Limulus* sperm: the false-discharge reaction. J. Cell Biol. **93**, 324–337.

DEBEC, A., SZOLLOSI, A., SZOLLOSI, D., 1982: A *Drosophila melanogaster* cell line lacking centriole. Biol. Cell. **44**, 133–138.

DEBUS, E., FLUGGE, G., WEBER, K., OSBORN, M., 1982: A monoclonal antibody specific for the 200 K polypeptide of the neurofilament triplet. EMBO J. **1**, 41–45.

DEERY, W. J., MEANS, A. R., BRINKLEY, B. R., 1984: Calmodulin-microtubule association in cultured mammalian cells. J. Cell Biol. **98**, 904–910.

DEITERS, 1865: Untersuchungen über Gehirn und Rückenmark des Menschen und der Säugetiere. Braunschweig: F. Vieweg.

DELLAGI, K., BROUET, J. C., PERREAU, J., PAULIN, D., 1982: Human monoclonal IgM with autoantibody activity against intermediate filaments. Proc. Nat. Acad. Sci. U.S.A. **79**, 446–450.

DENTLER, W. L., 1981: Microtubule-membrane interactions in cilia and flagella. Int. Rev. Cytol. **72**, 1–47.

— GRANETT, S., ROSENBAUM, J. L., 1975: Ultrastructural localization of the high molecular weight proteins associated with *in vitro*-assembled brain microtubules. J. Cell. Biol. **65**, 237–241.

— — WITMAN, G. B., ROSENBAUM, J. L., 1974: Directionality of brain microtubule assembly *in vitro*. Proc. Nat. Acad. Sci. U.S.A. **71**, 1710–1714.

— PRATT, M. M., STEPHENS, R. E., 1980: Microtubule-membrane interactions in cilia. Photochemical crosslinking of bridge structures and the identification of a membrane-associated dynein-like ATPase. J. Cell Biol. **84**, 381–403.

DETRICH, H. W., WILLIAMS, R. C., 1978: Reversible dissociation of the dimer of tubulin from bovine brain. Biochemistry **17**, 3900–3907.

— — WILSON, L., 1982: Effect of colchicine binding on the reversible dissociation of the tubulin dimer. Biochemistry **21**, 2392–2400.

DODGE, J. T., MITCHELL, C., HANAHAN, D. J., 1963: Preparation and chemical characteristics of hemoglobin-free ghosts of human erythrocytes. Arch. Biochem. Biophys. **100**, 119–130.

DOLKEN, G., LEISNER, E., PETTE, D., 1975: Immunofluorescent localization of glycogenolytic and glycolytic enzyme proteins and of malate dehydrogenase isoenzymes in cross-striated skeletal muscle and heart of the rabbit. Histochemistry **43**, 113–121.

DRENCKHAHN, D., GROESCHEL-STEWART, U., 1980: Localization of myosin, actin, and tropomyosin in rat intestinal epithelium: Immunohistochemical studies at the light and electron microscope level. J. Cell Biol. **86**, 475–482.

— KELLNER, J., MANNHERZ, H. G., GROESCHEL-STEWART, U., KENDRICK-JONES, J., SCHOLEY, J., 1982: Absence of myosin-like immunoreactivity in stereocilia of cochlear hair cells. Nature **300**, 531–532.

— ZINKE, K., SCHAUER, U., APPELL, K. C., LOW P. S., 1984: Identification of immunoreactive forms of human erythrocyte band 3 in nonerythroid cells. Eur. J. Cell Biol. **34**, 144–150.

DROCHMANS, P., FREUDENSTEIN, C., WANSON, J., LAURENT, L., KEENAN, T. W., STADLER, J., LELOUP, R., FRANKE, W. W., 1978: Structure and biochemical composition of desmosomes and tonofilaments isolated from calf muzzle epidermis. J. Cell Biol. **79**, 427–433.

DUERR, A., PALLAS, D., SOLOMON, F., 1981: Molecular analysis of cytoplasmic microtubules *in situ*: identification of both widespread and specific proteins. Cell **24**, 203–211.

DUHAIMAN, A. S., BAMBURG, J. R., 1984: Isolation of brain α-actinin. Its characterization and a comparison of its properties with those of muscle α-actinins. Biochemistry **23**, 1600–1608.

DUJARDIN, F., 1835: Recherches sur les organismes inférieurs III. Sur les prétendus estomacs des animalcules infusoires et sur une substance appelée sarcode. Ann. Sci. Nat. **4**, 343–377.

DULBECCO, R., AALLEN, R., OKADA, S., BOWMAN, M., 1980: Functional changes of intermediate filaments in fibroblastic cells revealed by a monoclonal antibody. Proc. Nat. Acad. Sci. U.S.A. **80**, 1915–1918.

DURICA, D. S., SCHLOSS, J. A., CRAIN, W. R., 1980: Organization of actin gene sequences in the sea urchin: molecular cloning of an intron-containing DNA sequence coding for a cytoplasmic actin. Proc. Nat. Acad. Sci. U.S.A. **77**, 5683–5687.

DUSTIN, P., 1984: Microtubules, 2nd ed. Berlin-Heidelberg-New York-Tokyo: Springer.

ECKERT, B. S., CAPUTI, S. E., WARREN, R. H., 1984: Dynamics of keratin filaments and the intermediate filament distribution center during shape change in PtK1 cells. Cell Motil. **4**, 169–181.

— DALEY, R. A., PARYSEK, L. M., 1982: *In vivo* disruption of the cytokeratin cytoskeleton in cultured epithelial cells by microinjection of antikeratin: evidence for the presence of an intermediate-filament-organizing center. Cold Spring Harbor Symp. Quant. Biol. **46**, 403–412.

— KOONS, S. J., SCHANTZ, A. W., ZOBEL, C. R., 1980: Association of creatine phosphokinase with the cytoskeleton of cultured mammalian cells. J. Cell Biol. **86**, 1–5.

Edds, K. T., 1977: Dynamic aspects of filopodial formation by reorganization of microfilaments. J. Cell Biol. **73**, 479–491.

Edelman, G. M., 1976: Surface modulation in cell recognition and cell growth. Science **192**, 218–226.

— Yahara, I., Wang, J. L., 1973: Receptor mobility and receptor cytoplasmic interactions in lymphocytes. Proc. Nat. Acad. Sci. U.S.A. **70**, 1442–1446.

Egelman, E. H., Francis, N., DeRosier, D. J., 1982: F-actin is a helix with a random variable twist. Nature **298**, 131–135.

— — — 1983: Helical disorder and the filament structure of F-actin are elucidated by the angle-layered aggregate. J. Mol. Biol. **166**, 605–629.

— Padron, R., 1984: X-ray diffraction evidence that actin is a 100 A filament. Nature **307**, 56–58.

Eichenlaub-Ritter, U., Tucker, J. B., 1984: microtubules with more than 13 protofilaments in the dividing nuclei of ciliates. Nature **307**, 60–62.

Eipper, B. A., 1974: Rat brain tubulin and protein kinase activity. J. Biol. Chem. **249**, 1398–1406.

Elbaum, D., Mimms, L. T., Branton, D., 1984: Modulation of actin polymerization by the spectrin-band 4.1 complex. Biochemistry **23**, 4813–4816.

Ellisman, M. H., Porter, K. R., 1980: Microtrabecular structure of the axoplasmic matrix: visualization of cross-linking structures and their distribution. J. Cell Biol. **87**, 464–479.

Elzinga, M., Collins, J. H., Kuehl, W. M., Adelstein, R. S., 1973: Complete amino acid sequence of actin of rabbit skeletal muscle. Proc. Nat. Acad. Sci. U.S.A. **70**, 2687–2691.

— Phelan, J. J., 1984: F-actin is intermolecularly crosslinked by N,N'-p-phenylene-dimaleimide through lysine-191 and cysteine-374. Proc. Nat. Acad. Sci. U.S.A. **81**, 6599–6602.

Eng, L. F., Vanderhaeghen, J. J., Bignami, A., Gertl, B., 1971: An acidic protein isolated from fibrous astrocytes. Brain Res. **28**, 351–354.

Erickson, H. P., 1974: Microtubule surface lattice and subunit structure and observations on reassembly. J. Cell Biol. **60**, 153–167.

— Voter, W. A., 1976: Polycation-induced assembly of purified tubulin. Proc. Nat. Acad. Sci. U.S.A. **73**, 2813–2817.

Euteneuer, U., Jackson, W. T., McIntosh, J. R., 1982: Polarity of spindle microtubules in *Haemanthus* endosperm. J. Cell Biol. **94**, 644–653.

— McIntosh, J. R., 1981 a: Polarity of some motility-related microtubules. Proc. Nat. Acad. Sci. U.S.A. **78**, 372–376.

— — 1981 b: Structural polarity of kinetochore microtubules in PtK 1 cells. J. Cell Biol. **89**, 338–345.

Evans, L., Mitchison, T., Kirschner, M., 1985: Influence of the centrosome on the structure of nucleated microtubules, J. Cell Biol. **100**, 1185–1191.

Evans, R. M., 1984: Peptide mapping of phosphorylated vimentin. J. Biol. Chem. **259**, 5372–5375.

Evans, R. R., Robson, R. M., Stromer, M. H., 1984: Properties of smooth muscle vinculin. J. Biol. Chem. **259**, 3916–3924.

262   References

FALKNER, F.-G., SAUMWEBER, H., BIESSMANN, H., 1981: Two *Drosophila melanogaster* proteins related to intermediate filament proteins of vertebrate cells. J. Cell Biol. **91**, 175–183.

FAURE-FREMIET, E., 1962: Le genre *Paranassula* Kahl (Ciliate Cyrtophorina). Cah. Biol. Mar. **3**, 61–77.

FAWCETT, D. W., 1959: Electron microscopic observations on the marginal band of nucleated erythrocytes. Anat. Rec. **133**, 379.

— 1975: The mammalian spermatozoon. Dev. Biol. **44**, 394–436.

— PORTER, K. R., 1954: A study of the fine structure of ciliated epithelia. J. Morphol. **94**, 221–228.

FEBVRE, J., FEBVRE-CHEVALIER, C., 1982: Motility processes in *Acantharia* (*Protozoa*). Biol. Cell. **44**, 283–304.

FECHHEIMER, M., TAYLOR, D. L., 1984: Isolation and characterization of a 30,000-dalton calcium-sensitive actin cross-linking protein from *Dictyostelium discoideum*. J. Biol. Chem. **259**, 4514–4520.

FERAMISCO, J. R., 1979: Microinjection of fluorescently labeled alpha-actinin into living fibroblasts. Proc. Nat. Acad. Sci. U.S.A. **76**, 3967–3971.

— BLOSE, S. H., 1980: Distribution of fluorescently labeled α-actinin in living and fixed fibroblasts. J. Cell Biol. **86**, 608–615.

— SMART, J. E., BURRIDGE, K., HELFMAN, D. M., THOMAS, G. P., 1982: Co-existence of vinculin and a vinculin-like protein of higher molecular weight in smooth muscle. J. Biol. Chem. **257**, 11024–11031.

FEY, E. G., WAN, K. M., PENMAN, S., 1984: Epithelial cytoskeletal framework and nuclear matrix-intermediate filament scaffold: Three-dimensional organization and protein composition. J. Cell Biol. **98**, 1973–1984.

FIELD, D. J., COLLINS, R. A., LEE, J. C., 1984: Heterogeneity of vertebrate brain tubulins. Proc. Nat. Acad. Sci. U.S.A. **81**, 4041–4045.

FIRTEL, R. A., BENDER, W., DAVIDSON, N., 1978: Multiple, heterogeneous actin genes in *Dictyostelium*. Cell **15**, 789–800.

— TIMM, R., KIMMEL, A. R., MCKEOWN, M., 1979: Unusual nucleotide sequences at the 5'end of actin genes in *Dictyostelium discoideum*. Proc. Nat. Acad. Sci. U.S.A. **76**, 6206–6210.

FLANAGAN, J., KOCH, G. L. E., 1978: Crosslinked surface immunoglobulin attaches to actin, Nature **273**, 278–281.

FLANAGAN, M. D., LIN, S., 1980: Cytochalasins block actin filament elongation by binding to high affinity sites associated with F-actin. J. Biol. Chem. **255**, 835–838.

FLEMMING, W., 1897: Zur Kenntnis der Zelle und ihrer Lebenserscheinungen. Arch. Mikros. Anat. **16**, 302–436.

FLORY, P. J., 1941: Molecular size distribution in three-dimensional polymers I. Gelation: J. Am. Chem. Soc. **63**, 3083–3091.

FOWLER, V. M., BENNETT, V., 1984: Erythrocyte membrane tropomyosin. Purification and properties. J. Biol. Chem **259**, 5978–5989.

— DAVIS, J., BENNETT, V., 1985: Human erythrocyte myosin: identification and purification. J. Cell Biol. **100**, 47–55.

— LUNA, E. J., HARGREAVES, W. R., TAYLOR, D. L., BRANTON, D., 1981: Spectrin promotes the association of F-actin with the cytoplasmic surface of the human erythrocyte membrane. J. Cell Biol. **88**, 388–395.

— POLLARD, H. B., 1982: Chromaffin granule membrane-F-actin interactions are calcium sensitive. Nature **295**, 336–339.

— TAYLOR, D. L., 1980: Spectrin plus band 4.1 cross-link actin. Regulation by micromolar calcium. J. Cell Biol. **85**, 361–376.

FOWLER, W. E., AEBI, U., 1983: A consistent picture of the actin filament related to the orientation of the actin molecule. J. Cell Biol. **97**, 264–269.

FOX, J. E. B., BOYLES, J. K., REYNOLDS, C. C., PHILLIPS, D. R., 1984: Actin filament content and organization in unstimulated platelets. J. Cell Biol. **98**, 1985–1991.

FRANK, E. D., WAREN, L., 1981: Aortic smooth muscle cells contain vimentin instead of desmin. Proc. Nat. Acad. Sci. U.S.A. **78**, 3020–3024.

FRANKE, W. W., DENK, H., KALT, R., SCHMID, E., 1981 a: Biochemical and immunological identification of cytokeratin proteins in hepatocytes of mammalian liver tissue. Exp. Cell Res. **131**, 299–318.

— GRUND, C., OSBORN, M., WEBER, K., 1978 c: The intermediate-sized filaments in rat kangeroo PtK 2 cells. I. Morphology *in situ*. Cytobiologie **17**, 365–391.

— MOLL, R., SCHILLER, D. L., SCHMID, E., KARTENBECK, J., MÜLLER, H., 1982 c: Desmoplakins of epithelial and myocardial desmosomes are immunologically and biochemically related. Differentiation **23**, 115–127.

— SCHILLER, D., GRUND, C., 1982 a: Protofilamentous and annular structures as intermediates during reconstitution of cytokeratin filaments *in vitro*. Biol. Cell **46**, 257–268.

— SCHILLER, D. L., HATZFELD, M., WINTER, S., 1983: Protein complexes of intermediate-sized filaments: melting of cytokeratin complexes in urea reveals different polypeptide separation characteristics. Proc. Nat. Acad. Sci. U.S.A. **80**, 7113–7117.

— — MOLL, R., WINTER, S., SCHMID, E., ENGELBRECHT, I., 1981 b: Diversity of cytokeratins. Differentiation specific expression of cytokeratin polypeptides in epithelial cells and tissues. J. Mol. Biol. **153**, 933–959.

— SCHMID, E., FREUDENSTEIN, C., APPELHANS, B., OSBORN, M., WEBER, K., 1980: Intermediate-sized filaments of the prekeratin type in myoepithelial cells. J. Cell Biol. **84**, 633–654.

— — GRUND, C., 1982 b: Intermediate filament proteins in non-filamentous structures: transient disintegration and inclusion of subunit proteins in granular aggregates. Cell **30**, 103–113.

— — — MÜLLER, H., ENGELBRECHT, I., MOLL, R., STADLER, J., JARASCH, E.-D., 1981 c: Antibodies to high molecular weight polypeptides of desmosomes: specific localization of a class of junctional proteins in cells and tissues. Differentiation **20**, 217–240.

— — MITTNACHT, S., GRUND, C., JORCANO, J. L., 1984: Integration of different keratins into the same filament system after microinjection of mRNA for epidermal keratins into kidney epithelial cells. Cell **36**, 813–825.

— — OSBORN, M., WEBER, K., 1978 a: Different intermediate-sized filaments distinguished by immunofluorescence microscopy. Proc. Nat. Acad. Sci. U.S.A. **75**, 5034–5038.

FRANKE, W. W., SCHMID, E., WEBER, K., OSBORN, M., 1979 a: HeLa cells contain intermediate-sized filaments of the prekeratin type. Exp. Cell Res. **118**, 95–109.

— — WINTER, S., OSBORN, M., WEBER, K., 1979 b: Widespread occurrence of intermediate-sized filaments of the vimentin-type in cultured cells from diverse vertebrates. Exp. Cell Res. **123**, 25–46.

— WEBER, K., OSBORN, M., SCHMID, E., FREUDENSTEIN, C., 1978 b: Antibody to prekeratin. Decoration of tonofilament-like arrays in various cells of epithelial character. Exp. Cell Res. **116**, 429–445.

FRANKEL, F. R., 1976: Organization and energy-dependent growth of microtubules in cells. Proc. Nat. Acad. Sci. U.S.A. **73**, 2798–2802.

FRANZ, J. K., GALL, L., WILLIAMS, M. A., PICHERAL, B., FRANKE, W. W., 1983: Intermediate-size filaments in a germ cell: expression of cytokeratins in oocytes and eggs of the frog *Xenopus*. Proc. Nat. Acad. Sci. U.S.A. **80**, 6254–6258.

FRANZ, V., 1939: Struktur und Mechanismus der Melanophoren. Z. Zellforsch. mikr. Anat. **30**, 194–234.

FRIEDEN, C., 1983: Polymerization of actin: mechanism of the Mg-induced process at pH 8 and 20 °C. Proc. Nat. Acad. Sci. U.S.A. **80**, 6513–6517.

FUCHS, E., COPPOCK, S. M., GREEN, H., CLEVELAND, D. W., 1981: Two distinct classes of keratin genes and their evolutionary significance. Cell **27**, 75–84.

— GREEN, H., 1978: The expression of keratin genes in epidermis and cultured epidermal cells. Cell **15**, 887–897.

— — 1979: Multiple keratins of cultured human epidermal cells are translated from different mRNA molcules. Cell **17**, 573–582.

— — 1980: Changes in keratin gene expression during terminal differentiation of the keratinocyte. Cell **19**, 1033–1042.

— HANUKOGLU, I., 1983: Unraveling the structure of the intermediate filaments. Cell **34**, 332–334.

— MARCHUK, D., 1983: Type I and type II keratins have evolved from lower eukaryotes to form the epidermal intermediate filaments in mammalian skin. Proc. Nat. Acad. Sci. U.S.A. **80**, 5857–5861.

FÜCHTBAUER, A., JOCKUSCH, B. M., MARUTA, H., KILIMANN, M. W., ISENBERG, G., 1983: Disruption of microfilament organization after injection of F-actin capping proteins into living tissue culture cells. Nature **304**, 361–364.

FUJIWARA, K., POLLARD, T. D., 1976: Fluorescent antibody localization of myosin in the cytoplasm, cleavage furrow, and mitotic spindle of human cells. J. Cell Biol. **71**, 848–875.

FUKADA, E., DATE, M., 1963: Viscoelastic properties of collagen solutions in dilute hydrochloric acid. Biorheology **1**, 101–109.

FUKUI, Y., 1978: Intranuclear actin bundles induced by dimethyl sulfoxide in interphase nucleus of *Dictyostelium*. J. Cell Biol. **76**, 146–157.

— KATSAMARU, H., 1980: Dynamics of nuclear actin bundle induction by dimethyl sulfoxide and factors affecting its development. J. Cell. Biol. **84**, 131–140.

FULTON, A. B., 1982: How crowded is the cytoplasm? Cell **30**, 345–347.

— PRIVES, J., FARMER, S. R., PENMAN, S., 1981: Developmental reorganization of the skeletal frame work and its surface lamina in fusing muscle cells. J. Cell Biol. **91**, 103–112.

— WAN, K. M., PENMAN, S., 1980: The spatial distribution of poly-ribosomes in 3T3 cells and the associated assembly of proteins into the skeletal frame work. Cell **20**, 849–857.

FULTON, C., KOWIT, J. D., 1975: Programmed synthesis of flagellar tubulin during cell differentiation in *Naegleria*. Ann. N.Y. Acad. Sci. **253**, 318–332.

— SIMPSON, P. A., 1976: Selective synthesis and utilization of flagellar tubulin. The multitubulin hypothesis. In: Cell Motility (GOLDMANN, R., POLLARD, T., ROSENBAUM, J., eds.), pp. 987–1005. Cold Spring Harbor.

FURCHT, L. T., MOSHER, D. F., WENDELSCHAFER-CRABB, G., 1978: Effects of cell density and transformation on the formation of a fibronectin extracellular filamentous matrix on human fibroblasts. Cancer Res. **38**, 4618–4623.

FYRBERG, E. A., BOND, B. J., HERSHEY, N. D., MIXTER, K. S., DAVIDSON, N., 1981: The actin genes of *Drosophila*: protein coding regions are highly conserved but intron positions are not. Cell **24**, 107–116.

— KINDLE, K. L., DAVIDSON, N., SODJA, A., 1980: The actin genes of *Drosophila*: a dispersed multigene family. Cell **25**, 107–116.

GABBIANI, G., CHAPONNIER, C., ZUMBE, A., VASSALLI, P., 1977: Actin and tubulin co-cap with surface immunoglobulin in mouse B lymphocytes. Nature **269**, 697–698.

GALL, L., PICHERAL, B., GOUNON, P., 1983: Cytochemical evidence for the presence of intermediate filaments and microfilaments in the egg of *Xenopus laevis*. Biol. Cell **47**, 331–342.

GALVIN, N. J., STOCKHAUSEN, D., MEYERS-HUTCHINS, B. L., FRAZIER, W. A., 1984: Association of the cyclic AMP chemotaxis receptor with the detergent-insoluble cytoskeleton of *Dictyostelium discoideum*. J. Cell Biol. **98**, 584–595.

GAMBETTI, P., AUTILIO-GAMBETTI, L., PAPASOZOMENOS, S. C., 1981: Bodian's silver method stains neurofilament polypeptides. Science **213**, 1521–1522.

GARD, D. L., BELL, P. B., LAZARIDES, E., 1979: Co-existence of desmin and the fibroblastic intermediate filament subunit in muscle and nonmuscle cells: identification and comparative peptide analysis. Proc. Nat. Acad. Sci. U.S.A. **76**, 3894–3898.

— LAZARIDES, E., 1980: The synthesis and distribution of desmin and vimentin during myogenesis *in vitro*. Cell **19**, 262–275.

— — 1982 a: Cyclic AMP-modulated phosphorylation of intermediate filament proteins in cultured avian myogenic cells. Mol. Cell. Biol. **2**, 1104–1114.

— — 1982 b: Analysis of desmin and vimentin phosphopeptides in cultured avian myogenic cells and their modulation by 8-bromo-adenosine 3′,5′-cyclic monophosphate. Proc. Nat. Acad. Sci. U.S.A. **79**, 6912–6916.

GASSNER, D., SHRAIDEH, Z., WOHLFAHRT-BOTTERMANN, K. E., 1983: An extraction resistant tensile protein in the protoplasmic matrix of *Physarum*. Cell Biol. Int. Rep. **7**, 905–909.

GAWLITTA, W., OSBORN, M., WEBER, K., 1981: Coiling of intermediate filaments induced by microinjection of a vimentin-specific antibody does not interfere with locomotion in mitosis. Eur. J. Cell Biol. **26**, 83–90.

GEIGER, B., 1979: A 130 K protein from chicken gizzard: Its localization at the termini of microfilament bundles in cultured chicken cells. Cell **187**, 193–205.

GEIGER, B., 1982: Microheterogeneity of avian and mammalian vinculin. Distinctive subcellular distribution of different isovinculins. J. Mol. Biol. **159**, 685–701.

— 1983: Membrane-cytoskeleton interaction. Biochim. Biophys. Acta **737**, 305–341.

— AVNUR, Z., KREIS, T. E., SCHLESSINGER, J., 1984: The dynamics of cytoskeletal organization in areas of cell contact. In: Cell and Muscle Motility, Vol. 5 (SHAY, J. W., ed.), pp. 195–234. New York: Plenum.

— — SCHLESSINGER, J., 1982: Restricted mobility of membrane constituents in cell substrate focal contacts of chicken fibroblasts. J. Cell Biol. **93**, 495–500.

— SCHMIDT, E., FRANKE, W. W., 1983: Spatial distribution of proteins specific for desmosomes and adhaerens junctions in epithelial cells demonstrated by double immunofluorescence microscopy. Differentiation **23**, 189–205.

— SINGER, S. J., 1979: The participation of α-actinin in the capping of cell membrane components. Cell **16**, 213–222.

— — 1980: Association of microtubules and intermediate filaments in chicken gizzard cells as detected by double immunofluorescence. Proc. Nat. Acad. Sci. U.S.A. **77**, 4769–4773.

— TOKUYASU, K. T., DUTTON, A. H., SINGER, S. J., 1980: Vinculin, an intracellular protein localized at specialized sites where microfilament bundles terminate at cell membranes. Proc. Nat. Acad. Sci. U.S.A. **77**, 4127–4131.

— — SINGER, S. J., 1979: Immunocytochemical localization of α-actinin in intestinal epithelial cells. Proc. Nat. Acad. Sci. U.S.A. **76**, 2833–2837.

GEISLER, N., FISCHER, S., VANDERKERCKHOVE, J., PLESSMANN, U., WEBER, K., 1984: Hybrid character of a large neurofilament protein (NF-M): intermediate filament type sequence followed by a long and acidic carboxy-terminal extension. EMBO J. **3**, 2701–2706.

— KAUFMANN, E., FISCHER, S., PLESSMANN, U., WEBER, K., 1983: Neurofilament architecture combines structural principles of intermediate filaments with carboxyterminal extensions increasing in size between triplet proteins. EMBO J. **2**, 1295–1302.

— — WEBER, K., 1982 a: Proteinchemical characterization of three structurally distinct domains along the protofilament unit of desmin 10 nm filaments. Cell **30**, 277–286.

— PLESSMANN, U., WEBER, K., 1982 b: Related sequences in neurofilaments and non-neuronal intermediate filaments. Nature **296**, 448–450.

— — — 1985: The complete amino acid sequence of the major mammalian neurofilament protein (NF-L). FEBS Lett. **182**, 475–481.

— VANDERKERCKHOVE, J., VAN DAMME, J., PLESSMANN, U., WEBER, K., 1985: Protein-chemical characterization of NF-H, the largest mammalian neurofilament component; intermediate filament-type sequences followed by a unique carboxy-terminal extension. EMBO J. **4**, 57–63.

— WEBER, K., 1980: Purification of smooth-muscle desmin and a protein-chemical comparison of desmin from chicken gizzard and hog stomach. Eur. J. Biochem. **111**, 425–433.

— — 1981 a: Comparison of the proteins of two immunological distinct intermediate-sized filaments by amino acid sequence analysis: desmin and vimentin. Proc. Nat. Acad. Sci. U.S.A. **78**, 120–124.

— — 1981 b: Self-assembly *in vitro* of the 68,000 molecular weight component of the mammalian neurofilament triplet proteins into intermediate-sized filaments. J. Mol. Biol. **151**, 565–571.

— — 1982: The amino acid sequence of chicken muscle desmin provides a common structural model for intermediate filament proteins including the wool α-keratins. EMBO J. **1**, 1649–1656.

GEORGE, H. J., MISRA, L., FIELD, D. J., LEE, J. C., 1981: Polymorphism of brain tubulin. Biochemistry **20**, 2402–2409.

GERKE, V., WEBER, K., 1984: Identity of p 36 K phosphorylated upon Rous sarcoma virus transformation with a protein purified from brush borders: calium-dependent binding to non-erythroid spectrin and F-actin. EMBO J. **3**, 227–233.

GEUENS, G., DE BRABANDER, M., NUYDENS, R., DE MEY, J., 1983: The interaction between microtubules and intermediate filaments in cultured cells treated with taxol and nocodazole. Cell Biol. Int. Rep. **7**, 35–47.

GIBBINS, J. R., 1982: Epithelial cell motility: the effect of 2-deoxyglucose on cell migration, ATP production, and the structure of the cytoplasmic ground substance in lammellipodia of epithelial cells in culture. Cell Motil. **2**, 25–46.

GIBBONS, I. R., 1981: Cilia and flagella of eukaryotes. J. Cell Biol. **91**, 107s–124s.

— COSSON, M. P., EVANS, J. A., GIBBONS, B. H., HOUCK, B., MARTINSON, K., 1978: Potent inhibition of dynein adenosine triphosphatase and of the motility of cilia and sperm flagella by vanadate. Proc. Nat. Acad. Sci. U.S.A. **75**, 2220–2224.

GIFFARD, R. G., WEEDS, A. G., SPUDICH, J. A., 1984: Ca-dependent binding of severin to actin: a one-to-one complex is formed. J. Cell Biol. **98**, 1796–1803.

GIGI, O., GEIGER, B., ESHHAR, Z., MOLL, R., SCHMID, E., WINTER, S., SCHILLER, D. I., FRANKE, W. W., 1982: Detection of a cytokeratin determinant common to diverse epithelial cells by a broadly crossreacting monoclonal antibody. EMBO J. **1**, 1429–1437.

GILBERT, D. S., NEWBY, B. J., ANDERTON, B. H., 1975: Neurofilament disguise, destruction and discipline. Nature **256**, 586–589.

GILLIS, J. M., O'BRIEN, E. J., 1975: The effect of calcium ions on the structure of reconstituted muscle thin filaments. J. Mol. Biol. **99**, 445–459.

GILMARTIN, M. E., CULBERTSON, V. B., FREEDBERG, I. M., 1980: Phosphorylation of epidermal keratins. J. Invest. Dermatol. **75**, 211–216.

— MITCHELL, J., VIDRICH, A., FREEDBERG, I. M., 1984: Dual regulation of intermediate filament phosphorylation. J. Cell Biol. **98**, 1144–1149.

GINGELL, D., 1981: The interpretation of interference-reflection images of spread cells: significant contribution from the peripheral cytoplasm. J. Cell Sci. **49**, 237–270.

GLACY, S. D., 1983: Subcellular distribution of rhodamine-actin microinjected into living fibroblastic cells. J. Cell Biol. **97**, 1207–1213.

GLENNEY, J. R., GEISLER, N., KAULFUS, P., WEBER, K., 1981 a: Demonstration of at least two different actin binding sites in villin, a calcium regulated modulator protein of F-actin organization. J. Biol. Chem. **256**, 8156–8161.

GLENNEY, J. R., GLENNEY, P., 1983: Fodrin is the general spectrin-like protein found in most cells whereas spectrin and the TW protein have a restricted distribution. Cell **34**, 503–512.

— — 1984: The microvillus 110 K cytoskeletal protein is an integral membrane protein. Cell **37**, 743–751.

— — OSBORN, M., WEBER, K., 1982 a: A F-actin- and calmodulin-binding protein from isolated intestinal brush borders has a morphology related to spectrin. Cell **28**, 843–854.

— — WEBER, K., 1982 c: Erythroid spectrin, brain fodrin, and intestinal brush border proteins (TW-260/240) are related molecules containing a common calmodulin-binding subunit bound to a variant cell type-specific subunit. Proc. Nat. Acad. Sci. U.S.A. **79**, 4002–4005.

— — — 1982 d: F-actin-binding and cross-linking properties of porcine brain fodrin, a spectrin-related molecule. J. Biol. Chem. **257**, 9781–9787.

— — — 1983: The spectrin-related molecule, TW-260/240, cross-links the actin bundles of the microvillus rootlets in the brush borders of intestinal epithelial cells. J. Cell Biol. **96**, 1491–1496.

— KAULFUS, P., MATSUDAIRA, P., WEBER, K., 1981 b: F-actin binding and bundling properties of fimbrin, a major cytoskeletal protein of microvillus core filaments. J. Biol. Chem. **256**, 9283–9289.

— — WEBER, K., 1981 c: F-actin assembly modulated by villin. Ca-dependent nucleation and capping of the barbed end. Cell **24**, 471–480.

— OSBORN, M., WEBER, K., 1982 b: The intracellular localization of the microvillus 110 K protein, a component considered to be involved in side-on membrane attachment of F-actin. Exp. Cell Res. **138**, 199–208.

— WEBER, K., 1980: Calmodulin-binding proteins of the microfilaments present in isolated brush borders and microvilli of intestinal epithelial cells. J. Biol. Chem. **255**, 10551–10554.

— — 1981: Calcium control of microfilaments: uncoupling of the F-actin-severing and -bundling activity of villin by limited proteolysis *in vitro*. Proc. Nat. Acad. Sci. U.S.A. **78**, 2810–2814.

GODSAVE, S. F., ANDERTON, B. H., HEASMAN, J., WYLIE, C. C., 1984: Oocytes and early embryos of *Xenopus laevis* contain intermediate filaments which react with anti-mammalian vimentin antibodies. J. Embryol. exp. Morph. **83**, 169–187.

GOLDBERG, D. J., HARRIS, D. A., LUBIT, B. W., SCHWARZ, J. H., 1980: Analysis of the mechanism of fast axonal transport by intracellular injection of potentially inhibitory macromolecules: Evidence for a possible role of actin filaments. Proc. Nat. Acad. Sci. U.S.A. **77**, 7448–7452.

GOLDMAN, J. E., SCHAUMBURG, H. M., NORTON, W. T., 1978: Isolation and characterization of glial filaments from human brain. J. Cell Biol. **78**, 426–440.

GOLDMAN, R. D., FOLLET, E., 1969: The structure of the major cell processes of isolated BHK-21 fibroblasts. Exp. Cell Res. **57**, 159–173.

— KNIPE, D. M., 1972: Functions of cytoplasmic fibers in nonmuscle cell motility. Cold Spring Harbor Symp. Quant. Biol. **37**, 523–534.

— POLLACK, R., HOPKINS, N. H., 1973: Preservation of normal behavior by enucleated cells in culture. Proc. Nat. Acad. Sci. U.S.A. **70**, 750–754.

GOLDSTEIN, J. L., ANDERSON, R. G. W., BROWN, M. S., 1979: Coated pits, coated vesicles, and receptor-mediated endocytosis. Nature **279**, 679–685.

GOODENOUGH, U., HEUSER, J., 1984: Structural comparison of purified dynein proteins with *in situ* dynein arms. J. Mol. Biol. **180**, 1083–1118.

GOODENOUGH, U. W., HEUSER, J. E., 1982: Substructure of the outer dynein arm. J. Cell Biol. **95**, 798–815.

GOODLOE-HOLLAND, C. M., LUNA, E. J., 1984: A membrane cytoskeleton from *Dictyostelium discoideum*. III. Plasma membrane fragments bind predominantly to the sides of actin filaments. J. Cell. Biol, **99**, 71–78.

GOODMAN, S. R., SHIFFER, K., 1983: The spectrin membrane skeleton of normal and abnormal human erythrocytes. J. Am. Phys. Soc. **244**, C121–C141.

— WEIDNER, S. A., 1980: Binding of spectrin α-β tetramers to human erythrocyte membranes. J. Biol. Chem. **255**, 8082–8086.

— ZAGON, I. S., KULIKOWSKI, R. R., 1981: Identification of a spectrin-like protein in noneythroid cells. Proc. Nat. Acad. Sci. U.S.A. **78**, 7570–7574.

GORBSKY, G., STEINBERG; M., 1981: Isolation of the intercellular glycoproteins of desmosomes. J. Cell Biol. **90**, 243–248.

GORDON, W. E., BUSHNELL, A., BURRIDGE, K., 1978: Characterization of the intermediate (10 nm) filaments of cultured cells using an autoimmune rabbit antiserum. Cell **13**, 249–261.

GOULD, R. R., BORISY, G. G., 1977: The pericentriolar material in Chinese hamster cells nucleates microtubule formation. J. Cell Biol. **73**, 601–615.

GOZES, I., SWEADNER, K. J., 1981: Multiple brain tubulin forms are expressed by a single neuron. Nature **294**, 477–480.

GRANGER, B. L., LAZARIDES, E., 1978: The exitence of an insoluble z-disc scaffold in chicken skeletal muscle. Cell **15**, 1253–1268.

— — 1979: Desmin and vimentin co-exist at the periphery of the myofibril z-disc. Cell **18**, 1053–1063.

— — 1982: Structural associations of synemin and vimentin filaments in avian erythrocytes revealed by immunoelectron microscopy. Cell **30**, 263–275.

GRANT, N. J., ORIOL-AUDIT, C., DICKENS, M. J., 1983: Supramolecular forms of actin induced by polyamines: an electron microscopic study. Eur. J. Cell Biol. **30**, 67–73.

GRATZER, W. B., 1981: The red cell membrane and its cytoskeleton. Biochem. J. **198**, 1–8.

— 1983: The cytoskeleton of the red blood cell. In: Muscle and Nonmuscle Motility (STRACHER, A., ed.), Vol. 2, pp. 37–125. New York: Academic Press.

GRAY, E. G., 1959: Electron microscopy of neuroglial fibrils of the cerebral cortex. J. Biophys. Biochem. Cytol. **6**, 121–122.

GREEN, K. J., GOLDMAN, R. D., 1983: The effects of taxol on cytoskeletal components in cultured fibroblasts and epithelial cells. Cell Motility **3**, 283–305.

GREENBERG, M. E., EDELMAN, G. M., 1983: The 34-kd pp 60 src substrate is located at the inner face of the plasma membrane. Cell **33**, 767–779.

GRIFFITH, L. M., POLLARD, T. D., 1978: Evidence for actin filament microtubule interaction mediated by microtubule-associated proteins. J. Cell Biol. **78**, 958–965.

GRIFFITH, L. M., POLLARD, T. D., 1982 a: Cross-linking of actin filament networks by self-association and actin-binding macromolecules. J. Biol. Chem. **257**, 9135–9142.
— — 1982 b: The interaction of actin filaments which microtubules and microtubule-associated proteins. J. Biol. Chem. **257**, 9143–9151.

GRIMSTONE, A. V., KLUG, A., 1966: Observation on the substructure of flagellar fibers. J. Cell Sci. **1**, 351–362.

GRINELL, F., 1980: Visualization of cell-substratum adhesion plaques by antibody exclusion. Cell Biol. Int. Rep. **4**, 1031–1036.

— FELD, M. K., 1979: Initial adhesion of human fibroblasts in serum-free medium: possible role of secreted fibronectin. Cell **17**, 117–129.

GROESCHEL-STEWART, U., 1980: Immunocytochemistry of cytoplasmic contractile proteins. Int. Rev. Cytol. **65**, 193–254.

GRUMET, M., LIN, S., 1980: A platelet inhibitor protein with cytochalasin-like activity against actin polymerization *in vitro*. Cell **21**, 439–445.

GUATELLI, J. C., PORTER, K. R., ANDERSON, K. L., BOGGS, D. P., 1982: Ultrastructure of the cytoplasmic and nuclear matrices of human lymphocytes observed using high voltage electron microscopy of embedment-free sections. Biol. Cell. **43**, 69–80.

GUNDERSEN, G. G., KALNOSKI, M. H., BULINSKY, J. C., 1984: Distinct populations of microtubules: tryosinated and nontyrosinated alpha tubulin are distributed differently *in vivo*. Cell **38**, 779–789.

GUNNING, B. E. S., HARDHAM, A. R., 1982: Microtubules. An. Rev. Plant Physiol. **33**, 651–698.

GUNNING, P., PONTE, P., KEDES, L., HICKEY, R. J., SKOULTCHI, A. I., 1984: Expression of human cardiac actin in mouse L cells: A sarcomeric actin associates with a nonmuscle cytoskeleton. Cell **37**, 709–715.

HAIMO, L. T., TELZER, B. R., ROSENBAUM, J. L., 1979: Dynein binds to and crossbridges cytoplasmic microtubules. Proc. Nat. Acad. Sci. U.S.A. **76**, 5759–5763.

HAMEL, E., LIN, C. M., 1984: Separation of active tubulin and microtubule-associated proteins by ultra-centrifugation and isolation of a component causing the formation of microtubule bundles. Biochemistry **23**, 4173–4184.

HAND, S. C., SAMERO, G. N., 1984: Influence of osmolytes, thin filaments, and solubility state on elasmobranch phosphofructokinase *in vitro*. J. Exp. Zool. **231**, 297–302.

HANUKOGLU, I., FUCHS, E., 1982: The cDNA sequence of a human epidermal keratin: divergence of sequence but conservation of structure among inter-mediate filament proteins. Cell **31**, 243–255.

— — 1983: The cDNA sequence of a type II cytoskeletal keratin reveals constant and variable structural domains among keratins. Cell **33**, 915–924.

HARDING, C. R., SCOTT, I. R., 1983: Histidine-rich proteins (filaggrins): structural and functional heterogeneity during epidermal differentiation. J. Mol. Biol. **170**, 651–673.

HARGREAVES, W. R., GIEDD, K. N., VERKLEIJ, A., BRANTON, D., 1980: Reassociation of ankyrin with band 3 in erythrocyte membranes and in lipid vesicles. J. Biol. Chem. **255**, 11965–11972.

HARRIS, A. K., 1976: Recycling of dissolved plasma membrane components as an explanation of the capping phenomenon. Nature **263**, 781–783.

— STOPAK, D., WILD, P., 1981: Fibroblast traction as a mechanism for collagen morphogenesis. Nature **290**, 249–251.

HARRIS, D. A., SCHWARTZ, J. H., 1981: Characterization of brevin, a serum protein that shortens actin filaments. Proc. Nat. Acad. Sci. U.S.A. **78**, 6798–6802.

HARRIS, H. E., GOOCH, J., 1981: An actin depolymerizing protein from pig plasma. FEBS Lett. **123**, 49–53.

— WEEDS, A. G., 1983: Plasma actin depolymerizing factor has both calcium-dependent and calcium-independent effects on actin. Biochemistry **22**, 2728–2741.

HARTSHORNE, D. J., 1982: Phosphorylation of myosin and the regulation of smooth-muscle actomyosin. In: Cell and Muscle Motility, Vol. II (DOWBEN, R. M., SHAY, J. W., eds.), pp. 188–220. New York: Plenum.

— GORECKA, A., 1980: The biochemistry of the contractile proteins of smooth muscle. In: Handbook of Physiology, Section II, Vol. 2 (BOHR, D. F., SOMLYO, A. P., SPARKS, H. V., eds.), pp. 93–120. Bethesda: Am. Physiology Soc.

HARTWIG, J. H., STOSSEL, T. P., 1975: Isolation and properties of actin, myosin, and a new actin-binding protein in rabbit alveolar macrophages. J. Biol. Chem. **250**, 5696–5705.

— — 1979: Cytochalasin B and the structure of actin gels. J. Mol. Biol. **134**, 539–553.

— — 1981: Structure of macrophage actin-binding protein molecules in solution and interacting with actin filaments. J. Mol. Biol. **145**, 563–581.

— TYLER, J., STOSSEL, T. P., 1980: Actin binding protein promotes the bipolar and perpendicular branching of actin filaments. J. Cell Biol. **87**, 841–848.

HASEGAWA, T., TAKAHASHI, S., HAYASHI, H., HATANO, S., 1980: Fragmin: a calcium ion sensitive regulatory factor on the formation of actin filaments. Biochemistry **19**, 2677–2683.

HAUSSMANN, K., MULISH, M., 1981: Das Epiplasma des Ciliaten *Pseudomicrothorax dubius*, ein Cytoskelett. Arch. Protistenk. **124**, 410–416.

HAY, E. D., 1981: Cell Biology of Extracellular Matrix. New York: Plenum.

— 1983: Cell and extracellular matrix: their organization and mutual dependence. Modern Cell Biol. **2**, 509–548.

HAYASHI, M., 1979: Depolymerization of brain microtubules by skeletal muscle myosin. J. Biochem. **85**, 691–698.

HEATH, I. B., SEAGULL, R. W., 1982: Oriented cellulose fibrils and the cytoskeleton: a critical comparison of models. In: The Cytoskeleton in Plant Growth and Development (LLOYD, C. W., ed.), pp. 163–182. London: Academic Press.

HEATH, J. P., 1983: Direct evidence for microfilament-mediated capping of the surface receptors on crawling fibroblasts. Nature **302**, 532–534.

— DUNN, G. A., 1978: Cell to substratum contacts of chick fibroblasts and their relation to the microfilament system: A correlated interference-reflexion and high-voltage electron microscope study. J. Cell Sci. **29**, 197–212.

HEGGENESS, M. H., ASH, J. F., SINGER, S. J., 1978 a: Transmembrane linkage of fibronectin to intracellular actin-containing filaments in cultured human fibroblasts. Ann. N. Y. Acad. Sci. **312**, 414–417.

Heggeness, M. H., Simon, M., Singer, S. J., 1978 b: Association of mitochondria with microtubules in cultured cells. Proc. Nat. Acad. Sci. U.S.A. 75, 3863–3866.

Heidemann, S. R., McIntosh, J. R., 1980: Visualization of the structural polarity of microtubules. Nature 286, 517–519.

— Sander, G., Kirschner, M. W., 1977: Evidence for a functional role of RNA in centrioles. Cell 10, 337–350.

Heidenhain, M., 1911: Plasma und Zelle, Vol. 2. Jena: G. Fischer.

Heilbrunn, L. V., 1926: The absolute viscosity of protoplasm. J. exp. Zool. 44, 255–278.

Helenius, A., Simons, K., 1975: Solubilization of membranes by detergents. Biochem. Biophys. Acta 415, 29–79.

Hellewell, S. B., Taylor, D. L., 1979: The contractile basis of amoeboid movement. VI. The solation-contraction coupling hypothesis. J. Cell Biol. 83, 633–648.

Henderson, D., Geisler, N., Weber, K., 1982: A periodic ultrastructure in intermediate filaments. J. Mol. Biol. 155, 173–176.

— Weber, K., 1979: Three-dimensional organization of microfilaments and microtubules in the cytoskeleton. Exp. Cell Res. 124, 301–316.

— Weber, K., 1980: Immunoelectron microscopic studies of intermediate filaments in cultured cells. Exp. Cell Res. 129, 441–453.

Henson-Stiennon, J.-A., 1965: Morphogenese de la cellule musculaire striée, etudiée au microscope éléctronique. I. Formation des structures fibrillaires. J. Microscopie 4, 657–678.

Hepler, P. K., Palewitz, B. A., 1974: Microtubules and microfilaments. Ann. Rev. Plant Physiol. 25, 309–362.

Herrmann, H., Pytela, R., Dalton, J. M., Wiche, G., 1984: Structural homology of microtubule-associated proteins 1 and 2 demonstrated by peptide mapping and immunoreactivity. J. Biol. Chem. 259, 612–617.

— Wiche, G., 1983: Specific in situ phosphorylation of plectin in detergent-resistent cytokeletons from cultured Chinese hamster ovary cells. J. Biol. Chem. 258, 14610–14618.

Hertwig, O., 1981: Über pathologische Veränderungen des Kernteilungsprozesses infolge experimenteller Eingriffe. Intern. Beitr. Wiss. Med. 1.

Herzog, W., Weber, K., 1977: In vitro assembly of pure tubulin into microtubules in the absence of microtubule-associated proteins and glycerol. Proc. Nat. Acad. Sci. U.S.A. 74, 1860–1864.

Hesterberg, L. K., Weber, K., 1983 a: Ligand-induced conformational changes in villin, a calcium-controlled actin-modulating protein. J. Biol. Chem. 258, 359–364.

— — 1983 b: Demonstration of three distinct calcium-binding sites in villin, a modulator of actin assembly. J. Biol. Chem. 258, 365–369.

Heuser, J. E., Kirschner, M. W., 1980: Filament organization revealed in platinum replicas of freeze-dried cytoskeletons. J. Cell. Biol. 86, 212–234.

Highsmith, S., Cooke, R., 1983: Evidence for actomyosin conformational changes involved in tension generation. In: Cell and Muscle Motility, Vol. IV (Dowben, R. M., Shay, J. W., eds.), pp. 207–238. New York: Plenum.

HILL, T. L., 1981: Microfilament or microtubule assembly or disassembly against a force. Proc. Nat. Acad. Sci. U.S.A. **78**, 5613–5617.

— 1984: Introductory analysis of the GTP-cap phase-change kinetics at the end of a microtubule. Proc. Nat. Acad. Sci. U.S.A. **81**, 6728–6732.

— CHEN, Y.-D., 1984: Phase changes at the end of a microtubule with a GTP cap. Proc. Nat. Acad. Sci. U.S.A. **81**, 5772–5776.

— KIRSCHNER, M. W., 1982: Bioenergetics and kinetics of microtubule and actin filament assembly-disassembly. Int. Rev. Cytol. **78**, 1–125.

HILLER, G., WEBER, K., 1977: Spectrin is absent in various tissue culture cells. Nature **266**, 181–183.

HIMES, R. H., KERSEY, R. N., HELLER-BETTINGER, I., SAMSON, F. E., 1976: Action of the Vinca alkaloids vincristine, vinblastine, and descetyl vinblastine amide on microtubules *in vitro*. Cancer Res. **36**, 3798–3802.

HINSSEN, H., 1972: Actin in isolated ground plasm of *Physarum polycephalum*. Cytobiologie **5**, 146–156.

— 1981 a: An actin-modulating protein from *Physarum pholycephalum*. I. Isolation and purification. Eur. J. Cell Biol. **23**, 225–233.

— 1981 b: An actin-modulating protein from *Physarum polycephalum*. II. Ca-dependence and other properties. Eur. J. Cell Biol. **23**, 234–240.

— D'HAESE, J., SMALL, J. V., SOBIESZEK, A., 1978: Mode of filament assembly of myosins from muscle and nonmuscle cells. J. Ultrastruct. Res. **64**, 282–302.

— SMALL, J. V., SOBIESZEK, A., 1984: A Ca-dependent actin modulator from vertebrate smooth muscle. FEBS Lett. **166**, 90–95.

HIROKAWA, N., 1982: Cross-linker system between neurofilaments, microtubules, and membranous organelles in frog axons revealed by the quick-freeze, deep-etching method. J. Cell Biol. **94**, 129–142.

— CHENEY, R. E., WILLARD, M., 1983: Location of a protein of the fodrin-spectrin-TW 260/240 family in the mouse intestinal brush border. Cell **32**, 953–965.

— GLICKSMAN, M. A., WILLORD, M. B., 1984: Organization of mammalian neurofilament polypeptides within the neuronal cytoskeleton. J. Cell Biol. **98**, 1523–1536.

— HEUSER, J. E., 1981: Quick-freeze, deep-etch visualization of the cytoskeleton beneath surface differentiations of intestinal epithelial cells. J. Cell Biol. **91**, 399–409.

— TILNEY, L. G., 1982: Interaction between actin filaments and between actin filaments and membranes in quick-frozen and deeply etched hair cells of the chick ear. J. Cell Biol. **95**, 249–261.

— — FUJIWARA, K., HEUSER, J. E., 1982: Organization of actin, myosin, and intermediate filaments in the brush border of intestinal epithelial cells. J. Cell Biol. **94**, 425–443.

HISANAGA, S., SAKAI, H., 1983: Cytoplasmic dynein of the sea urchin egg. II. purification, characterization and interaction with microtubules and Ca-calmodulin. J. Biochem. **93**, 87–98.

HITCHCOCK, S. E., CARLSSON, L., LINDBERG, U., 1976: Depolymerization of F-actin by deoxyribonuclease I. Cell **7**, 531–542.

Hoessli, C., Rungger-Braendle, E., Jokusch, B. M., Gabbiani, G., 1980: Lymphocyte α-actinin. Relationship to cell membrane and co-capping with surface receptors. J. Cell Biol. **84**, 305–314.

Hoffmann, P. N., Lasek, R. J., 1975: The slow component of axonal transport. Identification of major structural polypeptides of the axon and their generality among mammalian neurons. J. Cell Biol. **66**, 351–366.

Hoffmann-Berling, H., 1954: Adenosintriphosphat als Betriebsstoff von Zellbewegungen. Biochim. Biophys. Acta **14**, 182–194.

— Weber, H. H., 1953: Vergleich der Motilität von Zellmodellen und Muskelmodellen. Biochim. Biophys. Acta **10**, 629–638.

Hoglund, A. S., Karlsson, R., Arro, E., Fredrikson, B., Linberg, U., 1980: Visualization of the peripheral weave of microfilaments in glia cells. J. Muscle Res. Cell. Motil. **1**, 127–140.

Hollenbeck, P. J., Suprynowicz, F., Cande, W. Z., 1984: Cytoplasmic dynein-like ATPase cross-links microtubules in an ATP-sensitive manner. J. Cell Biol. **99**, 1251–1258.

Holtzer, H., Bennett, G. S., Tapscott, S. J., Croop, J. M., Toyama, Y., 1982: Intermediate-size filaments: changes in synthesis and distribution in cells of the myogenic and neurogenic lineages. Cold Spring Harbor Symp. Quant Biol. **46**, 317–329.

Horowitz, S. B., Miller, D. S., 1984: Solvent properties of ground substance studied by cryomicrodissection and intracellular reference-phase techniques. J. Cell Biol. **99**, 172s–179s.

Horwitz, B., Kupfer, B., Eshar, Z., Geiger, B., 1981: Reorganization of arrays of prekeratin filaments during mitosis. Exp. Cell Res. **134**, 281–290.

Horwitz, S. B., Parness, J., Schiff, P. B., Manfredi, J. J., 1982: Taxol: a new probe for studying the structure and function of microtubules. Cold Spring Harbor Symp. Quant. Biol. **46**, 219–226.

Hosoya, H., Mabuchi, I., 1984: A 45,000-mol-wt protein-actin complex from unfertilized sea urchin egg affects assembly properties of actin. J. Cell Biol. **99**, 994–1001.

— — Sakai, H., 1982: Actin modulating proteins in the sea urchin egg. I. Analysis of G-actin-binding proteins by DNAse I-affinity chromatography and purification of a 17,000 molecular weight component. J. Biochem. **92**, 1853–1862.

Howe, C. L., Mooseker, M. S., Graves, T. A., 1980: Brush border calmodulin: A major component of the isolated microvillus core. J. Cell Biol. **85**, 916–923.

Huang, B., Pitelka, D. R., 1973: The contractile process in the ciliate *Stentor coeruleus*. I. The role of microtubules and filaments. J. Cell Biol. **57**, 704–728.

Huang, C.-K., Hill, J. M. Bormann, B.-J., Mackin, W. M., Becker, E. L., 1984: Chemotactic factors induced vimentin phosphorylation in rabbit peritoneal neutrophils. J. Biol. Chem. **259**, 1386–1389.

Huber, G., Matus, A., 1984: Immunocytochemical localization of microtubule-associated protein 1 in rat cerebellum using monoclonal antibodies. J. Cell Biol. **98**, 777–781.

Hull, B. E., Staehelin, L. A., 1979: The terminal web, a reevaluation of its structure and function. J. Cell Biol. **81**, 67–82.

Hunt, R. C., Ellar, D. J., 1974: Isolation of the plasma membrane of a trypanosomatid flagellate: general characterization and lipid composition. Biochim. biophys. Acta **339**, 173–189.

HUSAIN, A., HOWLETT, G. J., SAWYER, W. H., 1984: The interaction of calmodulin with human and avian spectrin. Biochem. biophys. Res. Comm. **122**, 1194–1200.

HUXLEY, A. F., 1957: Muscle structure and theories of contraction. Progr. Bioph. Mol. Biol. **7**, 255–318.

— 1980: Reflections on muscle. Princeton, N J: Princeton University Press.

HUXLEY, H. E., 1963: Electron microscopic studies on the structure of natural and synthetic protein filaments from striated muscle. J. Mol. Biol. **7**, 281–308.

— BRAY, D., WEEDS, A. G., 1982: Molecular Biology of Cell Locomotion. London: The Royal Society.

— SIMMONS, R. M., FARUQI, A. R., KRESS, M., BORDAS, J., KOCH, M. H. J., 1983: Changes in the x-ray reflections from contracting muscle during rapid mechanical transients and their structural implications. J. Mol. Biol. **169**, 469–506.

HYAMS, J., 1984: A delicate molecular bouquet. Nature **302**, 292.

HYAMS, J. S., BORISY, G. G., 1978: Nucleation of microtubules *in vitro* by isolated spindle pole bodies of the yeast *Saccharomyces cerevisiae*. J. Cell. Biol. **78**, 401–414.

HYNES, R. O., 1973: Alteration of cell-surface proteins by viral transformation and by proteolysis. Proc. Nat. Acad. Sci. U.S.A. **70**, 3170–3174.

— 1976: Cell surface proteins and malignant transformation. Biochim. Biophys. Acta **458**, 73–107.

— 1981: Relationships between fibronectin and the cytoskeleton. In: Cytoskeletal Elements and Plasma Membrane Organization (POSTE, G., NICHOLSON, G. L., eds.), pp. 99–137. Amsterdam: Elsevier.

— DESTREE, A. T., 1978 a: 10 nm filaments in normal and transformed cells. Cell **13**, 151–163.

— — 1978 b: Relationship between fibronectin (LETS protein) and actin. Cell **15**, 875–886.

— — WAGNER, D. D., 1982: Relationship between microfilaments, cell-substratum adhesion, and fibronectin. Cold Spring Harbor Symp. Quant Biol. **46**, 659–670.

INOUE, S., SATO, H., 1967: Cell motility by labile association of molecules. J. Gen. Physiol. **50**, 259–292.

IP, W., DANTO, S. I., FISHMAN, D. A., 1983: Detection of desmin-containing intermediate filaments in cultured muscle and nonmuscle cells by immunoelectron microscopy. J. Cell Biol. **96**, 401–408.

— FISHMAN, D. A., 1979: High resolution scanning electron microscopy of isolated and *in situ* cytoskeletal elements. J. Cell Biol. **83**, 249–254.

ISENBERG, G., AEBI, U., POLLARD, T. D., 1980: An actin-binding protein from *Acanthamoeba* regulates actin filament polymerization and interactions. Nature **288**, 455–459.

— OHNHEISER, R., MARUTA, H., 1983: "cap 90", a 90-kDa Ca-dependent F-actin-capping protein from vertebrate brain. FEBS Lett. **163**, 225–231.

ISHIKAWA, H., 1984: Fine structure of skeletal muscle. Cell Muscle Motil. **4**, 1–84.
— BISCHOFF, R., HOLTZER, H., 1968: Mitosis and intermediate-sized filaments in developing skeletal muscle. J. Cell Biol. **38**, 538–555.
ISHIKAWA, M., MUROFUSHI, H., SAKAI, H., 1983: Bundling of microtubules *in vitro* by fodrin J. Biochem. **94**, 1209–1217.
ISHIZAKI, Y., TASHIRO, T., KUROKAWA, M., 1983: A calcium-activated protease which preferentially degrades the 160-kDa component of the neurofilament triplet. Eur. J. Biochem. **131**, 41–45.
ITO, S., WERTH, D. K., RICHERT, N. D., PASTAN, I., 1983: Vinculin phophorylation by the src kinase. J. Biol. Chem. **258**, 14626–14631.
IZANT, J. G., MCINTOSH, J. R., 1980: Microtubule-associated proteins: a monoclonal antibody to MAP 2 binds to differentiated neurons. Proc. Nat. Acad. Sci. U.S.A. **77**, 4741–4745.
— WEATHERBEE, J. A., MCINTOSH, J. R., 1982: A microtubule-associated protein in the mitotic spindle and the interphase nucleus. Nature **295**, 248–250.
IZZARD, C. S., LOCHNER, L. R., 1976: Cell-to-substrate contacts in living fibroblasts: An interference reflexion study with an evaluation of the technique. J. Cell Sci. **21**, 129–159.

JACKSON, B. W., GRUND, C., SCHMID, E., BURKI, K., FRANKE, W. W., ILMENSEE, K., 1980: Formation of cytoskeletal elements during mouse embryogenesis. Differentiation **17**, 161–179.
JACOBS, M., 1984: Tubulin nucleotide reactions and their role in microtubule assembly and dissociation. Ann. N.Y. Acad. Sci. **253**, 562–572.
— 1979: Tubulin and nucleotides. In: Microtubules (ROBERTS, K., HYAMS, J. S., eds.), pp. 255–277. New York: Academic Press.
JACOBSON, K., WOJCIESZYN, J., 1984: The translational mobility of substances within the cytoplasmic matrix. Proc. Nat. Acad. Sci. U.S.A. **81**, 6747–6751.
JARLFORS, U., SMITH, D. S., 1969: Associations between synaptic vesicles and neurotubules. Nature **224**, 710–711.
JEFFERY, W. R., 1984: Spatial distribution of messenger RNA in the cytoskeletal framework of ascidian eggs. Dev. Biol. **103**, 482–492.
JENNINGS, M. L., NICKNISH, J. S., 1984: Erythrocyte band 3 protein: evidence for multiple membrane crossing segments in the 17,000 dalton chymotryptic fragment. Biochemistry **23**, 6432–6436.
JOHNSON, K. A., BORISY, G. G., 1977: Kinetic analysis of microtubule self-assembly *in vitro*. J. Mol. Biol. **117**, 1–31.
— WALL, J. S., 1983: Structure and molecular weight of the dynein ATPase. J. Cell Biol. **96**, 669–678.
JOHNSON, P. J., FORAN, D. R., MOORE, G. P., 1983: Organization and evolution of the actin gene family in sea urchins. Mol. Cell. Biol. **3**, 1824–1833.
JOCKUSCH, B. M., ISENBERG, G., 1981: Interaction of α-actinin and vinculin with actin: Opposite effects on filament network formation. Proc. Nat. Acad. Sci. U.S.A. **78**, 3005–3009.
JONES, J. C. R., GOLDMAN, A. E., STEINERT, P. M., YUSPA, S., GOLDMAN, R. D., 1982: Dynamic aspects of the supramolecular organization of intermediate filament networks in cultured epidermal cells. Cell Motil. **2**, 197–213.

JONES, S. M., WILLIAMS, R. C., 1982: Phosphate content of mammalian neurofilaments. J. Biol. Chem. **257**, 9902–9905.

JORCANO, J. L., RIEGER, M., FRANZ, J. K., SCHILLER, D. L., MOLL, R., FRANKE, W. W., 1984: Identification of two types of keratin polypeptides within the acidic cytokeratin subfamily I. J. Mol. Biol. **179**, 257–281.

JULIEN, J.-P., MUSHYNSKI, W. E., 1982: Multiple phosphorylation sites in mammalian neurofilament polypeptides. J. Biol. Chem. **257**, 10467–10470.

JUNG, G., HELM, R. M., CAROTHERS CARRAWAY, C. A., CARRAWAY, K. L., 1984: Mechanism of concanavalin A-induced anchorage of the major cell surface glycoproteins to the submembrane cytoskeleton in 13762 ascites mammary adenocarcinoma cells. J. Cell Biol. **98**, 179–187.

KAMIYA, N., 1981: Physical and chemical basis of cytoplasmic streaming. Ann. Rev. Plant. Physiol. **32**, 205–236.

KARADSHEH, N. S., UYEDA, K., 1977: Changes in allosteric properties of phosphofructokinase bound to erythrocyte membranes. J. Biol. Chem. **252**, 7418–7423.

KARR, T. L., KRISTOFFERSON, D., PURICH, D. L., 1980 a: Mechanism of microtubule depolymerization. J. Biol. Chem. **255**, 8560–8566.

— — — 1980 b: Calcium ion induces endwise depolymerization of bovine brain microtubules. J. Biol. chem. **255**, 11853–11856.

KARSENTI, E., KOBAYASHI, S., MITCHISON, T., KIRSCHNER, M., 1984: Role of the centrosome in organizing the interphase microtubule array: properties of cytoplasts containing or lacking centrosomes. J. Cell Biol. **98**, 1763–1776.

KARTENBECK, J., FRANKE, W. W., MOSER, J. G., STOFFELS, U., 1983: Specific attachment of desmin filaments to desmosomal plaques in cardiac myocytes. EMBO J. **2**, 735–742.

— SCHMID, E., FRANKE, W. W., GEIGER, B., 1982: Different modes of internalization of proteins associated with adhaerens junctions and desmosomes: experimental separation of lateral contacts induces endocytosis of desmosomal plaque material. EMBO J. **6**, 725–732.

— SCHWECHHEIMER, K., MOLL, R., FRANKE, W. W., 1984: Attachment of vimentin filaments to desmosomal plaques in human meningiomal cells and arachnoidal tissue. J. Cell Biol. **98**, 1072–1081.

KEITH, A. D., 1979: The Aqueous Cytoplasm. New York: Marcel Dekker.

KEITH, C., DI PAOLA, M., MAXFIELD, F. R., SHELANSKI, M. L., 1983: Microinjection of Ca^{++}-calmodulin causes a localized depolymerization of microtubules. J. Cell Biol. **97**, 1918–1924.

KELLER, T. C. S., MOOSEKER, M. S., 1982: Ca-calmodulin-dependent phosphorylation of myosin, and its role in brush border contraction *in vitro*. J. Cell Biol. **95**, 943–959.

KELLEY, D. E., 1968: Myofibrillogenesis and Z-band differentiation. Anat. Rec. **163**, 403–426.

KELLY, D. E., 1966: Fine structure of desmosomes, hemi-desmosomes and an adepidermal layer in developing newt epidermis. J. Cell Biol. **28**, 51–72.

KEMPNER, E. S., MILLER, J. H., 1968: The molecular biology of *Euglena gracilius* V. Enzyme localization. Exp. Cell Res. **51**, 150–156.

KENDRICK-JONES, J., CANDE, W. Z., TOOTH, P. J., SMITH, R. C., SCHOLEY, J. M., 1983: Studies on the effect of phosphorylation of the 20,000 Mr light chain of vertebrate smooth muscle. J. Mol. Biol. **165**, 139–162.

— SCHOLEY, J. M., 1981: Myosin-linked regulatory systems. J. Muscle Res. Cell Motil. **2**, 347–372.

— SZENT-GYÖRGYI, A. G., COHEN, C., 1971: Segments from vertebrate smooth muscle myosin rods. J. Mol. Biol. **59**, 527–529.

KIEHART, D. P., 1981: Studies on the *in vivo* sensitivity of spindle microtubules to calcium ions and evidence for a vesicular calcium-sequestering system. J. Cell Biol. **88**, 604–617.

— KAISER, D. A., POLLARD, T. D., 1984: Monoclonal antibodies demonstrate limited structural homology between myosin isozymes from *Acanthamoeba*. J. Cell Biol. **99**, 1002–1014.

— POLLARD, T. D., 1984: Inhibition of *Acanthamoeba* actomyosin-II ATPase activity and mechanochemical function by specific monoclonal antibodies. J. Cell Biol. **99**, 1024–1033.

KIM, H., BINDER, L. I., ROSENBAUM, J. L., 1979: The periodic association of MAP 2 with brain microtubules *in vitro*. J. Cell Biol. **80**, 266–276.

KIM, K. H., RHEINWALD, J. D., FUCHS, E. V., 1983: Tissue specificity of epithelial keratins: differential expression of mRNAs from two multigene families. Mol. Cell. Biol. **3**, 495–502.

KIRSCHNER, M. W., 1978: Microtubule assembly and nucleation. Int. Rev. Cytol. **54**, 1–71.

— 1980: Implications of treadmilling for the stability and polarity of actin and tubulin polymers *in vivo*. J. Cell Biol. **86**, 330–334.

— HONIG, L. S., WILLIAMS, R. C., 1975: Quantitative electron microscopy of microtubule assembly *in vitro*. J. Mol. Biol. **99**, 263–276.

KLYMKOWSKY, M. W., 1981: Intermediate filaments in 3T3 cells collapse after intracellular injection of a monoclonal anti-intermediate filament antibody. Nature **291**, 249–251.

— 1982: Vimentin and keratin intermediate filament systems in cultured PtK epithelial cells are interrelated. EMBO J. **1**, 161–165.

— MILLER, R. H., LANE, E. B., 1983: Morphology, behavior and interaction of cultured epithelial cells after the antibody-induced disruption of keratin filament organization. J. Cell Biol. **96**, 494–509.

KNAPP, L. W., O'GUIN, W. M., SAWYER, R. H., 1983: Rearrangement of the keratin cytoskeleton after combined treatment with microtubule and microfilament inhibitors. J. Cell Biol. **97**, 1788–1794.

KNULL, H. R., BRONSTEIN, W. W., DES JARDINS, P., NIEHAUS, W. G., 1980: Interaction of selected brain glycolytic enzymes with an F-actin-tropomyosin complex. J. Neurochem. **34**, 222–225.

KOBAYASHI, R., BRADLEY, W. A., BRYAN, J., FIELD, J. B., 1983: Identification and purification of calcium ion dependent modulators of actin polymerization from bovine thyroid. Biochemistry **22**, 2463–2469.

KÖLLIKER, A., 1849: Das Sonnenthierchen, *Actinophrys sol.* Z. wiss. Zool. **1**, 198–217.

KONDO, H., 1984: Reexamination of the reality or artifacts of the microtrabeculae. J. Ultrastruct. Res. **87**, 124–135.

— ISHIWATA, S., 1976: Unidirectional growth of F-actin. J. Biochem. **79**, 159–171.

KORN, E. D., 1982: Actin polymerization and its regulation by proteins from nonmuscle cells. Physiol. Rev. **62**, 672–737.

KRAUHS, E., LITTLE, M., KEMPF, T., HOFER-WARBINEK, R., ADE, W., PONSTINGL, H., 1981: Complete amino acid sequence of β-tubulin from porcine brain. Proc. Nat. Acad. Sci. U.S.A. **78**, 4156–4160.

KRAVIT, N. G., REGULA, C. S., BERLIN, R. D., 1984: A reevaluation of the structure of purified tubulin in solution: Evidence for the prevalence of oligomers over dimers at room temperature. J. Cell Biol. **99**, 188–198.

KREIS, T. E., GEIGER, B., SCHLESSINGER, J., 1982: Mobility of microinjected rhodamine actin within living chicken gizzard cells determined by fluorescence photobleaching recovery. Cell **29**, 835–845.

— WINTERHALTER, H., BIRCHMEIER, W., 1979: *In vivo* distribution and turnover of fluorescently labelled actin microinjected into human fibroblasts. Proc. Nat. Acad. Sci. U.S.A. **76**, 3814–3818.

KRETSINGER, R. H., 1980: Structure and evolution of calcium-modulated proteins. CRC Crit. Rev. Biochem. **8**, 119–174.

KRISHNAN, N., KAISERMAN-ABRANOF, I. R., LASEK, R. J., 1979: Helical substructure of neurofilaments isolated from *Myxicola* and squid giant axons. J. Cell Biol. **82**, 323–335.

KROMAYER, W., 1982: Die Protoplasmafaserung der Epithelzelle. Arch. Mikr. Anat. **39**.

KUCZMARKSI, E. R., SPUDICH, J. A., 1980: Regulation of myosin self-assembly: phosphorylation of *Dictyostelium* heavy chain inhibits thick filament formation. Proc. Nat. Acad. Sci. U.S.A. **77**, 7292–7296.

KUMAGAI, H., NISHIDA, E., 1979: Interactions between calcium dependent regulator protein of cyclic nucleotide phosphodiesterase and microtubule protein. II. Association of calcium dependent regulator protein with tubulin dimer. J. Biochem. **85**, 1267–1274.

— — SAKAI, H., 1982: The interaction between calmodulin and microtubule proteins. IV. Quantitative analysis of the binding between calmodulin and tubulin dimer. J. Biochem. **91**, 1329–1336.

— SAKAI, H., 1983: A porcine brain protein (35 K protein) which bundles microtubules and its identification as glyceraldehyde 3-phosphate dehydrogenase. J. Biochem. **93**, 1259–1269.

KUMAR, N., FLAVIN, M., 1981: Preferential action of a brain detyrosinolating carboxypeptidase on polymerized tubulin. J. Biol. Chem. **256**, 7678–7686.

KUPFER, A., LOUVARD, D., SINGER, S. J., 1982: Polarization of the Golgi apparatus and the microtubule organizing center in cultured fibroblasts at the edge of an experimental wound. Proc. Nat. Acad. Sci. U.S.A. **79**, 2603–2607.

KUPKE, D. W., BEAMS, J. W., 1977: Solid-like character of virus solutions. Proc. Nat. Acad. Sci. U.S.A. **74**, 1993–1996.

KURIYAMA, R., BORISY, G. G., 1981: Microtubule-nucleating activity of centrosomes in Chinese hamster ovary cells is independent of the centriole cycle but coupled to the mitotic cycle. J. Cell Biol. **91**, 822–826.

KURKINEN, M., WARTIOVAARA, J., VAHERI, A., 1978: Cytochalasin B releases a major surface-associated glycoprotein, fibronectin, from cultured fibroblasts. Exp. Cell Res. **111**, 127–137.

KURTH, M. C., BRYAN, J., 1984: Platelet activation induces the formation of a stable gelsolin-actin complex from monomeric gelsolin. J. Biol. Chem. **259**, 7473–7479.

KUZNICKI, J., ALBANESI, J. P., COTE, G. P., KORN, E. D., 1983: Supramolecular regulation of the actin-activated ATPase activity of filaments of *Acanthamoeba* myosin II. J. Biol. Chem. **258**, 6011–6014.

— COTE, G. P., BOWERS, B., KORN, E. D., 1985: Filament formation and actin-activated ATPase activity are abolished by proteolytic removal of a small peptide from the tip of the tail of the heavy chain of *Acanthamoeba* myosin II. J. Biol. Chem. **260**, 1967–1972.

L'HERNAULT, S. W., ROSENBAUM, J. L., 1983: Chlamydomonas α-tubulin is posttranslationally modified in the flagella during flagellar regeneration. J. Cell Biol. **97**, 258–263.

LAL, A. A., BRENNER, S. L., KORN, E. D., 1984: Preparation and polymerization of skeletal muscle ADP-actin. J. Biol. Chem. **259**, 13061–13065.

— KORN, E. D., BRENNER, S. L., 1984: Rate constants for actin polymerization in ATP determined using cross-linked actin trimers as nuclei. J. Biol. Chem. **259**, 8794–8800.

LANE, E. B., GOODMAN, S. L., TREJDOSIEWICZ, L. K., 1982: Disruption of the keratin filament network during epithelial cell division. EMBO J. **1**, 1365–1372.

— HOGAN, B. L. M., KURKINEN, M., GARRELS, J. I., 1983: Co-expression of vimentin and cytokeratins in parietal endoderm cells of early mouse embryo. Nature **303**, 701–702.

LANGANGER, G., DE MEY, J., MOEREMANS, M., DANEELS, G., DE BRABANDER, M., SMALL, J. V., 1984: Ultrastructural localization of α-actinin and filamin in cultured cells with the immunogold staining (IGS) method. J. Cell Biol. **99**, 1324–1334.

LANGFORD, G. M., INOUE, S., 1979: Motility of the microtubular axostyle of *Pyrsonympha*. J. Cell Biol. **80**, 521–538.

LANGLEY, R. C., COHEN, C. M., 1984: Spectrin binds to intermediate filaments. J. Cell Biol. **99**, 303a.

LASEK, R. J., 1982: Translocation of the cytoskeleton in neurons and axonal growth. Phil. Trans. R. Soc. Lond. **299**, 319–327.

— BLACK, M. M., 1977: How do axons stop growing? In: Mechanisms, Regulation and Special Functions of Protein Synthesis in Brain (ROBERTS, S., LAJTHA, A., GISPEN, W. H., eds.), pp. 161–169. Amsterdam: Elsevier.

— GARNER, J. A., BRADY, S. T., 1984: Axonal transport of the cytoplasmic matrix. J. Cell Biol. **99**, 212s–221s.

— HOFFMAN, P. N., 1976: The neuronal cytoskeleton, axonal transport, and axonal growth. In: Cell Motility (GOLDMAN, R., POLLARD, T., ROSENBAUM, J., eds.), pp. 1021–1060. Cold Spring Harbor: Cold Spring Harbor Laboratory Press.

— KRISHNAN, N., KAISERMAN-ABRAMOF, I. R., 1979: Identification of the subunit proteins of 10 nm neurofilaments isolated from axoplasm of squid and *Myxicola* giant axons. J. Cell Biol. **82**, 336–346.

LAWSON, D., 1983: Epinemin: a new protein associated with vimentin filaments in non-neural cells. J. Cell Biol. **97**, 1891–1905.

LAZARIDES, E., 1975: Tropomyosin antibody: the specific localization of tropomyosin in non-muscle cells. J. Cell Biol. **65**, 549–560.

— 1976 a: Actin, α-actinin, and tropomyosin interaction in the structural organization of actin filaments in nonmuscle cells. J. Cell Biol. **68**, 202–219.

— 1976 b: Two general classes of cytoplasmic actin filaments in tissue culture cells: the role of tropomyosin. J. Supramol. Struct. **5**, 531–563.

— 1978: The distribution of desmin filaments in primary cultures of embryonic chick cardiac cells. Exp. Cell Res. **112**, 265–273.

— 1980: Intermediate filaments as mechanical integrators of cellular space. Nature **283**, 249–256.

— 1982 a: Intermediate filaments: a chemically heterogeneous, developmentally regulated class of proteins. Ann. Rev. Biochem. **51**, 219–250.

— 1982 b: Antibody production and immunofluorescent characterization of actin and contractile proteins. Methods Cell Biol. **24**, 313–332.

— BALZER, D. R., 1978: Specificity of desmin to avian and mammalian muscle cells. Cell **14**, 429–440.

— BURRIDGE, K., 1975: α-actinin: immunofluorescent localization of a muscle structural protein in nonmuscle cells. Cell **6**, 289–298.

— GRANGER, B. L., GARD, D. L., O'CONNOR, C. M., BRECKLER, J., PRICE, M., DANTO, S. I., 1982: Desmin- and vimentin-containing filaments and their role in the assembly of the Z disk in muscle cells. Cold Spring Harbor Symp. Quant. Biol. **46**, 351–378.

— HUBBARD, B. D., 1976: Immunological characterization of the subunit of the 100 nm filaments from muscle cells. Proc. Nat. Acad. Sci. U.S.A. **76**, 4344–4348.

— MOON, R. T., 1984: Assembly and topogenesis of the spectrin-based membrane skeleton in erythroid development. Cell **37**, 354–356.

— NELSON, W. J., 1982: Expression of spectrin in nonerythroid cells. Cell **31**, 505–508.

— — 1983: Erythrocyte and brain forms of spectrin in cerebellum: distinct membrane-cytoskeletal domains in neurons. Science **220**, 1295–1296.

— — KASAMATSU, T., 1984: Segregation of two spectrin forms in the chicken optic system: a mechanism for establishing restricted membrane-cytoskeletal domains in neurons. Cell **36**, 269–278.

— WEBER, K., 1974: Actin antibody: the specific visualization of actin filaments in nonmuscle cells. Proc. Nat. Acad. Sci. U.S.A. **71**, 2268–2272.

LEDBETTER, M. C., PORTER, K. R., 1963: A microtubule in plant fine structure. J. Cell Biol. **19**, 239–250.

LEE, J. J. L., SHOTT, R. J., ROSE III, S. J., THOMAS, T. L., BRITTEN, R. J., DAVIDSON, E. H., 1984: Sea urchin actin gene subtypes. J. Mol. Biol. **172**, 149–176.

LEE, V., WU, H. L., SCHLAEPFER, W. W., 1982: Monoclonal antibodies recognize individual neurofilament triplet proteins. Proc. Nat. Acad. Sci. U.S.A. **79**, 6089–6092.

LEE, Y. C., WOLFF, J., 1984 a: Calmodulin binds to both microtubule-associated protein 2 and τ proteins. J. Biol. Chem. **259**, 1226–1230.

— — 1984 b: The calmodulin-binding domain on microtubule-associated protein 2. J. Biol. Chem. **259**, 8041–8044.

LEES, A., HADDAD, J. G., LIN, S., 1984: Brevin and vitamin D binding protein: comparison of the effects of two serum proteins on actin assembly and disassembly. Biochemistry **23**, 3038–3047.

LEFEBVRE, P. A., SILFLOW, C. D., WIEBEN, E. D., ROSENBAUM, J. L., 1980: Increased levels of mRNAs for tubulin and other flagellar proteins after amputation or shortening of *Chlamydomonas* flagella. Cell **20**, 469–477.

LEHNERT, M. E., JORCANO, J. L., ZENTGRAF, H., BLESSING, M., FRANZ, J. K., FRANKE, W. W., 1984: Characterization of bovine keratin genes: similarities of exon patterns in genes coding for different keratins. EMBO J. **3**, 3279–3287.

LEHRER, S. S., 1981: Damage to actin filaments by glutaraldehyde: protection by tropomyosin. J. Cell Biol. **90**, 459–466.

LEHTO, V. P., 1983: 140,000 dalton surface glycoprotein. A plasma membrane component of the detergent-resistant cytoskeletal preparation of cultured human fibroblasts. Exp. Cell Res. **143**, 271–286.

— VARTIO, T., BADLEY, R. A., VIRTANEN, I., 1983 a: Characterization of a detergent-resistant surface lamina in cultured human fibroblasts. Exp. Cell Res. **143**, 287–294.

— — VIRTANEN, I., 1980: Enrichment of a 140 K surface glycoprotein in adherent, detergent-resistant cytoskeletons of cultured human fibroblasts. Biochem. Biophys. Res. Comm. **95**, 909–916.

— VIRTANEN, I., KURKI, P., 1978: Intermediate filaments anchor the nuclei in nuclear monolayers of cultured human fibroblasts. Nature **272**, 175–177.

— — PAASIVUO, R., RALSTON, R., ALITALO, K., 1983 b: The p 36 K substrate of tyrosine-specific protein kinases colocalizes with non-erythroid α-spectrin antigen, p 230, in surface lamina of cultured fibroblasts. EMBO J. **2**, 1701–1705.

LENGSFELD, A. M., LOW, I., WIELAND, T., DANCKER, P., HASSELBACH, W., 1974: Interaction of phalloidin with actin. Proc. Nat. Acad. Sci. U.S.A. **71**, 2803–2807.

LENK, R., RANSOM, L., KAUFMAN, Y., PENMAN, S., 1977: A cytoskeletal structure with associated polyribosomes obtained from HeLa cells. Cell **10**, 67–78.

LETERRIER, J.-F., LIEM, R. K. H., SHELANSKI, M. C., 1982: Interactions between neurofilaments and microtubule-associated proteins: a possible mechanism for intraorganellar bridging. J. Cell Biol. **95**, 982–986.

LETO, T. L., MARCHESI, V. T., 1984: A structural model of human erythrocyte protein 4.1. J. Biol. Chem. **259**, 4603–4608.

LEVINE, J., WILLARD, M., 1983: Redistribution of fodrin (a component of the cortical cytoplasm) accompanying capping of cell surface molecules. Proc. Nat. Acad. Sci. U.S.A. **80**, 191–195.

LEWIS, S. A., BALCAREK, J. M., KREK, V., SHELANSKI, M., COWAN, N. J., 1984: Sequence of a cDNA clone encoding mouse glial fibrillary acidic protein: structural conservation of intermediate filaments. Proc. Nat. Acad. Sci. U.S.A. **81**, 2743–2746.

LIEM, R. K. H., HUTCHISON, S. B., 1982: Purification of individual components of the neurofilament triplet: filament assembly from the 70,000 dalton subunit. Biochemistry 21, 3221–3226.

— YEN, S.-H., SOLOMON, G. D., SHELANSKI, M. L., 1978: Intermediate filaments in nervous tissue. J. Cell Biol. 79, 637–645.

LIN, D. C., LIN, S., 1979: Actin polymerization induced by a motility related high affinity complex from human erythrocyte membranes. Proc. Nat. Acad. Sci. U.S.A. 76, 2345–2349.

LIN, J. J. C., 1981: Monoclonal antibodies against myofibrillar components of rat skeletal muscle decorate the intermediate filaments of cultured cells. Proc. Nat. Acad. Sci. U.S.A. 78, 2335–2339.

— FERAMISCO, J. R., 1981: Disruption of the *in vivo* distribution of the intermediate filaments in fibroblasts through the microinjection of a specific monoclonal antibody. Cell 24, 185–193.

— MATSUMURA, F., YAMASHIRO-MATSUMURA, S., 1984: Tropomyosin-enriched and alpha-actinin-enriched microfilaments isolated from chicken embryo fibroblasts by monoclonal antibodies. J. Cell Biol. 98, 116–127.

— QUEALLY, S. A., 1982: A monoclonal antibody that recognizes Golgi-associated protein of cultured fibroblast cells. J. Cell Biol. 92, 108–112.

LIN, W., FUNG, B., SHYAMALA, M., KASAMATSU, H., 1981: Identification of antigenically related polypeptides at centrioles and basal bodies. Proc. Nat. Acad. Sci. U.S.A. 78, 2373–2377.

LINCK, R. W., AMOS, L. A., 1974: The hands of helical lattices in flagellar doublet microtubules. J. Cell Sci. 14, 551–559.

LINDWALL, G., COLE, R. D., 1984: The purification of tau protein and the occurrence of two phosphorylation states of tau in brain. J. Biol. Chem. 259, 12241–12245.

LIU, S. C., PALEK, J., 1980: Spectrin tetramer-dimer equilibrium and the stability of erythrocyte membrane cytoskeletons. Nature 285, 586–588.

— WINDISCH, P., KIM, S., PALEK, J., 1984: Oligomeric states of spectrin in normal erythrocyte membranes: biochemical and electron microscopic studies. Cell 37, 587–594.

LLOYD, C. W., 1982: The cytoskeleton in plant growth and development. London: Academic Press.

— 1983: Helical microtubular arrays in onion root hairs. Nature 305, 311–313.

LOCKWOOD, A., PENNINGROTH, S. M., KIRSCHNER, M. W., 1975: Function of GTP in microtubule formation. Fed. Proc. 34, 540.

LOEWY, A. G., 1952: An actomyosin-like substance from the plasmodium of a myxomycete. J. Cell. Comp. Physiol. 40, 127–156.

— WILSON, F. J., TAGGART, N. M., GREENE, E. A., FRASCA, P., KAUFMAN, H. S., SORRELL, M. J., 1983: A covalently cross-linked matrix in skeletal muscle fibers. Cell Motil. 3, 463–483.

LOFTUS, J. C., CHOATE, J., ALBRECHT, R. M., 1984: Platelet activation and cytoskeletal reorganization: high voltage electron microscopic examinations of intact and triton-extracted whole mounts. J. Cell Biol. 98, 2019–2025.

LOHMANN, S. M., DECAMILLI, P., EINIG, I., WALTER, U., 1984: High-affinity binding of the regulatory subunit (R II) of cAMP-dependent protein kinase to microtubule-associated and other cellular proteins. Proc. Nat. Acad. Sci. U.S.A. **81**, 6723–6727.

LONSDALE-ECCLES, J. D., TELLER, D. C., DALE, B. A., 1982: Characterization of a phosphorylated form of the intermediate filament-aggregating protein filaggrin. Biochemistry **21**, 5940–5948.

LOOR, F., 1981: Plasma membrane and cell cortex interactions in lymphocyte functions. Adv. Immunol. **30**, 1–120.

LOPATA, M. A., HAVERCROFT, J. C., CHOW, L. T., CLEVELAND, D. W., 1983: Four unique genes required for β-tubulin expression in vertebrates. Cell **32**, 713–724.

LOWEY, S., SLAYTER, H. S., WEEDS, A. G., BAKER, H., 1969: Substructure of the myosin molecule. I. Subfragments of myosin by enzymic degradation. J. Mol. Biol. **42**, 1–29.

LU, R. C., ELZINGA, M., 1977: Partial amino acid sequence of brain actin and its homology with muscle actin. Biochemistry **16**, 5801–5806.

LUBY, K. J., PORTER, K. R., 1980: The control of pigment migration in isolated erythrophores of *Holocentrus ascensionis* (Osbeck). I. Energy requirements. Cell **21**, 13–23.

LUBY-PHELPS, K., PORTER, K. R., 1982: The control of pigment migration in isolated erythrophores of *Holocentrus ascensionis* (Osbeck). II. The role of Ca^{++}. Cell **29**, 441–450.

LUCK, D. J., 1984: Genetic and biochemical dissection of the eukaryotic flagellum. J. Cell Biol. **98**, 789–794.

LUDFORD, R. J., 1936: The action of toxic substances upon the division of normal and malignant cells *in vitro* and *in vivo*. Arch. Exp. Zellforsch. **18**, 411–441.

LUDUENA, R. F., 1979: Biochemistry of tubulin. In: Microtubules (ROBERTS, K., HYAMS, J. S., eds.), pp. 65–116. London: Academic Press.

— WOODWARD, D. O., 1975: α- and β-tubulin: separation and partial sequence analysis. Ann. N.Y. Acad. Sci. **253**, 272–283.

LUNA, E. J., FOWLER, V. M., SWANSON, J., BRANTON, D., TAYLOR, D. L., 1981: A membrane cytoskeleton from *Dictyostelium discoideum*. I. Identification and partial characterization of an actin-binding activity. J. Cell Biol. **88**, 396–409.

— GOODLOE-HOLLAND, C. M., INGALLS, H. M., 1984: A membrane cytoskeleton from *Dictyostelium discoideum*. II. Integral proteins mediate the binding of plasma membranes to F-actin affinity beads. J. Cell Biol. **99**, 58–70.

LYNCH, G., BAUDRY, M., 1984: The biochemistry of memory: a new and specific hypothesis. Science **224**, 1057–1063.

LYNLEY, A. M., DALE, B. A., 1983: The characterization of human epidermal filaggrin, a histidine-rich, keratin filament-aggregating protein. Biochim. Biophys. Acta **744**, 28–35.

MABUCHI, I., 1983: An actin-depolymerizing protein (depactin) from starfish oocytes: properties and interaction with actin. J. Cell Biol. **97**, 1612–1621.

MacLEAN-FLETCHER, S. D., POLLARD, T. D., 1980a: Viscometric analysis of the gelation of *Acanthamoeba* extracts and purification of two gelation factors. J. Cell Biol. **85**, 414–428.

— — 1980b: Mechanism of action of cytochalasin B on actin. Cell **20**, 329–341.

MacRae, T. H., 1984: Bundling of bovine and brine shrimp (*Artemia*) microtubules *in vitro*. Cell Biol. Int. Rep. **8**, 423–431.

Magin, T. M., Jarcano, J. L., Franke, W. W., 1983: Translational products of mRNA coding for non-epidermal cytokeratins. EMBO J. **2**, 1387–1392.

Maher, P., Singer, S. J., 1983: A 200-kd protein isolated from the fascia adherens membrane domains of chicken cardic muscle cells is detected immunologically in fibroblast focal adhesions. Cell Motil. **3**, 419–429.

Malinoff, H. L., Wicha, M. S., 1983: Isolation of a cell surface receptor protein for laminin from murine fibrosarcoma cells. J. Cell Biol. **96**, 1475–1479.

Mandelkow, E., Mandelkow, E.-M., 1983: Structure of tubulin rings studied by X-ray scattering using synchrotron radiation. J. Mol. Biol. **167**, 179–196.

Mandelkow, E.-M., Mandelkow, E., 1979: Junctions between microtubule walls. J. Mol. Biol. **129**, 135–148.

— — 1985: Unstained microtubules studied by cryo-electron microscopy. J. Mol. Biol. **181**, 123–135.

Mangeat, P. H., Burridge, K., 1984 a: Immunoprecipitation of nonerythrocyte spectrin within live cells following microinjection of specific antibodies: relation to cytoskeletal structures. J. Cell Biol. **98**, 1363–1377.

— — 1984 b: Actin-membrane interaction in fibroblasts: What proteins are involved in this association. J. Cell Biol. **99**, 95 s–103 s.

Manton, I., Clarke, B., 1952: An electron microscope study of the spermatozoids of *Sphagnum*. J. Exptl. Botany **3**, 265–275.

Marchesi, V. T., 1983: The red cell membrane skeleton: recent progress. Blood **61**, 1–11.

— Furthmayr, H., Tomita, M., 1976: The red cell membrane. Ann. Rev. Biochem. **45**, 667–697.

— Steers, E., 1968: Selective solubilization of a protein component of red cell membrane. Science **159**, 203–204.

Marchiso, P. C. Capasso, O., Nitsch, L., Cancedda, R., Gionti, E., 1984: Cytoskeleton and adhesion patterns of cultured chick embryo chondrocytes during cell spreading and Rous sarcoma virus transformation. Exp. Cell Res. **151**, 332–343.

Marchuk, D., McCrohon, S., Fuchs, E., 1984: Remarkable conservation of structure among intermediate filament genes. Cell **39**, 491–498.

Marcum, J. M., Dedman, J. R., Brinkley, B. R., Means, A. R., 1978: Control of microtubule assembly-disassembly by calcium-dependent regulator protein. Proc. Nat. Acad. Sci. U.S.A. **75**, 3771–3775.

Margolis, R. L., 1983: Calcium and microtubules. In: Calcium and Cell Function. Vol. IV (Cheung, W. Y., ed.), pp. 313–335. New York: Academic Press.

— Wilson, L., 1977: Addition of colchicine-tubulin complex to microtubule ends: the mechanism of substoichiometric colchicine poisoning. Proc. Nat. Acad. Sci. U.S.A. **74**, 3466–3470.

— — 1978: Opposite end assembly and disassembly of microtubules at steady state *in vitro*. Cell **13**, 1–8.

— — 1979: Regulation of the microtubule steady state *in vitro* by ATP. Cell **18**, 673–679.

MARGOLIS, R. L., WILSON, L., 1981: Microtubule treadmills–possible molecular machinery. Nature **293**, 705–711.

— — KIEFER, B. J., 1978: Mitotic mechanisms based on intrinsic microtubule behaviour. Nature **272**, 450–452.

MARO, B., BORNENS, M., 1982: Reorganization of the HeLa cell cytoskeleton induced by an uncoupler of oxydative phosphorylation. Nature **295**, 334–336.

— SAURON, M.-E., PAULIN, D., BORNENS, M., 1983: Further evidence for interaction between microtubules and vimentin filaments: taxol and cold effects. Biol. Cell **47**, 243–246.

MARSLAND, D. A., 1938: The effects of high hydrostatic pressure upon cell division in Arbacia eggs. J. Cell. Comp. Physiol. **12**, 57–70.

MARTIN, F., GABRION, J., CAVADORE, J. C., 1981: Thyroid myosin filament assembly-disassembly is controlled by myosin light chain phosphorylation-dephosphorylation. FEBS Lett. **131**, 235–238.

MARUTA, H., GADASI, H., COLLINS, J. H., KORN, E. D., 1979: Multiple forms of *Acanthamoeba* myosin-I. J. Biol. Chem. **254**, 3624–3630.

— ISENBERG, G., 1983: Ca-dependent actin-binding phosphoprotein in *Physarum polycephalum*. J. Biol. Chem. **258**, 10151–10158.

— — SCHRECKENBACH, T., HALLMANN, R., RISSE, G., SHIBAYAMA, T., HESSE, J., 1983: Ca-dependent actin-binding phosphoprotein in *Physarum polycephalum*. J. Biol. Chem. **258**, 10144–10150.

— KNOERZER, W., HINSSEN, H., ISENBERG, G., 1984: Regulation of actin polymerization by non-polymerizable actin-like proteins. Nature **312**, 424–426.

— KORN, E. D., 1977: Purification from *Acanthamoeba castellani* of proteins that induce gelation and syneresis of F-actin. J. Biol. Chem. **252**, 399–402.

MARUYAMA, K., 1976: Actinins, regulatory proteins of muscle. Adv. Biophys. **9**, 157–185.

— ABE, S., ISHII, T., 1975: Dynamic viscoelasticity study of the effect of beta-actinin on interaction between F-actin and heavy-meromyosin. J. Biochem. **77**, 131–136.

— HARTWIG, J. H., STOSSEL, T. P., 1980: Cytochalasin B and the structure of actin gels. II. Further evidence for the splitting of F-actin by cytochalasin B. Biochim. Biophys. Acta **626**, 494–500.

— KIMURA, S., KURODA, M., HAYDA, S., 1977: Connectin, an elastic protein of muscle: its abundance in cardic myofibrils. J. Biochem. **82**, 347–355.

— — YOSHIDOMI, H., SAWADA, H., KIKUCHI, M., 1984 a: Molecular size and shape of β-connectin, an elastic protein of striated muscle. J. Biochem. **95**, 1423–1433.

— NATORI, R., NONOMURA, Y., 1976: New elastic protein from muscle. Nature **262**, 58–60.

— SAWADA, H., KIMURA, S., OHASHI, K., UMAZUME, Y., 1984: Connectin filaments in stretched skinned fibers of frog skeletal muscle. J. Cell Biol. **99**, 1391–1397.

MASTERS, C. J., 1981: Interactions between soluble enzymes and subcellular structure. C.R.C. Crit. Rev. Biochem. **11**, 105–144.

MASTRO, A. M., BABICH, M. A., TAYLOR, W. D., KEITH, A. D., 1984: Diffusion of a small molecule in the cytoplasm of mammalian cells. Proc. Nat. Acad. Sci. U.S.A. **81**, 3414–3418.

— KEITH, A. D., 1984: Diffusion in the aqueous compartment. J. Cell Biol. **99**, 180s–187s.

MASUROVSKY, E. B., PETERSON, E. R., CRAIN, S. M., HOROWITZ, S. B., 1981: Microtubule arrays in taxol-treated mouse dorsal root ganglion-spinal cord cultures. Brain Res. **217**, 392–398.

MATSUDAIRA, P. T., BURGESS, D. R., 1979: Identification and organization of the components in the isolated microvillus cytoskeleton. J. Cell Biol. **83**, 667–673.

— — 1982 a: Partial reconstitution of the microvillus core bundle: Characterization of villin as a Ca-dependent, actin-bundling/depolymerizing protein. J. Cell Biol. **92**, 648–656.

— — 1982 b: Organization of the cross-filaments in intestinal microvilli. J. Cell Biol. **92**, 657–664.

— MANDELKOW, E., RENNER, W., HESTERBERG, L. K., WEBER, K., 1983: Role of fimbrin and villin in determining the interfilament distances of actin bundles. Nature **301**, 209–214.

MATUS, A., BERNHARDT, R., HUGH-JONES, T., 1981: High-molecular weight microtubule-associated proteins are preferentially associated with dendritic microtubules in brain. Proc. Nat. Acad. Sci. U.S.A. **78**, 3010–3014.

MAUNOURY, R., 1978: Localisation immunocytochimique de la centroshere de cellules tumorales humaines par utilisation d'anticorps naturels de Lapin. C.R. Acad. Sc. Paris **286**, 503–506.

MAUPIN, P., POLLARD, T. D., 1983: Improved preservation and staining of HeLa cell actin filaments, clathrin-coated membranes, and other cytoplasmic structures by tannic acid-glutaraldehyde-saponin fixation. J. Cell Biol. **96**, 51–62.

MAUPIN-SZAMIER, P., POLLARD, T. D., 1978: Actin filament destruction by osmium tetroxide. J. Cell Biol. **77**, 837–852.

MAUTNER, V., HYNES, R. O., 1977: Surface distribution of LETS protein in relation to the cytoskeleton of normal and transformed cells. J. Cell Biol. **75**, 743–758.

MAZIA, D., 1984: Centrosomes and mitotic poles. Exp. Cell Res. **153**, 1–15.

— DAN, K., 1952: The isolation and biochemical characterization of the mitotic apparatus of dividing cells. Proc. Nat. Acad. Sci. U.S.A. **38**, 826–838.

McCONKEY, E. H., 1982: Molecular evolution, intracellular organization, and the quinary structure of proteins. Proc. Nat. Acad. Sci. U.S.A. **79**, 3236–3240.

McDONALD, K., 1984: Osmium-ferricyanide fixation improves microfilament preservation and membrane visualization in a variety of animal cell types. J. Ultrastruct. Res. **86**, 107–118.

McGILL, M., BRINKLEY, B. R., 1975: Human chromosomes and centrioles as nucleating sites for the *in vitro* assembly of microtubules from bovine brain tubulin. J. Cell Biol. **67**, 189–199.

McINTOSH, J. R., 1973: The axostyle of *Saccinobaculus*.I. Motion of the microtubule bundle and a structural analysis of straight and bent axostyles. J. Cell Biol. **56**, 324–339.

— 1983: The centrosome as an organizer of the cytoskeleton. In: Spatial Organization of Eukaryotic Cells (McINTOSH, J. R., ed.), pp. 115–142. New York: Alan Liss.

McIntosh, J. R., Euteneuer, U., 1984: Tubulin hooks are probes for microtubule polarity: an analysis of the method and an evaluation of data on microtubule polarity in the mitotic spindle. J. Cell Biol. **98**, 525–533.

— Hepler, P. K., van Wie, D. G., 1969: Model for mitosis. Nature **224**, 659–663.

McKeithan, T. W., Lefebvre, P. A., Silflow, C. D., Rosenbaum, J. L., 1983: Tubulin heterogeneity in *Polytomella* and *Chlamydomonas*. Evidence for a precursor of flagellar alpha tubulin. J. Cell Biol. **96**, 1056–1063.

— Rosenbaum, J. L., 1981: Multiple forms of tubulin in the cytoskeletal and flagellar microtubules of *Polytomella*. J. Cell Biol. **91**, 352–360.

— — 1984: The biochemistry of microtubules. In: Cell and Muscle Motility (Shay, J. W., ed.), pp. 255–288. New York: Plenum.

McKeown, M., Firtel, R. A., 1981: Evidence for subfamilies of actin genes in *Dictyostelium discoideum* as determined by comparison of 3'end sequences. J. Mol. Biol. **151**, 593–606.

McNiven, M. A., Wang, M., Porter, K. R., 1984: Microtubule polarity and the direction of pigment transport reverse simultaneously in surgically severed melanophore arms. Cell **37**, 753–765.

McNutt, N. W., Culp, L. A., Black, P. H., 1973: Contact-inhibited revertant cell lines isolated from SV-40-transformed cells. IV. Microfilament distribution and cell shape. J. Cell Biol. **56**, 413–428.

Means, A. R., Dedman, J. R., 1980: Calmodulin- an intracellular calcium receptor. Nature **285**, 73–77.

Meek, R. L., Lonsdale-Eccles, J. D., Dale, B. A., 1983: Epidermal filaggrin is synthesized on a large messenger ribonucleic acid as a high-molecular-weight precursor. Biochemistry **22**, 4867–4871.

Mehrabian, M., Bame, K. J., Rome, L. H., 1984: Interaction of liver lysosomal membranes with actin. J. Cell Biol. **99**, 680–685.

Mendelson, R. A., 1983: Muscle crossbridges in disarray. Nature **303**, 131.

Mescher, M., Jose, M., Balk, S., 1981: Actin-containing matrix associated with the plasma membrane of the murine tumor and lymphoid cells. Nature **289**, 139–144.

Meves, F., 1911: Gesammelte Studien an den roten Blutkörperchen der Amphibien. Arch. Mikrosk. Anat. **77**, 465–540.

Micko, S., Schlaepfer, W. W., 1978: Protein composition of axons and myelin from rat and human peripheral nerves. J. Neurochem. **30**, 1041–1049.

Milam, L., Erickson, H. P., 1982: Visualization of a 21 nm axial periodicity in shadowed keratin filaments and neurofilaments. J. Cell Biol. **94**, 592–595.

Miller, M., Solomon, F., 1984: Kinetics and intermediates of marginal band reformation: evidence for peripheral determinants of microtubule organization. J. Cell Biol. **99**, 70 s–75 s.

Miller, P., Walter, U., Theurkauf, W. E., Vallee, R. B., de Camilli, P., 1982: Frozen tissue sections as an experimental system to reveal specific binding sites for the regulatory subunit of type II cAMP-dependent protein kinase in neurons. Proc. Nat. Acad. Sci. U.S.A. **79**, 5562–5566.

Milstone, L. M., McGuire, J., 1981: Different polypeptides from the intermediate filaments in bovine hoof and esophageal epithelium and in aortic endothelium. J. Cell Biol. **88**, 312–316.

MIMURA, N., ASANO, A., 1979: Ca-sensitive gelation of actin filaments by a new protein factor. Nature **282**, 44–48.

MINAMI, Y., ENDO, S., SAKAI, H., 1984: Participation of 200 K or 150 K subunit of neurofilament in construction of the filament core with 70 K subunit and promotion of tubulin polymerization by incorporated 200 K subunit. J. Biochem. **96**, 1481–1490.

— MUROFUSHI, H., SAKAI, H., 1982: Interaction of tubulin with neurofilaments-formation of networks by neurofilament-dependent tubulin polymerization. J. Biochem. **92**, 889–898.

— SAKAI, H., 1983: Network formation by neurofilament-induced polymerization of tubulin: 200 K subunit of neurofilament triplet promotes nucleation of tubulin polymerization and enhances microtubule assembly. J. Biochem. **94**, 2023–2033.

MINTON, A. P., 1983: The effect of volume occupancy upon the thermodynamic activity of proteins: some biochemical consequences. Mol. Cell Biochem. **55**, 119–140.

— WILF, J., 1981: Effect of macromolecular crowding upon the structure and function of an enzyme: glyceraldehyde-3-phosphate dehydrogenase. Biochemistry **20**, 4821–4826.

MINTY, A. J., ALONSO, S., GUENET, J.-L., BUCKINGHAM, M. E., 1983: Number and organization of actin-related sequences in the mouse genome. J. Mol. Biol. **167**, 77–101.

MITCHELL, D. R., WARNER, F. D., 1981: Binding of dynein 21s ATPase to microtubules. J. Biol. Chem. **256**, 12535–12544.

MITCHISON, T., KIRSCHNER, M., 1984 a: Microtubule assembly nucleated by isolated centrosomes. Nature **312**, 232–237.

— — 1984 b: Dynamic instability of microtubule growth. Nature **312**, 237–242.

MOCKRIN, S. C., KORN, E. D., 1983: Kinetics of polymerization and ATP hydrolysis by covalently cross-linked actin dimer. J. Biol. Chem. **258**, 3215–3221.

MOHRI, H., 1968: Amino acid composition of "tubulin" constituting microtubules of sperm flagella. Nature **217**, 1053–1054.

MOLL, R., FRANKE, W. W., SCHILLER, D. L., GEIGER, B., KREPLER, R., 1982: The catalog of human cytokeratins: Patterns of expression in normal epithelia, tumors and cultured cells. Cell **31**, 11–24.

MOON, H. W., WISNIEWSKI, T., MERZ, P., DE MARTIN, J., WISNIEWSKI, H. M., 1981: Partial purification of neurofilament subunits from bovine brains and studies on neurofilament assembly. J. Cell Biol. **89**, 560–567.

MOON, R. T., LAZARIDES, E., 1983: Canavanine inhibits vimentin assembly but not its synthesis in chicken embryo erythroid cells. J. Cell Biol. **97**, 1309–1314.

MOORE, P., HUXLEY, H. E., DEROSIER, D. J., 1970: Three-dimensional reconstruction of F-actin, thin filaments and decorated thin filaments. J. Mol. Biol. **50**, 279–295.

MOOSEKER, M. S., 1983: Actin binding proteins of the brush border. Cell **35**, 11–13.

— GRAVES, T. A., WHARTON, K. A., FALCO, N., HOWE, C. L., 1980: Regulation of microvillus structure: Calcium-dependent solation and cross-linking of actin filaments in the microvilli of intestinal epithelial cells. J. Cell Biol. **87**, 809–822.

MOOSEKER, M. S., POLLARD, T. D., FUJIWARA, K., 1978: Characterization and localization of myosin in the brush border of intestinal epithelial cells. J. Cell Biol. **79**, 444–453.

— — WHARTON, K. A., 1982: Nucleated polymerization of actin from the membrane-associated ends of microvillar filaments in the intestinal brush border. J. Cell Biol. **95**, 223–233.

— TILNEY, L. G., 1973: Isolation and reactivation of the axostyle. Evidence for a dynein-like ATPase in the axostyle. J. Cell Biol. **56**, 13–26.

— — 1975: Organization of an actin filament-membrane complex: Filament polarity and membrane attachment in the microvilli of intestinal epithelial cells. J. Cell Biol. **67**. 725–743.

MORKEY, D. R., BOUCK, G. B., 1977: Mastigoneme attachment in *Ochromonas*. J. Ultrastruct. Res. **59**, 173–177.

MORRIS, A., TANNENBAUM, J., 1980: Cytochalasin D does not produce net depolymerization of actin filaments in HEp-2 cells. Nature **287**, 638–640.

MORRIS, N. R., LAI, M. H., OAKLEY, B. E., 1979: Identification of a gene for α-tubulin in *Aspergillus nidulans*. Cell **16**, 437–442.

MORROW, J. S., MARCHESI, V. T., 1981: Self-assembly of spectrin oligomers *in vitro*: a basis for a dynamic cytoskeleton. J. Cell Biol. **88**, 463–468.

— SPEICHER, D. W., KNOWLES, W. J., HSU, D. J., MARCHESI, V. T., 1980: Identification of functional domains of human erythrocyte spectrin. Proc. Nat. Acad. Sci. U.S.A. **77**, 6592–6596.

MOSE-LARSEN, P., BRAVO, R., FEY, S. J., SMALL, J. V., CELIS, J. E., 1982: Putative association of mitochondria with a subpopulation of intermediate-sized filaments in cultured human skin fibroblasts. Cell **31**, 681–692.

MOSKALEWSKI, S., THYBERG, J., LOHMANDER, S., FRIBERG, U., 1975: Influence of colchicine and vinblastine on the Golgi complex and matrix deposition in chondrocyte aggregates: an ultrastructural study. Exp. Cell Res. **95**, 440–454.

MOSS, D. J., 1983: Cytoskeleton-associated glycoproteins from chicken sympathetic neurons and chicken embryo brain. Eur. J. Biochem. **135**, 291–297.

MUELLER, H., FRANKE, W. W., 1983: Biochemical and immunological characterization of desmoplakins I and II, the major polypeptides of the desmosomal plaque. J. Mol. Biol. **163**, 647–671.

MUKHERJEE, T. M., STAEHELIN, L. A., 1971: The fine structural organization of the brush border of intestinal epithelial cells. J. Cell Sci. **8**, 573–594.

MUROFUSHI, H., MINAMI, Y., MATSUMOTO, G., SAKAI, H., 1983: Bundling of microtubules *in vitro* by a high molecular weight protein prepared from the squid axon. J. Biochem. **93**, 639–650.

MURPHY, D. B., BORISY, G. G., 1975: Association of high molecular weight proteins with microtubules and their role in microtubule assembly *in vitro*. Proc. Nat. Acad. Sci. U.S.A. **72**, 2696–2700.

— HIEBSCH, R. R., WALLIS, K. T., 1983 a: Identity and origin of the ATPase activity associated with neuronal microtubules. I. The ATPase activity is associated with membrane vesicles. J. Cell Biol. **96**, 1298–1305.

— JOHNSON, K. A., BORISY, G. G., 1977: Role of tubulin-associated proteins in microtubule nucleation and elongation. J. Mol. Biol. **117**, 33–52.

— WALLIS, K. T., HIEBSCH, R. R., 1983 b: Identity and origin of the ATPase activity associated with neuronal microtubules. II. Identification of a 50,000-

dalton polypeptide with ATPase activity similar to F-1 ATPase from mito-chondria. J. Cell Biol. **96**, 1306–1315.

MURRAY, J. M., 1984 a: Three-dimensional structure of a membrane-microtubule complex. J. Cell Biol. **98**, 283–295.

— 1984 b: Disassembly and reconstitution of a membrane-microtubule complex. J. Cell Biol. **98**, 1481–1487.

MURTHY, A. S. N., FLAVIN, M., 1983: Microtubule assembly using the microtubule-associated protein MAP-2 prepared in defined states of phosphorylation with protein kinase and phosphatase. Eur. J. Biochem. **137**, 37–46.

NAKAOKA, Y. KASAI, M., 1969: Behaviour of sonicated actin polymers: adenosine triphosphate splitting and polymerization. J. Mol. Biol. **44**, 319–332.

NATH, J., FLAVIN, M., GALLIN, J. I., 1982: Tubulin tyrosinolation in human polymorphonuclear leukocytes: studies in normal subjects and in patients with the Chediak-Higashi syndromes. J. Cell Biol. **95**, 519–526.

— WHITLOCK, J., FLAVIN, M., 1978: Tyrosylation of tubulin in synchronized HeLa cells. J. Cell Biol. **79**, 294 a.

NELSON, G. A., ANDREWS, M. L., KARNOVSKY, M. J., 1982: Participation of calmodulin in immunoglobulin capping. J. Cell Biol. **95**, 771–780.

NELSON, W. J., COLACO, C. A. L. S., LAZARIDES, E., 1983: Involvement of spectrin in cell-surface receptor capping in lymphocytes. Proc. Nat. Acad. Sci. U.S.A. **80**, 1626–1630.

— LAZARIDES, E., 1983 a: Expression of the β subunit of spectrin in nonerythroid cells. Proc. Nat. Acad. Sci. U.S.A. **80**, 363–367.

— — 1983 b: Switching of subunit composition of muscle spectrin during myogenesis *in vitro*. Nature **304**, 364–368.

— — 1984 a: Goblin (ankyrin) in striated muscle: Identification of the potential membrane receptor for erythroid spectrin in muscle cells. Proc. Nat. Acad. Sci. U.S.A. **81**, 3292–3296.

— — 1984 b: The patterns of expression of two ankyrin isoforms demonstrate distinct steps in the assembly of the membrane skeleton in neuronal mor-phogenesis. Cell **39**, 309–320.

— TRAUB, P., 1981: Properties of a Ca-activated protease specific for the intermediate-sized filament protein vimentin in Ehrlich ascites tumor cells. Eur. J. Biochem. **116**, 51–57.

— — 1982: Intermediate (10 nm) filament protein and the Ca-activated proteinase specific for vimentin and desmin in the cells from fish to man: an example for evolutionary conservation. J. Cell Sci. **57**, 25–49.

NEMHAUSER, I., ORNBERG, R., COHEN, W. D., 1980: Marginal bands in blood cells of invertebrates. J. Ultrastruct. Res. **70**, 308–317.

NERMUT, M. V., 1981: Visualization of the "membrane skeleton" in human erythrocytes by freeze-etching. Eur. J. Cell Biol. **25**, 265–271.

NEYFAKH, A. A., TINT, I. S., SVITKINA, T. M., BERSHADSKY, A. D., GELFAND, V. I., 1983: Visualization of cellular focal contacts using a monoclonal antibody to 80 kD protein adsorbed on the substratum. Exp. Cell Res. **149**, 387–396.

NG, R., ABELSON, J., 1980: Isolation and sequence of the gene for actin in *Saccharomyces cerevisiae*. Proc. Nat. Acad. Sci. U.S.A. **77**, 3912–3916.

NGAI, P. K., WALSH, M. P., 1984: Inhibition of smooth muscle actin-activated myosin Mg-ATPase activity by caldesmon. J. Biol. Chem. **259**, 13656–13659.

NIEDERMANN, R., AMREIN, P. C., HARTWIG, J., 1983: Three-dimensional structure of actin filaments and of an actin gel made with actin binding protein. J. Cell Biol. **96**, 1400–1413.

— POLLARD, T. D., 1975: Human platelet myosin. II. *In vitro* assembly and structure of myosin filaments. J. Cell Biol. **67**, 72–92.

NIELSEN, P., GOELZ, S., TRACHSEL, H., 1983: The role of the cytoskeleton in eukaryotic protein synthesis. Cell Biol. Int. Rep. **7**, 245–254.

NIGG, E. A., CHERRY, R. J., 1980: Anchorage of a band 3 population at the erythrocyte cytoplasmic membrane surface: protein rotational diffusion measurements. Proc. Nat. Acad. Sci. U.S.A. **77**, 4702–4706.

— SEFTON, B. M., HUNTER, T., WALTER, G., SINGER, S. J., 1982: Immunofluorescent localization of the transforming protein of Rous sarcoma virus with antibodies against a synthetic src peptide. Proc. Nat. Acad. Sci. U.S.A. **79**, 5322–5326.

NISHIDA, E., KUWAKI, T., SAKAI, H., 1981: Phosphorylation of microtubule-associated proteins (MAPs) and pH of the medium control interaction between MAPs and actin filaments. J. Biochem. **90**, 575–578.

— MACKAWA, S., SAKAI, H., 1984: Cofilin, a protein in porcine brain that binds to actin filaments and inhibits their interactions with myosin and tropomyosin. Biochemistry **23**, 5307–5313.

— OHTA, Y., SAKAI, H., 1983: The regulation of actin polymerization by the 88 K protein/actin complex and cytochalasin B. J. Biochem. **94**, 1671–1683.

— SAKAI, H., 1977: Calcium-sensitivity of the microtubule reassembly system. Difference between crude brain extract and purified microtubule proteins. J. Biochem. **82**, 303–306.

NISHIKAWA, M., DE LANEROLLE, P., LINCOLN, T. M., ADELSTEIN, R. S., 1984: Phosphorylation of mammalian myosin light chain kinases by the catalytic subunit of cyclic AMP-dependent protein kinase. J. Biol. Chem. **259**, 8429–8436.

NUMATA, O., WATANABE, Y., 1982: *In vitro* assembly and disassembly of 14 nm filament from *Tetrahymena pyriformis*. J. Biochem. **91**, 1563–1573.

NUNNALLY, M. H., POWELL, L. D., CRAIG, S. W., 1981: Reconstitution and regulation of actin gel-sol transformation with purified filamin and villin. J. Biol. Chem. **256**, 2083–2086.

— STULL, J. T., 1984: Mammalian skeletal muscle light chain kinases. J. Biol. Chem. **259**, 1776–1780.

NYSTROM, L.-E., LINDBERG, I., KENDRICK-JONES, J., JAKES, R., 1979: The amino acid sequence of profilin from calf spleen. FEBS Lett. **101**, 161–165.

O'CONNOR, C. M., GARD, D. L., LAZARIDES, E., 1981: Phosphorylation of intermediate filament proteins by cAMP-dependent protein kinases. Cell **23**, 135–143.

OCKLEFORD, C. D., TUCKER, J. B., 1973: Growth, breakdown, repair and rapid contraction of microtubular axopodia in the heliozoan *Actinophrys sol*. J. Ultrastruct. Res. **44**, 369–387.

OESCH, B., BIRCHMEIER, W., 1982: New surface component of fibroblast's focal contacts identified by a monoclonal antibody. Cell **31**, 671–679.

OHANIAN, V., WOLFE, L. C., JOHN, K. M., PINDER, J. C., LUX, S. E., 1984: Analysis of the ternary interaction of the red cell membrane skeletal proteins spectrin, actin, and 4.1. Biochemistry **23**, 4416–4420.

OHTA, Y. ENDO, S., NISHIDA, E., MUROFUSHI, H., SAKAI, H., 1984: An 18 K protein from ascites hepatoma cell depolymerizes actin filaments rapidly J. Biochem. **96**, 1547–1558.

OLIVER, J. M., ALBERTINI, D. S., BERLIN, R. D., 1976: Effects of glutathione-oxidizing agents on microtubule assembly and microtubule-dependent surface properties of human neutrophils. J. Cell Biol. **71**, 921–932.

— BERLIN, R. D., 1982: Mechanisms that regulate the structural and functional architecture of cell surfaces. Int. Rev. Cytol. **74**, 55–93.

— LALCHANDAI, R., BECKER, E. L., 1977: Actin redistribution during concanavalin A cap formation in rabbit neutrophils. J. Reticuloendothel. Soc. **21**, 359–364.

OLMSTED, J. B., BORISY, G. G., 1975: Ionic and nucleotide requirements for microtubule polymerization *in vitro*. Biochemistry **14**, 2996–3005.

— MARCUM, J. M., JOHNSON, K. A., ALLEN, C., BORISY, G. G., 1974: Microtubule assembly: some possible regulatory mechanisms. J. Supramol. Struct. **2**, 429–450.

OOSAWA, F., KASAI, M., 1962: A theory of linear and helical aggregation of macromolecules. J. Mol. Biol. **4**, 10–21.

OSBORN, M., BORN, T., KOITSCH, H., WEBER, K., 1978: Stereo immunofluorescence microscopy: I. Three-dimensional arrangement of microfilaments, microtubules, and tonofilaments. Cell **14**, 477–488.

— FRANKE, W. W., WEBER, K., 1977: Visualization of a system of filaments 7–10 nm thick in cultured cells of an epithelioid line (PtK 2) by immunofluorescence microscopy. Proc. Nat. Acad. Sci. U.S.A. **74**, 2490–2494.

— — — 1980: Direct demonstration of the presence of two immunologically distinct intermediate filament systems in the same cell by double immunofluorescence microscopy. Exp. Cell Res. **125**, 37–46.

— GEISLER, N., SHAW, G., WEBER, K., 1982 Intermediate filaments. Cold Spring Harbor Symp. Quant. Biol. **46**, 413–429.

— WEBER, K., 1976: Cytoplasmic microtubules in tissue culture cells appear to grow from an organizing structure towards the plasma membrane. Proc. Nat. Acad. Sci. U.S.A. **73**, 867–871.

— — 1977: The detergent-resistant cytoskeleton of tissue culture cells includes the nucleus and the microfilament bundles. Exp. Cell Res. **106**, 339–349.

— — 1980: Dimethylsulfoxide and the ionophore A 23187 affect the arrangement of actin and induce nuclear actin paracrystals in PtK 2 cells. Exp. Cell Res. **129**, 103–114.

— — 1982 a: Intermediate filaments: cell-type-specific markers in differentiation and pathology. Cell **31**, 303–306.

— — 1982 b: Immunofluorescence and immunocytochemical procedures with affinity-purified antibodies: tubulin-containing structures. Methods Cell Biol. **24**, 98–132.

Osborn, M., Weber, K., 1983: Tumor diagnosis by intermediate filament typing: a novel tool for surgical pathology. Lab. Invest. **48**, 372–394.

—— 1984: Actin paracrystal induction by forskolin and by db-cAMP in CHO cells. Exp. Cell Res. **150**, 408–418.

— Webster, R. E., Weber, K., 1978: Individual microtubules viewed by immunofluorescence and electron-microscopy in the same PtK 2 cell. J. Cell Biol. **77**, R 27–R 34.

Oster, G. F., 1984: On the crawling of cells. J. Embryol. Exp. Morph. **83**, 329–364.

Ostlund, R. E., Leung, J. T., Kipnis, D. M., 1977: Muscle actin filaments bind pituitary secretory granules in vitro. J. Cell Biol. **73**, 78–87.

Otto, J. J., 1983: Detection of vinculin-binding proteins with an 125-I-vinculin gel overlay technique. J. Cell Biol. **97**, 1283–1287.

— Kane, R. E., Bryan, J., 1979: Formation of filapodia in coelomocytes: localization of fascin, a 58,000-dalton actin cross-linking protein. Cell **17**, 285–293.

Overton, J., 1974: Cell junctions and their development. Prog. Surf. Membr. Sci. **8**, 161–208.

Owaribe, K., Masuda, H., 1982: Isolation and characterization of circumferential microfilaments bundles from retinal pigmented epithelial cells. J. Cell Biol. **95**, 310–315.

Ozaki, K., Hatano, S., 1984: Mechanism of regulation of actin polymerization by Physarum profilin. J. Cell Biol. **98**, 1919–1925.

— Sugino, H., Hasegawa, T., Takahashi, S., Hatano, S., 1983: Isolation and characterization of Physarum profilin. J. biochem. **93**, 295–298.

Paetau, A., Virtanen, I., Stenman, S., Kurki, P., Linder, E., Vaheri, A., Westermark, B., Dahl, D., Haltia, M., 1979: Glial fibrillary acidic protein and intermediate filaments in human glioma cells. Acta Neuropathol. **47**, 71–74.

Pagh, K., Maruta, H., Claviez, M., Gerisch, G., 1984: Localization of two phosphorylation sites adjacent to a region important for polymerization on the tail of Dictyostelium myosin. EMBO J. **3**, 3271–3278.

Painter, R., Ginsberg, M., 1982: Concanavalin A induces interactions between surface glycoproteins and platelet cytoskeleton. J. Cell Biol. **92**, 565–573.

Palay, S. L., 1956: Synapses in the central nervous system. J. Biophys. Biochem. Cytol., Suppl. **1**, 193–201.

Pallas, D., Solomon, F., 1982: Cytoplasmic microtubule-associated proteins: phosphorylation at novel sites is correlated with their incorporation into assembled microtubules. Cell **30**, 407–414.

Pant, H. C., Terakawa, S., Gainer, H., 1979: A calcium activated protease in squid axoplam. J. Neurochem. **32**, 99–102.

Pantaloni, D., Carlier, M.-F., Simon, C., Batelier, G., 1981: Mechanism of tubulin assembly: role of rings in the nucleation process and of associated proteins in the stabilization of microtubules. Biochemistry **20**, 3709–3716.

Pardee, J. D., Spudich, J. A., 1982: Mechanism K^+-induced actin assembly. J. Cell Biol. **93**, 648–654.

Pardo, J. V., d'Angelo Siliciano, J., Craig, S. W., 1983: A vinculin-containing cortical lattice in skeletal muscle: transverse lattice elements ("costamers") mark sites of attachment between myofibrils and sarcolemma. Proc. Nat. Acad. Sci. U.S.A. **80**, 1008–1012.

PATO, M. D., ADELSTEIN, R. S., 1980: Dephosphorylation of the 20,000-dalton light chain of myosin by two different phosphatases from smooth muscle. J. Biol. Chem. **255**, 6535–6541.

PAWELETZ, N., 1981: Membranes in the mitotic apparatus. Cell Biol. Int. Rep. **5**, 323–336.

— MAZIA, D., FINZE, E. M., 1984: The centrosome cycle in the mitotic cycle of sea urchin eggs. Exp. Cell Res. **152**, 47–65.

PAYNE, M. R., RUDNIK, S. E., 1984: Tropomyosin as a modulator of microfilaments. TIBS **9**, 361–363.

PEARL, M., FISHKIND, D., MOOSEKER, M., KEENE, D., KELLER, T., 1984: Studies on the spectrin-like protein from the intestinal brush border, TW 260/240, and characterization of its interaction with the cytoskeleton and actin. J. Cell Biol. **98**, 66–78.

PECK, R. K., 1977: The ultrastructure of the somatic cortex of *Pseudomicrothorax dubius*: structure and function of the epiplasm in ciliated *Protozoa*. J. Cell Sci. **25**, 367–385.

PENNINGROTH, S. M., CLEVELAND, D. W., KIRSCHNER, M. W., 1976: *In vitro* studies of the regulation of microtubule assembly. In: Cell Motility (GOLDMAN, R., POLLARD, T., ROSENBAUM, J., eds.), pp. 1233–1258. Cold Spring Harbor Laboratory.

PERNICE, B., 1889: Sulla cariocinesi delle cellule epiteliali e dell' endotelio dei vasi della mucosa dello stomaco et dell' intestino, nelle studio della gastroenterite sperimentale (nell'avvelenamento per colchico). Sicilia Med. **1**, 265–279.

PETERS, A., VAUGHN, J. E., 1967: Microtubules and filaments in axons and astrocytes of early postnatal rat optic nerves. J. Cell Biol. **32**, 113–119.

PETRUCCI, T. C., THOMAS, C., BRAY, D., 1983: Isolation of a Ca-dependent actin-fragmenting protein from brain, spinal cord, and cultured neurons. J. Neurochem. **40**, 1507–1516.

PETZELT, C., 1984: Localization of an intracellular membrane-bound Ca^{++}-ATPase in PtK-cells using immunofluorescence techniques. Eur. J. Cell Biol. **33**, 55–59.

PHAIRE-WASHINGTON, L., SILVERSTEIN, S. C., WANG, E., 1980: Phorbol myristate acetate stimulates microtubule and 10 nm filament extension and lysosome redistribution in mouse macrophages. J. Cell Biol. **86**, 641–655.

PICKETT-HEAPS, J. D., 1969: The evolution of the mitotic apparatus: an attempt at comparative cytology in dividing plant cells. Cytobios **1**, 257–280.

PIERSON, G. B., BURTON, P. R., HIMES, R. H., 1978: Alterations in number of protofilaments in microtubules assembled *in vitro*. J. Cell Biol. **76**, 223–228.

PINDER, J. C., GRATZER, W. B., 1983: Structure and dynamic states of actin in the erythrocyte. J. Cell Biol. **96**, 768–775.

— UNGEWICKELL, E., CALVERT, R., MORRIS, E., GRATZER, W. B., 1979: Polymerization of G-actin by spectrin preparations: identification of the active constituent. FEBS Lett. **104**, 396–400.

PIPERNO, G., 1984: Monoclonal antibodies to dynein subunits reveal the existence of cytoplasmic antigens in sea urchin eggs. J. Cell Biol. **98**, 1842–1850.

— HUANG, B., RAMANIS, Z., LUCK, D. J. L., 1981: Radial spokes of *Chlamydomonas* flagella: polypeptide composition and phosphorylation of stalk components. J. Cell Biol. **88**, 73–79.

PIRAS, R., PIRAS, M. M., 1975: Changes in microtubule phosphorylation during cell cycle of HeLa cells. Proc. Nat. Acad. Sci. U.S.A. **72**, 1161–1165.

PITELKA, D., 1969: Fibrillar systems in protozoa. In: Research in Protozoology (CHEN, T. T., ed.), Vol. 3, pp. 279–388. New York: Pergamon Press.

POLLACK, G. H., SUGI, H., 1984: Contractile Mechanisms in Muscle. New York: Plenum.

POLLACK, R., OSBORN, M., WEBER, K., 1975: Patterns of organization of actin and myosin in normal and transformed cultured cells. Proc. Nat. Acad. Sci. U.S.A. **72**, 994–998.

— RIFKIN, D., 1975: Actin-containing cables within anchorage-dependent rat embryo cells are dissociated by plasmin and trypsin. Cell **6**, 495–506.

POLLARD, T. D., 1976: The role of actin in the temperature-dependent gelation and contraction of extracts of *Acanthamoeba*. J. Cell Biol. **68**, 579–601.

— 1981 a: Cytoplasmic contractile proteins. J. Cell Biol. **91**, 156 s–165 s.

— 1981 b: Purification of a calcium-sensitive actin gelation protein from *Acanthamoeba*. J. Biol. Chem. **256**, 7666–7670.

— 1982: Structure and polymerization of *Acanthamoeba* myosin II filaments. J. Cell Biol. **95**, 816–825.

— 1984: Polymerization of ADP-actin. J. Cell Biol. **99**, 769–777.

— AEBI, U., COOPER, J. A., ELZINGA, M., FOWLER, W. E., GRIFFITH, L. M., 1982: The mechanism of actin-filament assembly and cross-linking. In: Cell and Muscle Motility, Vol. 2 (DOWBEN, R. M., SHAY, J. W., eds.), pp. 15–44. New York: Plenum.

— COOPER, J. A., 1984: Quantitative analysis of the effect of *Acanthamoeba* profilin on actin filament nucleation and elongation. Biochemistry **23**, 6631–6641.

— — 1982: Methods to characterize actin filament networks. Methods Enzymol. **85**, 211–233.

— KORN, E. D., 1973 a: Electron microscope identification of actin associated with isolated amoeba plasma membranes. J. Biol. Chem. **248**, 448–450.

— — 1973 b: *Acanthamoeba* myosin. I. Isolation from *Acanthamoeba castellanii* of an enzyme similar to muscle myosin. J. Biol. Chem. **248**, 4682–4690.

— MOOSEKER, M. S., 1981: Direct measurement of actin polymerization rate constants by electron microscopy of actin filaments nucleated by isolated microvillus cores. J. Cell Biol. **88**, 654–659.

— SELDEN, C. S., MAUPIN, P., 1984: Interaction of actin filaments with microtubules. J. Cell Biol. **99**, 33 s–37 s.

— WEIHING, R. R., 1974: Actin and myosin and cell movement. CRC Crit. Rev. Biochem. **2**, 1–65.

PONSTINGL, H., KRAUHS, E., LITTLE, M., KEMPF, T., 1981: Complete amino acid sequence of α-tubulin from porcine brain. Proc. Nat. Acad. Sci. **78**, 2757–2761.

PONTE, P., GUNNING, P., BLAU, H., KEDES, L., 1983: Human actin genes are single copy for α-skeletal and α-cardiac actin but multicopy for β- and γ-cytoskeletal genes: 3′ untranslated regions are isotype specific but are conserved in evolution. Mol. Cell. Biol. **3**, 1783–1791.

PORTER, K. R., 1964: Cell fine structure and biosynthesis of extracellular macromolecules. Biophys. J. **4**, 167–196.

— 1966: Cytoplasmic microtubules and their functions. In: Ciba Foundation Symposium on Principles of Biomolecular Organization. London: Churchill.

— 1984: The cytomatrix: a short history of its study. J. Cell Biol. **99**, 3 s–12 s.

— BECKERLE, M., MCNIVEN, M. A., 1983: The cytoplasmic matrix. In: Spatial Organization of Eukarytic Cells. Modern Cell Biology (MCINTOSH, J. R., ed.) Vol. 2, pp. 259–302. New York: Alan R. Liss.

— BOGGS, D. P., ANDERSON, K. L., 1982: The distribution of water in the cytoplasm. Proc. EMSA **40**, 4–7.

— CLAUDE, A., FULLAM, E. F., 1945: A study of tissue culture cells by electron microscopy. J. Exp. Med. **81**, 233–246.

— MCNIVEN, M. A., 1982: The cytoplast: a unit structure in chromatophores. Cell **29**, 23–32.

PORTER, R., COLLINS, G. M., 1983: Brush Border Membranes. Ciba Found. Symp., 95. London: Pitman.

POSTLETHWAITE, A. E., SEYER, J. M., KANG, A. H., 1978: Chemotactic attraction of human fibroblasts to type I, II, and III collagen and collagen-derived peptides. Proc. Nat. Acad. Sci. U.S.A. **75**, 871–875.

POULSEN, F. R., LOWY, J., 1983: Small-angle X-ray scattering from myosin heads in relaxed and rigor frog skeletal muscle. Nature **303**, 146–152.

PRATT, M. M., 1980: The identification of a dynein ATPase in unfertilized sea urchin eggs. Dev. Biol. **74**, 364–378.

— 1984: ATPases in mitotic spindles. Int. Rev. Cytol. **87**, 83–105.

— OTTER, T., SALMON, T. D., 1980: Dynein-like Mg^{++}-ATPase in mitotic spindles isolated from sea urchin embryos (*Strongylocentrotus droebaciensis*). J. Cell Biol. **86**, 738–745.

PRICE, M., LAZARIDES, E., 1983: Expression of intermediate filament-associated proteins paranemin and synemin in chicken development. J. Cell. Biol. **97**, 1860–1874.

PRUSS, R. M., MIRSKY, R., RAFF, M. C., THORPE, R., DOWDING, A. J., ANDERTON, B. H., 1981: All classes of intermediate filaments share a common antigenic determinant defined by a monoclonal antibody. Cell **27**, 419–427.

PRYZWANKSI, K. B., SCHLIWA, M., PORTER, K. R., 1983: Comparison of the three-dimensional organization of unextracted and triton-extracted human neutrophilic polymorphonuclear leucocytes. Eur. J. Cell Biol. **30**, 112–125.

PUDNEY, J., SINGER, R. H., 1979: Electron microscopic visualization of filamentous reticulum in whole cultured presumptive chick myoblasts. Am. J. Anat. **156**, 321–336.

— — 1980: Intracellular filament bundles in whole mounts of chick and human myoblasts extracted with triton X-100. Tissue & Cell **12**, 595–612.

PURICH, C. L., KRISTOFFERSON, D., 1984: Microtubule assembly: a review of progress, principles, and perspectives. Adv. Protein Chem. **36**, 133–212.

PYTELA, R., PIERSCHBACHER, M. D., RUOSLAHTI, E., 1985: Identification and isolation of a 140 kD cell surface glycoprotein with properties expected of a fibronectin receptor. Cell **40**, 191–198.

PYTELA, T. WICHE, G., 1980: High molecular weight polypeptides (270,000–340,000) from cultured cells are related to hog brain microtubule-associated proteins but copurify with intermediate filaments. Proc. Nat. Acad. Sci. U.S.A. **77**, 4808–4812.

QUAX, W., EGBERTS, W. V., HENDRIKS, W., QUAX-JEUKEN, Y., BLOEMENDAL, H., 1983: The structure of the vimentin gene. Cell **35**, 215–223.

— VAN DEN HEUVEL, R., EGBERTS, W. V., QUAX-JEUKEN, Y., BLOEMENDAL, H., 1984: Intermediate filament cDNAs from BHK-21 cells: demonstration of distinct genes for desmin and vimentin in all vertebrate classes. Proc. Nat. Acad. Sci. U.S.A. **81**, 5970–5974.

QUINLAN, R. A., FRANKE, W. W., 1982: Heteropolymer filaments of vimentin and desmin in vascular smooth muscle tissue and cultured baby hamster kidney cells demonstrated by chemical cross-linking. Proc. Nat. Acad. Sci. U.S.A. **79**, 3452–3456.

— — 1983: Molecular interactions in intermediate-sized filaments revealed by chemical cross-linking. Eur. J. Cell Biol. **132**, 477–484.

RADKE, K., CARTER, C., MOSS, P., DEHAZYA, P., SCHLIWA, M., MARTIN, G. S., 1983: Membrane association of a 36,000 dalton substrate for tyrosine phophorylation in chicken embryo fibroblasts transformed by avian sarcoma virus. J. Cell Biol. **97**, 1601–1611.

RAFF, E. C., 1979: The control of microtubule assembly *in vivo*. Int. Rev. Cytol. **59**, 1–96.

— 1984: Genetics of microtubule systems. J. Cell Biol. **99**, 1–10.

RAFTERY, M. A., HUNKAPILLER, M. W., STRADER, C. D., HOOD, L. E., 1980: Acetylcholine receptor: complex of homologous subnits. Science **208**, 1454–1457.

RAINE, C. S., GHETTI, G., SHELANSKI, L. M., 1971: On the association between tubules and mitochondria in axons. Brain Res. **34**, 389–393.

RAMAEKERS, F. C. S., DUNI, I., DODEMONT, H. J., BENEDETTI, E. L., BLOEMENDAHL, H., 1982: Lenticular intermediate-sized filaments: biosynthesis and interaction with plasma membrane. Proc. Nat. Acad. Sci. U.S.A. **79**, 3208–3212.

— HAAG, D., KANT, A., MOESKER, O., JAP, P. H. K., VOOIJS, G. P., 1983: Coexpression of keratin- and vimentin-type intermediate filaments in human metastatic carcinoma cells. Proc. Nat. Acad. Sci. U.S.A. **80**, 2618–2622.

— PUTS, J. J. G., KANT, A., MOESKER, O., JAP, P. H. K., VOOIJS, G. P., 1982: Use of antibodies to intermediate filaments in the characterization of human tumors. Cold Spring Harbor Symp. Quant. Biol. **46**, 331–339.

RANSCHT, B., MOSS, D. J., THOMAS, C., 1984: A neuronal surface glycoprotein associated with the cytoskeleton. J. Cell Biol. **99**, 1803–1813.

RANVIER, L., 1875: Recherches sur les éléments du sang. Arch. Physiol. **2**, 1–15.

— 1882: Sur la structure des cellules du corps muqueux de Malpighi. Compt. Rend. Acad. Sci. **95**.

RASH, J. E., BIESELE, J. J., GEY, G. O., 1970: Three classes of filaments in cardiac differentiation. J. Ultrastruct. Res. 33, 408–435.

— SHAY, J. W., BIESELE, J. J., 1970: Preliminary biochemical investigations of the intermediate filaments. J. Ultrastruct. Res. 33, 399–409.

RAYBIN, D., FLAVIN, M., 1975: An enzyme tyrosylating α-tubulin and its role in microtubule assembly. Biochem. Biophys. Res. Commun. 65, 1088–1095.

— FLAVIN, M., 1977: Modification of tubulin by tyrosylation in cells and extracts and its effect on assembly in vitro. J. Cell Biol. 73, 492–504.

REBHUN, L. I., JEMIOLO, D., KELLER, T., BURGESS, W., KRETSINGER, R., 1980: Calcium, calmodulin, and control of assembly of brain and spindle microtubules. In: Microtubules and Microtubule Inhibitors (DE BRABANDER, M., DE MEY, J., eds.), pp. 242–252. Amsterdam: Elsevier.

REICHSTEIN, E., KORN, E. D., 1979: Acanthamoeba profilin. A protein of low molecular weight from Acanthamoeba castellani that inhibits actin nucleation. J. Biol. Chem. 254, 6174–6179.

REISLER, E., SMITH, C., SEEGAN, G., 1980: Myosin minifilaments. J. Mol. Biol. 143, 129–145.

REMAK, F., 1838: Über die Struktur des Nervensystems. Frorieps Neue Notizen 6.

— — 1844: Neurologische Erläuterungen. Müller's Archiv.

RICH, S. A., ESTES, 1976: Detection of conformational changes in actin by proteolytic digestion: evidence for a new monomeric species. J. Mol. Biol. 104, 777–792.

RIMME, A. H., 1981: The ultrastructure of the erythrocyte cytoskeleton at neutral and reduced pH. J. Ultrastruct. Res. 77, 199–209.

RIS, H., 1980: The cytoplasmic "microtrabecular lattice"–reality or artifact. EMSA Proc. 38, 812–813.

ROBBINS, E., GONATAS, N. K., 1964: Histochemical and ultrastructural studies on HeLa cell cultures exposed to spindle inhibitors, with special reference to the interphase cell. J. Histochem. Cytochem. 12, 704–711.

ROBBINS, E. G., JENTZSCH, G., MICALI, A., 1968: The centriole cycle in synchronized HeLa cells. J. Cell Biol. 36, 329–338.

ROBERTS, K., 1974: Cytoplasmic microtubules and their functions. Progr. Bioph. Mol. Biol. 28, 373–420.

— HYAMS, J. S., 1979: Microtubules. London: Academic Press.

ROBINSON, D. G., QUADER, H., 1982: The microtubule-microfibril syndrome. In: The cytoskeleton in plant growth and development (LLOYD, C. W., ed.), pp. 109–126. London: Academic Press.

RODEWALD, R., NEWMAN, S. B., KARNOVSKY, M. J., 1976: Contraction of isolated brush borders from intestinal epithelium. J. Cell Biol. 70, 541–554.

ROGALSKI, A. A., BERGMANN, J. E., SINGER, S. J., 1984: Effect of microtubule assembly status on the intracellular processing and surface expression of an integral protein of the plasma membrane. J. Cell Biol. 99, 1101–1109.

— SINGER, S. J., 1984a: Associations of elements of the Golgi apparatus with microtubules. J. Cell Biol. 99, 1092–1100.

— — 1984b: A 120 kD integral membrane protein associated with membrane-microfilament attachment regions in chick fibroblasts and gizzard smooth muscle cells. J. Cell Biol. 99, 296a.

ROHRSCHNEIDER, L., ROSOK, M. J., 1983: Transformation parameters and pp60src localization in cells infected with partial transformation mutants of Rous sarcoma virus. Mol. Cell Biol. **3**, 731–746.

ROOP, D. R., HAWLEY-NELSON, P., CHENG, C. K., YUSPA, S. H., 1983: Keratin gene expression in mouse epidermis and cultured epidermal cells. Proc. Nat. Acad. Sci. U.S.A. **80**, 716–720.

ROSENBERG, S., STRACHER, A., BURRIDGE, K., 1981: Isolation and characterization of a calciumsensitive α-actinin-like protein from human platelet cytoskeletons. J. Biol. Chem. **256**, 12986–12991.

ROSENFELD, A. C., ZACKROFF, R. V., WEISENBERG, R. C., 1976: Magnesium stimulation of calcium binding to tubulin and calcium induced depolymerization of microtubules. FEBS Lett. **65**, 144–147.

ROSLANSKY, P. F., CORNELL-BELL, A., RICE, R. V., ADELMAN, W. J., 1980: Polypeptide composition of squid neurofilaments. Proc. Nat. Acad. Sci. U.S.A. **77**, 404–408.

ROSOK, M. J., ROHRSCHNEIDER, L., 1983: Increased phosphorylation of vinculin does not occur during the release of stress fibers before mitosis in animal cells. Mol. Cell Biol. **3**, 475–479.

ROTH, L. E., 1958: A filamentous component of protozoal fibrillar systems. J. Ultrastruct. Res. **1**, 223–234.

— PIHLAJA, D. J., SHIGENAKA, Y., 1970: Microtubules in the heliozoan axopodium. I. The gradion hyptohesis of allosterism in structural proteins. J. Ultrastruct. Res. **30**, 7–37.

ROUAYRENC, J. F., TRAVERS, F., 1981: The first step in the polymerization of actin. Eur. J. Biochem. **116**, 73–77.

ROUTLEDGE, L. M., AMOS, W. B., YEW, F. F., WEIS-FOGH, T., 1976: New calcium-binding contractile proteins. In: Cell Motility (GOLDMAN, R., POLLARD, T. D., ROSENBAUM, J. L., eds.), pp. 93–113. Cold Spring Harbor: Cold Spring Harbor Laboratory.

RUEGER, D. C., HUSTON, J. S., DAHL, D., BIGNAMI, A., 1979: formation of 100 nm filaments from purified glial acidic protein *in vitro*. J. Mol. Biol. **135**, 53–68.

RUNGE, M. S., LAUE, T. M., YPHANTIS, D. A., LIFSICS, M. R., SAITO, A., ALTIN, M., 1981: ATP-induced formation of an associated complex between microtubules and neurofilaments. Proc. Nat. Acad. Sci. U.S.A. **78**, 1431–1435.

— WILLIAMS, R. C., 1982: Formation of an ATP-dependent microtubule-neurofilament complex *in vitro*. Cold Spring Harbor Symp. Quant. Biol. **46**. 483–492.

RUSKA, H., 1939: Übermikroskopische Bilder zu Strukturproblemen. Zool. Anz., Suppl. **12**, 295–302.

RUSSEL, D. G., GULL, K., 1984: Flagellar regeneration of the trypanosome *Crithidia fasciculata* involves post-translational modification of cytoplasmic alpha tubulin. Mol. Cell. Biol. **41**, 1182–1185.

SABATINI, D. D., BENSCH, K., BARRNETT, R. J., 1963: Cytochemistry and electron microscopy. J. Cell Biol. **17**, 19–58.

SAGERA, J., NAGATA, K., ISHIKAWA, Y., 1982: A cofactor protein required for actin activation of myosin Mg-ATPase activity in leukemic myeloblasts. J. Biochem. **92**, 1845–1851.

SAITO, K., HAMA, K., 1982: Structural diversity of microtubules in the supporting cells of the sensory epithelium of guinea pig organ of Corti. J. Electron Microsc. **31**, 278–281.

SAKAI, H., MOHRI, H., BORISY, G. G., 1982: Biological Functions of Microtubules and Related Structures. Tokyo: Academic Press.

SAKAKIBARA, K., MOMOI, T., UCHIDA, T., NAGAI, Y., 1981: Evidence for association of glycosphingolipid with a colchicine-sensitive microtubule-like cytoskeletal structure of cultured cells. Nature **293**, 76–79.

SALE, W. S., SATIR, P., 1977: Direction of active sliding of microtubules in *Tetrahymena* cilia. Proc. Nat. Acad. Sci. U.S.A. **74**, 2045–2049.

SALISBURY, J. L., BARON, A., SUREK, B., MELKONIAN, M., 1984: Striated flagellar roots: isolation and partial characterization of a calcium-modulated contractile organelle. J. Cell Biol. **99**, 962–970.

— CONDEELIS, J. S., MAIHLE, N. J., SATIR, P., 1981: Calmodulin localization during capping and receptor-mediated endocytosis. Nature **294**, 163–166.

— — SATIR, P., 1980: Role of coated vesicles, microfilaments, and calmodulin in receptor-mediated endocytosis by cultured B lymphoblastoid cells. J. Cell Biol. **87**, 132–141.

— FLOYD, G. L., 1978: Calcium-induced contraction of the rhizoplast of a quadriflagellate green alga. Science **202**, 975–976.

SALMON, E. D., LESLIE, R. J., SAXTON, W. M., KAROW, M. L., McINTOSH, J. R., 1984: Spindle microtubule dynamics in sea urchin embryos: analysis using a fluorescein-labeled tubulin and measurements of fluorescence redistribution after laser photobleaching. J. Cell Biol. **99**, 2165–2174.

SALTARELLI, D., PANTALONI, D., 1983: Copolymerization of tubulin-colchicine complex and unligated tubulin in a nonmicrotubular polymer. Biochemistry **22**, 4607–4614.

SANCHEZ, F., NATZLE, J. E., CLEVELAND, D. W., KIRSCHNER, M. W., McCARTHY, B. J., 1980: A dispersed multigene family encoding tubulin in *Drosophila melanogaster*. Cell **22**, 845–854.

SANDOVAL, I. V., COLACO, A. L. S., LAZARIDES, E., 1983: Purification of the intermediate filament-associated protein, synemin, from chicken smooth muscle. J. Biol. Chem. **258**, 2568–2576.

SANGER, J. M., SANGER, J. W., 1980: Banding and polarity of actin filaments in interphase and cleaving cells. J. Cell Biol. **86**, 568–575.

SANGER, J. W., SANGER, J. M., KREIS, T., JOCKUSCH, B. M., 1980: Reversible translocation of cytoplasmic actin into the nucleus caused by dimethylsulfoxide. Proc. Nat. Acad. Sci. U.S.A. **77**, 5268–5272.

SATIR, P., 1965: Studies on cilia. II. Examination of the distal region of the ciliary shaft and the role of the filaments in motility. J. Cell Biol. **26**, 805–834.

— WAIS-STEIDER, J., LEBDUSKA, S., NASR, A., AVOLIO, J., 1981: The mechanochemical cycle of the dynein arm. Cell Motility **1**, 303–327.

SATO, C., NISHIZAWA, K., NAKAMURA, H., KOMAGOE, Y., SHIMADA, K., UEDA, R., SUZUKI, S., 1983: Monoclonal antibody against microtubule associated protein-

1 produces immunofluorescent spots in the nucleus and centrosome of cultured mammalian cells. Cell Struct. Funct. **8**, 245–254.

SATTILARO, R. F., DENTLER, W. L., LeCLUYSE, E. L., 1981: Microtubule-associated proteins (MAPs) and the organization of actin filaments *in vitro*. J. Cell Biol. **90**, 467–473.

SAUK, J. J., KRUMWEIDE, M., COCKING-JOHNSON, D., WHITE, J. G., 1984: Reconstruction of cytokeratin filaments *in vitro*: Further evidence for the role of nonhelical peptides in filament assembly. J. Cell Biol. **99**, 1590–1597.

SCHEELE, R. B., BERGEN, L. G., BORISY, G. G., 1982: Control of the structural fidelity of microtubules by initiation sites. J. Mol. Biol. **154**, 485–500.

SCHIFF, P. B., FANT, J., HOROWITZ, S. B., 1979: Promotion of microtubule assembly *in vitro* by Taxol. Nature **277**, 665–667.

— HORWITZ, S. B., 1980: Taxol stabilizes microtubules in mouse fibroblast cells. Proc. Nat. Acad. Sci. U.S.A. **77**, 1561–1565.

SCHLAEPFER, W. W., FREEMAN, L. A., 1978: Neurofilament protein of rat peripheral nerve and spinal cord. J. Cell Biol. **78**, 653–662.

— HASLER, M. B., 1979: Characterization of the calcium-induced disruption of neurofilaments in rat peripheral nerve. Brain Res. **168**, 299–309.

— LYNCH, R. G., 1977: Immunofluorescence studies of neurofilaments in rat and human peripheral and central nervous system. J. Cell Biol. **74**, 241–250.

— MICKO, S., 1979: Calcium-dependent alterations of neurofilament proteins of rat peripheral nerve. J. Neurochem. **32**, 211–219.

SCHLESSINGER, J., SHECHTER, Y., WILLINGHAM, M. C., PASTAN, I., 1978: Direct visualization of binding, aggregation, and internalization of insulin and epidermal growth factor on living fibroblastic cells. Proc. Nat. Acad. Sci. U.S.A. **75**, 2659–2663.

SCHLIWA, M., 1975: Cytoarchitecture of surface layer cells of the teleost epidermis. J. Ultrastruct. Res. **52**, 377–386.

— 1976: The role of divalent cations in the regulation of microtubule assembly. J. Cell Biol. **70**, 527–540.

— 1981: Proteins associated with cytoplasmic actin. Cell **25**, 587–591.

— 1982: Action of cytochalasin D on cytoskeletal networks. J. Cell Biol. **92**, 79–91.

— 1984: Mechanisms of intracellular organelle transport. In: Cell and Muscle Motility, Vol. 5 (SHAY, J. W., ed.), pp. 1–82. New York: Plenum.

— EUTENEUER, U., BULINSKI, J. C., IZANT, J. G., 1981: Calcium lability of cytoplasmic microtubules and its modulation by microtubule–associated proteins. Proc. Nat. Acad. Sci. U.S.A. **78**, 1037–1041.

— — HERZOG, W., WEBER, K., 1979: Evidence for rapid structural and functional changes of the melanophore microtubule-organizing center upon pigment movements. J. Cell. Biol. **83**, 623–632.

— NAKAMURA, T., PORTER, K. R., EUTENEUER, U., 1984: A tumor promoter induces rapid and coordinated reorganization of actin and vinculin in cultured cells. J. Cell Biol. **99**, 1045–1059.

— PRYZWANSKY, K. B., VAN BLERKOM, J., 1982: Implications of cytoskeletal interactions for cellular architecture and behaviour. Phil. Trans. R. Soc. Lond. **B 299**, 199–205.

— van Blerkom, J., 1981: Structural interaction of cytoskeletal components. J. Cell Biol. **90**, 222–235.

Schloss, J. A., Goldman, R. D., 1979: Isolation of a high molecular weight actin-binding protein from baby hamster kidney (BHK-21) cells. Proc. Nat. Acad. Sci. U.S.A. **76**, 4484–4488.

— — 1980: Microfilaments and tropomyosin of cultured mammalian cells: isolation and characterization. J. Cell Biol. **87**, 633–642.

Schmid, E., Osborn, M., Rungger-Brandle, E., Gabbiani, G., Weber, K., Franke, W. W., 1982: Distribution of vimentin and desmin filaments in smooth muscle tissue of mammalian and avian aorta. Exp. Cell Res. **137**, 329–340.

Schmidt, W. J., 1937: Die Doppelbrechung von Karyoplasma, Zytoplasma und Metaplasma. (Protoplasma Monographien 11.) Berlin: Bornträger.

Schnapp, B. J., Reese, T. S., 1982: Cytoplasmic structure in rapid frozen axons. J. Cell Biol. **94**, 667–679.

Schneider, K. C., 1908: Histologisches Praktikum der Niere. Jena: G. Fischer.

Schnitzer, J., Franke, W. W., Schachner, M., 1981: Immunocytochemical demonstration of vimentin in astrocytes and ependymal cells of developing and adult mouse nervous system. J. Cell Biol. **90**, 435–447.

Scholey, J. M., Neighbors, B., McIntosh, J. R., Salmon, E. D., 1984: Isolation of microtubules and a dynein-like MgATPase from unfertilized sea urchin eggs. J. Biol. Chem. **259**, 6516–6525.

— Taylor, K. A., Kendrick-Jones, J., 1980: Regulation of non-muscle myosin assembly by calmodulin-dependent light chain kinase. Nature **287**, 233–235.

Schreiner, G. F., Fujiwara, K., Pollard, T. D., Unanue, E., 1977: Redistribution of myosin accompanying capping of surface immunoglobulin. J. Exp. Med. **145**, 1393–1398.

— Unanue, E. R., 1976: Membrane and cytoplasmic changes in lymphocytes induced by ligand-surface immunoglobulin interaction. Adv. Immunol. **24**, 37–165.

Schroeder, T. E., 1982: Interrelations between the cell surface and the cytoskeleton in cleaving sea urchin eggs. Cell Surface Rev. **7**, 170–216.

Schubert, D., Dorst, D., 1979: Self-association of band 3 protein from human erythrocyte membranes in aqueous solutions. Hoppe-Seyler's Z. Physiol. Chem. **360**, 1605–1618.

Schulmann, H., 1984: Phosphorylation of microtubule-associated proteins by a Ca^{++}/Calmodulin-dependent protein kinase. J. Cell Biol. **99**, 11–19.

Schultheiss, R., Mandelkow, E., 1983: Three-dimensional reconstruction of tubulin sheets and re-investigation of microtubule surface lattice. J. Mol. Biol. **170**, 471–496.

Schultze, M., 1868: Observationes de structures cellularum fibrarumque nervorum. Bonn: A. Marcus.

Sefton, B. M., Hunter, T., Ball, E. H., Singer, S. J., 1981: Vinculin: a cytoskeletal target of the transforming protein of Rous sarcoma virus. Cell **24**, 165–174.

Selden, L. A., Gershman, L. C., Estes, J. E., 1980: A proposed mechanism of action of cytochalasin D on muscle actin. Biochem. Biophys. Res. Comm. **95**, 1854–1860.

SELDEN, S. C., POLLARD, T. D., 1983: Phosphorylation of microtubule-associated proteins regulates their interaction with actin filaments. J. Biol. Chem. **258**, 7064–7074.

SHARP, G., OSBORN, M., WEBER, K., 1982 a: Occurrence of two different intermediate filament proteins in the same filament *in situ* within a human glioma cell line. Exp. Cell Res. **141**, 385–395.

SHARP, G. A., SHAW, G., WEBER, K., 1982 b: Immunoelectronmicroscopical localization of the three neurofilaments triplet proteins along neurofilaments of cultured dorsal root ganglion neurons. Exp. Cell Res. **137**, 403–413.

SHARPE, A. H., CHEN, L. B., MURPHY, J. R., FIELDS, B. N., 1980: Specific disruption of vimentin filament organization in monkey kidney CV-1 cells by diphtheria toxin, oxotoxin, and cycloheximide. Proc. Nat. Acad. Sci. U.S.A. **77**, 7267–7271.

SHAW, G., DEBUS, E., WEBER, K., 1984: The immunological relatedness of neurofilament proteins of higher vertebrates. Eur. J. Cell Biol. **34**, 130–136.

— WEBER, K., 1981: The distribution of the neurofilament triplet proteins within individual neurones. Exp. Cell Res. **136**, 119–125.

SHAY, J. W., PORTER, K. R., PRESCOTT, D. M., 1974: The surface morphology and fine structure of CHO (Chinese Hamster Ovary) cells following enucleation. Proc. Nat. Acad. Sci. U.S.A. **71**, 3059–3063.

SHECKET, G., LASEK, R. J., 1980: Preparation of neurofilament protein from guinea pig peripheral nerve and spinal cord. J. Neurochem. **35**, 1335–1344.

SHEETS, M. P., PAINTER, R. R., SINGER, S. J., 1976: Relationships of the spectrin complex of human erythrocyte membranes to the actomyosins of muscle cells. Biochemistry **15**, 4486–4492.

SHEIR-NEISS, G., LAI, M. H., MORRIS, N. R., 1978: Identification of a gene for β-tubulin in *Aspergillus nidulans*. Cell **15**, 638–647.

SHEN, B. W., JOSEPHS, R., STECK, T. L., 1984: Ultrastructure of unit fragments of the skeleton of the human erythrocyte membrane. J. Cell Biol. **99**, 810–821.

SHERLINE, P., LEE, Y. C., JACOBS, L. S., 1977: Binding of microtubules to pituitary secretory granules and secretory granule membranes. J. Cell Biol. **72**, 380–389.

— MASCARDO, R. N., 1982: Epidermal growth factor induces rapid centrosomal separation in HeLa and 3T3 cells. J. Cell Biol. **93**, 507–511.

— SHIAVONE, K., 1978: High molecular weight MAPs are part of the mitotic spindle. J. Cell Biol. **77**, R 9–R 12.

SHETERLINE, P., 1983: Mechanisms of cell motility. Molecular Aspects of Contractility. London: Academic Press.

— HOPKINS, C. R., 1981: Transmembrane linkage between surface glycoproteins and components of the cytoplasm in neutrophil leucocytes. J. Cell Biol. **90**, 743–754.

SHIFFER, K. A., GOODMAN, S. R., 1984: Protein 4.1: Its association with the human erythrocyte membrane Proc. Nat. Acad. Sci. U.S.A. **81**, 4404–4408.

SHOTTON, D. M., BURKE, B. E., BRANTON, D., 1979: Molecular structure of human erythrocyte spectrin: biophysical and electron microscopic studies. J. Mol. Biol. **131**, 303–329.

SHRIVER, K., ROHRSCHNEIDER, L., 1981: Organization of pp60src and selected cytoskeletal proteins within adhesion plaques and junctions of Rous sarcoma virus-transformed cells. J. Cell Biol. **89**, 525–535.

SIGEL, P., PETTE, D., 1969: Intracellular localization of glycogenolytic and glycolytic enzymes in white and red rabbit skeletal muscles. J. Histochem. Cytochem. **17**, 225–237.

SILFLOW, C. D., ROSENBAUM, J. L., 1981: Multiple alpha- und beta-tubulin genes in *Chlamydomonas* and regulation of tubulin mRNA levels after deflagellation. Cell **24**, 81–88.

SIMAN, R., BAUDRY, M., LYNCH, G., 1984: Brain fodrin: substrate for calpain I, an endogenous calcium-activated protease. Proc. Nat. Acad. Sci. U.S.A. **81**, 3572–3576.

SINGER, I. I., 1982: Association of fibronectin and vinculin with focal contacts and stress fibers in stationary hamster fibroblasts. J. Cell Biol. **92**, 398–408.

— KAWKA, D. W., KAZAZIS, D. M., CLARK, R. A. F., 1984: *In vivo* co-distribution of fibronectin and actin fibers in granulation tissue: immunofluorescence and electron microscope studies of the fibronexus at the myofibroblast surface. J. Cell Biol. **98**, 2091–2106.

SINGER, S. J., BALL, E. H., GEIGER, B., CHEN, W.-T., 1982: Immunolabeling studies of cytoskeletal associations in cultured cells. Cold Spring Harbor Symp. Quant. Biol. **46**, 303–316.

— NICHOLSON, G. L., 1972: The fluid mosaic model of the structure of cell membranes. Science **175**, 720–731.

SITU, R., LEE, E. C., McCOY, J. P., VARANI, J., 1984: Stimulation of murine tumor cell mobility by laminin. J. Cell Sci. **70**, 167–176.

SKERROW, C. J., MATOLTSY, A. G., 1974: Chemical characterization of isolated epidermal desmosomes. J. Cell Biol. **63**, 524–530.

SLAUTERBACK, D. B., 1963: Cytoplasmic microtubules. I. Hydra. J. Cell Biol. **18**, 367–388.

SLOBODA, R. D., ROSENBAUM, J. L., 1979: Decoration and stabilization of intact, smooth-walled microtubules with microtubule-associated protein. Biochemistry **18**, 48–55.

— Rudolph, S. A., ROSENBAUM, J. L., GREENGARD, P., 1975: Cyclic AMP-dependent endogenous phosphorylation of a microtubule-associated protein. Proc. Nat. Acad. Sci. U.S.A. **72**, 177–181.

SMALL, J. V., 1981: Organization of actin in the leading edge of cultured cells. J. Cell Biol. **91**, 695–705.

— CELIS, J. E., 1978: Direct visualization of the 10 nm (100 A) filament network in whole and enucleated cultured cells. J. Cell Sci. **31**, 393–409.

— ISENBERG, G. T., CELIS, J. E., 1978: Polarity of actin at the leading edge of cultured cells. Nature **272**, 638–639.

— SOBIESZEK, A., 1977: Studies on the function and composition of the 10 nm (100 A) filaments of vertebrate smooth muscle. J. Cell Sci. **23**, 243–268.

— — 1980: The contractile apparatus of smooth muscle. Int. Rev. Cytol. **64**, 241–306.

SMITH, D. S., 1971: On the significance of cross-bridges between microtubules and synaptic vesicles. Phil. Trans. R. Soc. London **B 261**, 365–405.

SMITH, D. S., JARLFORS, U., CAYER, M. L., 1977: Structural crossbridges between micro-tubules and mitochondria in central axons of an insect (*Periplaneta americana*). J. Cell Sci. **27**, 235–272.

SMITH, P. R., FOWLER, W. E., AEBI, U., 1984: Towards an alignment of the actin molecule within the actin filament. Ultramicroscopy **13**, 113–124.

— FOWLER, W. E., POLLARD, T. D., AEBI, U., 1983: Structure of the actin molecule determined from electron micrographs of crystalline actin sheets with a tentative alignment of the molecule in the actin filament. J. Mol. Biol. **167**, 641–660.

SNYDER, J. A., HAMILTON, B. T., MULLINS, J. M., 1982: Loss of mitotic centrosomal microtubule initiation capacity at the metaphase-anaphase transition. Eur. J. Cell Biol. **27**, 191–199.

— MCINTOSH, J. R., 1975: Initiation and growth of microtubules from mitotic centers in lysed mammalian cells. J. Cell Biol. **67**, 744–760.

SOBIESZEK, A., 1977: Ca^{++}-linked phosphorylation of a light chain of vertebrate smooth muscle myosin. Eur. J. Biochem. **73**, 477–485.

— SMALL, J. V., 1981: Effect of muscle and nonmuscle tropomyosins in recon-stituted skeletal muscle actomyosin. Eur. J. Biochem. **118**, 533–539.

SOBUE, K., FUJITA, M., MURAMOTO, Y., KAKUICHI, S., 1981: The calmodulin-binding protein in microtubules is tau factor. FEBS Lett. **132**, 137–143.

— KANDA, K., YAMAGAMI, K., KAKIUCHI, S., 1982: Ca^{++}- and calmodulin-dependent phosphorylation of calspectin (spectrin-like calmodulin-binding protein; fodrin) by protein kinase systems in synaptosomal cytosol and membranes. Biomed. Res. **3**, 561–570.

— MURAMOTO, Y., FUJITA, M., KAKIUCHI, S., 1981 a: Purification of a calmodulin-binding protein from chicken gizzard that interacts with F-actin. Proc. Nat. Acad. Sci. U.S.A. **78**, 5652–5655.

— — — — 1981 b: Calmodulin-binding protein of erythrocyte cytoskeleton. Biochem. Biophys. Res. Comm. **100**, 1063–1070.

SOIFER, D., 1975: The Biology of Cytoplasmic Microtubules. Ann. N.Y. Acad. Sci. **253**.

— MACK, K., 1984: Microtubules in the nervous system. Handbook of Neu-rochemistry **7**, 245–280.

SOLOMON, F., 1977: Binding sites for calcium on tubulin. Biochemistry **16**, 358–363.

— MAGENDANTZ, M., SALZMAN, A., 1979: Identification with cellular microtubules of one of the co-assembling microtubule-associated proteins. Cell **18**, 431–438.

SOMLYO, A. V., 1980: Ultrastructure of vascular smooth muscle. In: Handbook of Physiology, Section 2, The Cardiovascular System, Vol. II, Vascular Smooth Muscle (BOHR, D. F., SOMLYO, A. V., SPARKS, H. V., eds.), pp. 33–67. Bethesda: American Physiology Society.

SORIANO, P., SZABO, P., BERNARDI, G., 1982: The scattered distribution of actin genes in the mouse and human genomes. EMBO J. **1**, 579–583.

SORRANO, T., BELL, E., 1982: Cytostructural dynamics of spreading and translocat-ing cells. J. Cell Biol. **95**, 127–136.

SOUTHWICK, F. S., STOSSEL, T. P., 1981: Isolation of an inhibitor of actin polymerization from human polymorphonuclear leukocytes. J. Biol. Chem. **256**, 3030–3036.

— TATSUMI, N., STOSSEL, T. P., 1982: Acumentin, an actin-modulating protein of rabbit pulmonary macrophages. Biochemistry 21, 6321–6326.

SPEICHER, D. W., DAVIS, G., MARCHESI, V. T., 1983 b: Structure of human erythrocyte spectrin. II. The sequence of the α-1 domain. J. Biol. Chem. 258, 14938–14947.

— — YURCHENCO, P. D., MARCHESI, V. T., 1983 a: Structure of human erythrocyte spectrin. I. Isolation of the α-1 domain and its cyanogen bromide peptides. J. Biol. Chem. 258, 14931–14937.

— MARCHESI, V. T., 1984: Erythrocyte spectrin is comprised of many homologous triple helical segments. Nature 311, 177–179.

SPIEGEL, J. E., SCHULTZ BEARDSLEY, D., SOUTHWICK, F. S., LUX, S. E., 1984: An analogue of the erythroid membrane skeletal protein 4.1 in nonerythroid cells. J. Cell Biol. 99, 886–893.

SPOONER, B. S., ASH, J. F., WRENN, J. T., FRATER, R. B., WESSELLS, N. K., 1973: Heavy meromyosin binding to microfilaments involved in cell and morphogenetic movements. Tissue Cell 5, 37–46.

SPRUILL, W. A., STEINER, A. L., TRES, L. T., KIERSZENBAUM, A. L., 1983 a: Follicle-stimulating hormone-dependent phosphorylation of vimentin in cultures of rat Sertoli cells. Proc. Nat. Acad. Sci. U.S.A. 80, 993–997.

— ZYSK, J. R., TRES, L. T., KIERSZENBAUM, A. L., 1983 b: Calcium/calmodulin-dependent phosphorylation of vimentin in rat Sertoli cells. Proc. Nat. Acad. Sci. U.S.A. 80, 760–764.

SPUDICH, J. A., 1973: Effects of cytochalasin B on actin filaments. Cold Spring Harbor Symp. Quant. Biol. 37, 585–598.

— 1974: Biochemical and structural studies of actomyosin-like proteins from nonmuscle cells. II. Purification, properties, and membrane association of actin from amoeba of Dictyostelium discoideum. J. Biol. Chem. 249, 6013–6020.

— AMOS, L. A., 1979: Structure of actin filament bundles from microvilli of sea urchin eggs. J. Mol. Biol. 129, 319–331.

SQUIRE, J., 1981: The structural basis of muscle contraction. New York: Plenum.

STAEHELIN, A., 1974: Structure and function of intracellular junctions. Int. Rev. Cytol. 39, 191–283.

STAUFENBIEL, M., DEPPERT, W., 1982: Intermediate filament systems are collapsed onto the nuclear surface after isolation of nuclei from tissue culture cells. Exp. Cell Res. 138, 207–214.

STEARNS, M. E., BROWN, D. L., 1979: Purification of cytoplasmic tubulin and microtubule organizing center proteins functioning in microtubule initiation from the alga Polytomella. Proc. Nat. Acad. Sci. U.S.A. 76, 5745–5749.

STECK, T. L., 1972: Cross-linking the major proteins of the isolated erythrocyte membrane. J. Mol. Biol. 66, 295–305.

— 1974: Organization of proteins in the human red blood cell membrane. J. Cell Biol. 62, 1–19.

— 1978: The band 3 protein of the human red cell membrane: a review. J. Supramol. Struct. 8, 311–324.

— KOZIARZ, J. J., SINGH, M. K., REDDY, G., KOEHLER, H., 1978: Preparation and analysis of seven major, topographically defined fragments of band 3, the predominant transmembrane polypeptide of human erythrocyte membranes. Biochemistry **17**, 1216–1222.

STEINERT, P. M., 1978: Structure of the three-chain unit of the bovine epidermal keratin filament. J. Mol. Biol. **123**, 49–70.

— CANTIERI, J. S., TELLER, D. C., LONSDALE-ECCLES, J. D., DALE, B. A., 1981: Characterization of a class of cationic proteins that specifically interact with intermediate filaments. Proc. Nat. Acad. Sci. U.S.A. **78**, 4097–4101.

— IDLER, W. W., 1975: The polypeptide composition of bovine epidermal α-keratin. Biochem. **151**, 603–614.

— — AYNARDI-WHITMAN, M., ZACKROFF, R., GOLDMAN, R. D., 1982 b: Heterogeneity of intermediate filaments assembled *in vitro*. Cold Spring Harbor Symp. Quant. Biol. **46**, 465–474.

— — GOLDMAN, R. D., 1980: Intermediate filaments of baby hamster (BHK-21) cells and bovine epidermal keratinocytes have similar ultrastructures and subunit domain structures. Proc. Nat. Acad. Sci. U.S.A. **77**, 4534–4538.

— — ZIMMERMAN, S. B., 1976: Self-assembly of bovine epidermal keratin filaments *in vitro*. J. Mol. Biol. **108**, 547–567.

— RICE, R. H., ROOP, D. R., TRUS, B. L., STEVEN, A. C., 1983: Complete amino acid sequence of a mouse epidermal keratin subunit and implications for the structure of intermediate filaments. Nature **302**, 794–800.

— WANTZ, M. L., IDLER, W. W., 1982 a: O-phosphoserine content of intermediate filament subunits. Biochemistry **21**, 177–183.

STENDAHL, O. I., STOSSEL, T. P., 1980: Actin-binding protein amplifies actomyosin contraction, and gelsolin confers calcium control on the direction of the contraction. Biochem. biophys. Res. Comm. **92**, 675–681.

STEPHENS, N. L., 1984: Smooth muscle contraction. New York: Marcel Dekker.

STEPHENS, R. E., 1975 a: Structural chemistry of the axoneme: Evidence for chemically and functionally unique tubulin dimers of outer fibers. In: Molecules and Cell Movement (INOUE, S., STEPHENS, R. E., eds.), pp. 181–204. New York: Raven Press.

— 1975 b: The basal apparatus. Mass isolation from the molluscan ciliated gill epithelium and preliminary characterization of striated rootlets. J. Cell Biol. **64**, 408–424.

— 1978: Primary structural differences among tubulin subunits from flagella, cilia, and the cytoplasm. Biochemistry **14**, 2882–2891.

— EDDS, K. T., 1976: Microtubules: structure, chemistry, and function. Physiol. Rev. **56**, 709–777.

STERNBERGER, L. A., STERNBERGER, N. H., 1983: Monoclonal antibodies distinguish phosphorylated and nonphosphorylated forms of neurofilaments *in situ*. Proc. Nat. Acad. Sci. U.S.A. **80**, 6126–6130.

STERNLICHT, H., RINGEL, I., 1979: Colchicine inhibition of microtubule assembly via copolymer formation. J. Biol. Chem. **254**, 10540–10548.

STEVEN, A. C., WALL, J., HAINFELD, J., STEINERT, P. M., 1982: Structure of fibroblastic intermediate filaments: analysis by scanning transmission electron microscopy. Proc. Nat. Acad. Sci. U.S.A. **79**, 3101–3105.

STIDWILL, R. P., WYSOLMERSKI, T., BURGESS, D. R., 1984: The brush border cytoskeleton is not static: *in vivo* turnover of proteins. J. Cell Biol. **98**, 641–645.

STOPAK, D., HARRIS, A. K., 1982: Connective tissue morphogenesis by fibroblast traction. I Tissue culture observations. Dev. Biol. **90**, 383–398.

STOSSEL, T. P., 1982: The structure of cortical cytoplasm. Phil. Trans. R. Soc. Lond. **B 299**, 275–289.

— 1984: Contribution of actin to the structure of the cytoplasmic matrix. J. Cell Biol. **99**, 15 s–21 s.

— HARTWIG, J. H., 1976: Interactions of actin, myosin, and a new actin-binding protein of rabbit pulmonary macrophages. II. Role in cytoplasmic movement and phagocytosis. J. Cell Biol. **68**, 602–619.

STRAUB, F. B., 1942: Actin. Stud. Szeged. **2**, 3–15.

STREULI, C., PATEL, B., CRITCHLEY, D., 1981: The cholera toxin receptor ganglioside GM 1 remains associated with triton X-100 cytoskeletons of Balb/c 3T3 cells. Exp. Cell Res. **136**, 247–254.

SUCK, D., KABSCH, W., MANNHERZ, H. G., 1981: Three-dimensional structure of skeletal muscle actin and bovine pancreatic deoxyribonuclease I at 6 Å resolution. Proc. Nat. Acad. Sci. U.S.A. **78**, 4319–4323.

SUGINO, H., SAKABE, N., SAKABE, K., HATANO, S., OOSAWA, F., MIKAWA, T., EBASHI, S., 1979: Crystallization and preliminary crystallographic data of chicken gizzard G-actin-DNAse I complex. J. Biochem. **86**, 257–260.

SUGRUE, S. P., HAY, E. D., 1981: Response of basal epithelial cell surface and cytoskeleton to solubilized extracellular matrix molecules. J. Cell Biol. **91**, 45–54.

SULLIVAN, K. F., CLEVELAND, D. W., 1984: Sequence of a highly divergent β tubulin gene reveals regional heterogeneity in the β tubulin polypeptide. J. Cell Biol. **99**, 1754–1760.

— WILSON, L., 1984: Development and biochemical analysis of chick brain tubulin heterogeneity. J. Neurochem. **42**, 1363–1371.

SUMMERHAYES, I. C., WONG, D., CHEN, L. B., 1983: Effect of microtubules and intermediate filaments on mitochondrial distribution. J. Cell Sci. **61**, 87–105.

SUMMERS, K., GIBBONS, I. R., 1971: Adenosine triphosphate-induced sliding of tubules in trypsin-treated flagella of sea urchin sperm. Proc. Nat. Acad. Sci. U.S.A. **68**, 3092–3096.

— KIRSCHNER, M. W., 1979: Characteristics of the polar assembly and disassembly of microtubules observed *in vitro* by darkfield light microscopy. J. Cell Biol. **83**, 205–217.

SUN, T.-T., GREEN, H., 1977: Cultured epithelial cells of cornea, conjunctive and skin: absence of marked intrinsic divergence of their differentiated states. Nature **269**, 489–493.

— — 1978 a: Immunofluorescence staining of keratin fibers in cultured cells. Cell **14**, 469–476.

— — 1978 b: Keratin filaments of cultured human epidermal cells. Formation of intermolecular disulphide bonds during terminal differentiation. J. Biol. Chem. **253**, 2053–2060.

— SHIH, C., GREEN, H., 1979: Keratin cytoskeletons in epithelial cells of internal organs. Proc. Nat. Acad. Sci. U.S.A. **76**, 3813–2817.

SUNDQUIST, K. G., EHRUST, A., 1976: Cytoskeletal control of surface membrane motility. Nature **264**, 226–231.

SUPRENANT, K. A., DENTLER, W. L., 1982: Association between endocrine pancreatic secretory granules and *in vitro*-assembled microtubules is dependent upon microtubule-associated protein. J. Cell Biol. **93**, 164–174.

SUTOH, K., 1984: Actin-actin and actin-deoxyribonuclease I contact sites in the actin sequence. Biochemistry **23**, 1942–1946.

— IWANE, M., MATSUZAKI, F., KIKUCHI, M., IKAI, A., 1984: Isolation and characterization of a high molecular weight actin-binding protein from *Physarum polycephalum* plasmodia. J. Cell. Biol. **98**, 1611–1618.

SUZUKI, H., OUISHI, H., TAKAHASHI, K., WATANABE, S., 1978: Structure and function of chicken gizzard myosin. J. Biochem. **84**, 1529–1542.

SVITKINA, T. M., SHEVELEV, A. A., BERSHADSKY, A. D., GELFAND, V. I., 1984: Cytoskeleton of mouse embryo fibroblasts. Electron microscopy of platinum replicas. Eur. J. Cell Biol. **34**, 64–74.

SWAN, J. A., SOLOMON, F., 1984: Reformation of the marginal band of avian erythrocyte *in vitro* using calf-brain tubulin: peripheral determinants of microtubule form. J. Cell Biol. **99**, 2108–2113.

SZOLLOSI, C., CALARCO, P., DONAHUE, R. P., 1972: Absence of centrioles in the first and second meiotic spindles of mouse oocytes. J. Cell Sci. **11**, 521–541.

TANNENBAUM, S. W., 1978: Cytochalasins: Biochemical and Cell Biological Aspects. Amsterdam: Elsevier.

TAPSCOTT, S. J., BENNETT, G. S., HOLTZER, H., 1981: Neuronal precursor cells in the chick neural tube express neurofilament proteins. Nature **292**, 836–838.

TARONE, G., FERRACINI, R., GALETTO, G., COMOGLIO, P., 1984: A cell surface integral membrane glycoprotein of 85,000 mol wt (gp 85) associated with triton X-100-insoluble cell skeleton. J. Cell Biol. **99**, 512–519.

TAYLOR, D. G., WILLIAMS, V. M., CRAWFORD, N., 1976: Platelet membrane actin: solubility and binding studies with ^{125}-I-labelled actin. Biochem. Soc. Trans. **4**, 156–160.

TAYLOR, D. L., CONDEELIS, J. S., 1979: Cytoplasmic structure and contractility in amoeboid cells. Int. Rev. Cytol. **56**, 57–144.

— FECHHEIMER, M., 1982: Cytoplasmic structure and contractility: the solation-contraction coupling hypothesis. Phil. Trans. R. Soc. Lond. **B 299**, 185–197.

TAYLOR, K. A., AMOS, L. A., 1981: A new model for the geometry of the binding of myosin crossbridges to muscle thin filaments. J. Mol. Biol. **147**, 297–324.

TAYLOR, R. B., DUFFUS, P. H., RAFF, M. C., DE PETRIS, S., 1971: Redistribution and pinocytosis of lymphocyte surface immunoglobulin molecules induced by anti-immunoglobulin antibody. Nat. New Biol. **233**, 225–229.

TEMMINK, J. H. M., SPIELE, H., 1980: Different cytoskeletal domains in murine fibroblasts. J. Cell Sci. **41**, 19–32.

THERIEN, H.-M., GRUDA, J., CARRIER, F., 1984: Interaction of filamentous actin with isolated liver plasma membranes. Eur. J. Cell Biol. **35**, 112–121.

THEURKAUF, W. E., VALLEE, R. B., 1983: Extensive cAMP dependent and cAMP independent phosphorylation of microtubule-associated protein 2. J. Biol. Chem. **258**, 7883–7886.

THOMPSON, W. C., ASAI, D. J., CARNEY, D. H., 1984: Heterogeneity among microtubules of the cytoplasmic microtubule complex detected by a monoclonal antibody to alpha tubulin. J. Cell Biol. **98**, 1017–1025.

THORSTENSSON, R., UTTER, G., NORBERG, R., 1982: Further characterization of the Ca-dependent F-actin-depolymerizing protein of human serum. Eur. J. Biochem. **126**, 11–16.

THYBERG, J., PIASEK, A., MOSKALEWSKI, S., 1980: Effects of colchicine on the Golgi complex and GERL of cultured rat peritoneal macrophages and epiphyseal chondrocytes. J. Cell Sci. **45**, 42–58.

TILNEY, L. G., 1968 a: Ordering of subcellular units: the assembly of microtubules and their role in the development of cell form. Develop. Biol., Suppl. **2**, 63–103.

— 1968 b: Studies on the microtubules in heliozoa. IV. The effect of colchicine on the formation and maintenance of the axopodia and the redevelopment of pattern in *Actinosphaerium nucleofilum* (BARRETT). J. Cell Sci. **3**, 549–562.

— 1975: Actin filaments in the acrosomal reaction of *Limulus* sperm. Motion generated by alterations in the packing of the filaments. J. Cell Biol. **64**, 289–310.

— 1978: Polymerization of actin. V. A new organelle, the actomere, that initiates the assembly of actin filaments in *Thyone* sperm. J. Cell Biol. **77**, 551–564.

— BONDER, E. M., COLUCCIO, L. M., MOOSEKER, M. S., 1983 b: Actin from *Thyone* sperm assembles on only one end of an actin filament: a behavior regulated by profilin. J. Cell Biol. **97**, 112–124.

— — DEROSIER, D. J., 1981: Actin filaments elongate from their membrane-associated ends. J. Cell Biol. **90**, 485–494.

— BRYAN, J., BUSH, D. J., FUJIWARA, K., MOOSEKER, M. S., MURPHY, D. B., 1973: Microtubules: Evidence for 13 protofilaments. J. Cell Biol. **59**, 267–275.

— BYERS, B., 1969: Studies on the microtubules in *Heliozoa*. V. Factors controlling the organization of microtubules in the axonemal pattern in *Echinosphaerium (Actinosphaerium) nucleofilum*. J. Cell Biol. **43**, 148–165.

— DEROSIER, D. J., MULROY, M. J., 1980: The organization of actin filaments in the stereocilia of chochlear hair cells. J. Cell Biol. **86**, 244–259.

— DETMERS, P., 1975: Actin in erythrocyte ghosts and its association with spectrin. J. Cell Biol. **66**, 508–520.

— EGELMAN, E. H., DEROSIER, D. J., SAUNDERS, J. C., 1983 a: Actin filaments, stereocilia, and hair cells of the bird cochlea. II. Packing of actin filaments in the stereocilia and in the cuticular plate and what happens to the organization when stereocilia are bent. J. Cell Biol. **96**, 822–834.

— INOUE, S., 1982: Acrosomal reaction of *Thyone* sperm. II. The kinetics and possible mechanism of acrosomal process elongation. J. Cell Biol. **93**, 820–827.

— PORTER, K. R., 1965: Studies on the microtubules in *Heliozoa*. I. The fine structure of *Actinosphaerium nucleofilum* (Barrett) with particular reference to the axial rod structure. Protoplasma **60**, 317–344.

— — 1967: Studies on the microtubules in *Heliozoa*. II. The effect of low temperature on these structures in the formation and maintenance of the axopodia. J. Cell Biol. **34**, 327–343.

TOBACMAN, L. S., BRENNER, S. L., KORN, E. D., 1983: Effect of *Acanthamoeba* profilin on the pre-steady state kinetics of actin polymerization and on the concentration of F-actin at steady state. J. Biol. Chem. **258**, 8806–8812.

Tobacman, L. S., Korn, E. D., 1983: The kinetics of actin nucleation and polymerization. J. Biol. Chem. **258**, 3207–3214.

Toh, B. H., Hard, G. C., 1977: Actin co-caps with concanavalin A receptors. Nature **269**, 695–697.

Tokuyasu, K. T., Dutton, A. H., Geiger, B., Singer, S. J., 1981: Ultrastructure of chicken cardiac muscle as studied by double immunolabeling in electron microscopy. Proc. Nat. Acad. Sci. U.S.A. **78**, 7619–7623.

— Maher, P. A., Singer, S. J., 1984: Distribution of vimentin and desmin in developing myotubes *in vivo*. I. Immunofluorescence study. J. Cell Biol. **98**, 1961–1972.

Tomasek, J. J., Hay, E. D., 1984: Analysis of the role of microfilaments and microtubules in acquisition of bipolarity and elongation of fibroblasts in hydrated collagen gels. J. Cell Biol. **99**, 536–549.

— — Fujiwara, K., 1982: Collagen modulates cell shape and cytoskeleton of embryonic corneal fibroblasts. Distribution of actin, α-actinin, and myosin. Devel. Biol. **92**, 107–122.

Tosney, K. W., Wessels, N. K., 1983: Neuronal motility: the ultrastructure of veils and microspikes correlates with their motile activities. J. Cell. Sci. **61**, 389–411.

Toyama, Y., Forry-Schaudies, S., Hoffman, B., Holtzer, H., 1982: Effects of taxol and colcemid on myofibrillogenesis. Proc. Nat. Acad. Sci. U.S.A. **79**, 6556–6560.

Traub, P., Nelson, W. J., 1981: Occurrence in various mammalian cells and tissues of the Ca-activated protease specific for the intermediate-sized filament proteins vimentin and desmin. Eur. J. Cell Biol. **26**, 61–67.

Traub, U. E., Nelson, W. J., Traub, P., 1983: Polyacrylamide gel electrophoretic screening of mammalian cells cultured *in vitro* for the presence of the intermediate filament protein vimentin. J. Cell Sci. **62**, 129–147.

Travo, P., Weber, K., Osborn, M., 1982: Co-existence of vimentin and desmin type intermediate filaments in a subpopulation of adult rat vascular smooth muscle cells growing in primary culture. Exp. Cell Res. **139**, 87–94.

Trinick, J., Knight, P., Whiting, A., 1984: Purification and properties of native titin. J. Mol. Biol. **180**, 331–356.

— Offer, G., 1979: Cross-linking of actin filaments by heavy meromyosin. J. Mol. Biol. **133**, 549–556.

Trotter, J. A., Adelstein, R. S., 1979: Macrophage myosin-regulation of actin-activated ATPase activity by phosphorylation of the 20,000-dalton light chain. J. Biol. Chem. **254**, 8781–8785.

— Foerder, B. A., Keller, J. M., 1978: Intracellular fibers in cultured cells: analysis by scanning and transmission electron microscopy and by SDS-polyacrylamide gel electrophoresis. J. Cell Sci. **31**, 369–392.

Tseng, P., C.-H., Runge, M. S., Cooper, J. A., Williams, R. C., Pollard, T. D., 1984: Physical, immunochemical, and functional properties of *Acanthamoeba* profilin. J. Cell Biol. **98**, 214–221.

Tsukita, S., Ishikawa, H., 1981: The cytoskeleton in myelinated axons: serial section study. Biomed. Res. **2**, 424–437.

— — Kurokawa, M., 1981: Isolation of 10 nm filaments from astrocytes in the mouse optic nerve. J. Cell Biol. **88**, 245–250.

— — — MORIMOTO, K., SOBUE, K., KAKIUCHI, S., 1983: Binding sites of calmodulin and actin on the brain spectrin, calspectin. J. Cell Biol. **97**, 574–578.

— TSUKITA, S., ISHIKAWA, H., 1980: Cytoskeletal network underlying the human erythrocyte membrane. Thin section electron microscopy. J. Cell Biol. **85**, 567–576.

— — — 1984: Bidirectional polymerization of G-actin on the human erythrocyte membrane. J. Cell Biol. **98**, 1102–1110.

TUCKER, J. B., 1968: Fine structure and function of the cytopharyngeal basket in the ciliate *Nassula*. J. Cell Sci. **3**, 493–514.

— 1970: Morphogenesis of a large microtubular organelle and its association with basal bodies in the ciliate *Nassula*. J. Cell Sci. **6**, 385–429.

— 1977: Shape and pattern specification during microtubule bundle assembly. Nature **266**, 22–26.

— 1979: Spatial organization of microtubules. In: Microtubules (ROBERTS, K., HYAMS, J. S., eds.), pp. 315–358. London: Academic Press.

— 1981: Cytoskeletal coordination and intercellular signalling during metazoan embryogenesis. J. Embryol. Exp. Morphol. **65**, 1–25.

— 1984: Spatial organization of microtubule-organizing centers and microtubules. J. Cell Biol. **99**, 55s–62s.

TUFFANELLI, D. L., McKEON, F., KLEINSMITH, D. K., BURNHAM, T. K., KIRSCHNER, M., 1983: Anticentrosome and anticentriole antibodies in the scleroderma spectrum. Arch. Derm. **119**, 560–566.

TURKSEN, K., AUBIN, J. E., KALNINS, V. I., 1982: Identification of a centriole-associated protein by antibodies present in normal rabbit sera. Nature **298**, 763–765.

TURNER, P. F., MARGOLIS, R. L., 1984: Taxol-induced bundling of brain-derived microtubules. J. Cell Biol. **99**, 940–946.

TUSZYNSKI, G. P., FRANK, E. D., DAMSKY, C. H., BUCK, C. A., WARREN, L., 1979: The detection of smooth muscle desmin-like protein in BHK-21/C-13 fibroblasts. J. Biol. Chem. **254**, 6138–6143.

TYDELL, M., BLACK, M. M., GARNER, J. A., LASEK, R. J., 1981: Axonal transport: each major rate component reflects the movement of distinct macromolecular complexes. Science **214**, 179–181.

TYLER, J., HARGREAVES, W., BRANTON, D., 1979: Purification of two spectrin-binding proteins: biochemical and electron microscopic evidence for site-specific reassociation between spectrin and bands 2.1 and 4.1. Proc. Nat. Acad. Sci. U.S.A. **76**, 5192–5196.

TYLER, J. M., ANDERSON, J. M., BRANTON, D., 1980: Structural comparison of several actin-binding molecules. J. Cell Biol. **85**, 489–495.

UEHARA, Y., CAMPBELL, G., BURNSTOCK, G., 1971: Cytoplasmic filaments in developing and adult vertebrate smooth muscle. J. Cell Biol. **50**, 484–497.

UNGEWICKELL, E., BENNETT, P. M., CALVERT, R., OHANIAN, V., GRATZER, W. B., 1979: *In vitro* formation of a complex between cytoskeletal proteins of the human erythrocyte. Nature **280**, 811–814.

— GRATZER, W., 1978: Self-association of human spectrin. A thermodynamic and kinetic study. Eur. J. Biochem. **88**, 379–385.

UNNA, P. G., 1895: Über die neueren Protoplasmathorien und das Spongioplasma. Dt. Medizinal-Ztg.

VAHERI, A., MOSHER, D. F., 1978: High molecular weight, cell surface-associated glycoprotein (fibronectin) lost in malignant transformation. Biochim. Biophys. Acta **516**, 1–25.

VALENZUELA, P., QUIROGA, M., ZALDIVAR, J., RUTTER, W. J., KIRSCHNER, M. W., CLEVELAND, D. W., 1981: Nucleotide and corresponding amino acid sequence encoded by α- and β-tubulin mRNAs. Nature **289**, 650–655.

VALLEE, R. B., 1982: A taxol-dependent procedure for the isolation of microtubule-associated proteins (MAPs). J. Cell Biol. **92**, 435–442.

— 1984: MAP 2 (microtubule-associated protein 2). In: Cell and Muscle Motility (SHAY, J. W., ed.), pp. 289–311. New York: Plenum.

— BLOOM, G. S., 1983: Isolation of sea urchin egg microtubules with taxol and identification of mitotic spindle microtubule-associated proteins with monoclonal antibodies. Proc. Nat. Acad. Sci. U.S.A. **80**, 6259–6263.

— BORISY, G. G., 1978: The non-tubulin component of microtubule protein oligomers. J. Biol. Chem. **253**, 2834–2845.

— DIBARTOLOMEIS, M. J., THEURKAUF, W. E., 1981: A protein kinase bound to the projection protion of MAP 2 (microtubule-associated protein 2). Cell Biol. **90**, 528–576.

VAN BAELEN, H., BOUILLON, R., DE MOOR, P., 1980: Vitamin D-binding protein (Gc-globulin) binds actin. J. Biol. Chem. **255**, 2270–2272.

VANDEKERCKHOVE, J., FRANKE, W. W., WEBER, K., 1981: Diversity of expression of non-muscle actin in *Amphibia*. J. Mol. Biol. **152**, 413–426.

— SANDOVAL, I. V., 1982: Purification and characterization of a new mammalian serum protein with the ability to inhibit actin polymerization and promote depolymerization of actin filaments. Biochemistry **21**, 3983–3991.

— WEBER, K., 1978 a: Actin amino acid sequences. Comparison of actins from calf thymus, bovine brain, and SV-40 transformed 3T3 cells with rabbit skeletal muscle actin. Eur. J. Biochem. **90**, 451–462.

— — 1978 b: At least six different actins are expressed in a higher mammal: an analysis based on the amino acid sequence of the aminoterminal peptide. J. Mol. Biol. **126**, 783–802.

— — 1978 c: The amino acid sequence of *Physarum* actin. Nature **276**, 720–721.

— — 1984: Chordate muscle actins differ distinctly from invertebrate muscle actins. J. Mol. Biol. **179**, 391–413.

VAUDAUX, P. E., 1976: Isolation and identification of specific cortical proteins in *Tetrahymena pyriformis* strain G.L. J. Protozool. **23**, 458–464.

VERKHOVSKY, A. B., SURGUCHEVA, I. G., GELFAND, V. I., 1984: Phalloidin and tropomyosin do not prevent actin filament shortening by the 90-kD protein-actin complex from brain. Biochem. Biophys. Res. Comm. **123**, 596–603.

— — — ROSENBLATT, V. A., 1981: G-actin-tubulin interaction. FEBS Lett. **135**, 290–294.

VIGUES, B., METENIER, G., SENAUD, J., 1984: The sub-surface cytoskeleton of the ciliate *Polyplastron multivesiculatum*: isolation and major protein components. Eur. J. Cell Biol. **35**, 336–342.

VIRTANEN, I., BADLEY, R. A., PAASIVUO, R., LEHTO, V. P., 1984: Distinct cytoskeletal domains revealed in sperm cells. J. Cell Biol. **99**, 1083–1091.
— LEHTO, V.-P., LEHTONEN, E., BADLEY, R. A., 1980: Organization of intermediate filaments in cultured fibroblasts upon disruption of microtubules by cold treatment. Eur. J. Cell Biol. **23**, 80–84.
VOLK, T., GEIGER, B., 1984: A 135 kD membrane protein of intercellular adherens junctions. EMBO J. **3**, 2249–2260.
VOTER, W. A., ERICKSON, H. P., 1982: Electron microscopy of MAP 2 (microtubule-associated protein 2). J. Ultrastruct. Res. **80**, 374–382.

WAGNER, P. D., 1984: Calcium-sensitive modulation of the actomyosin ATPase by fodrin. J. Biol. Chem. **259**, 6306–6310.
— GINIGER, E., 1980: Hydrolysis of ATP and reversible binding to F-actin by myosin heavy chains free of all light chains. Nature **292**, 560–562.
WAIS-STEIDER, C., EAGLES, P., GILBERT, D. S., HOPKINS, J. M., 1983: Structural similarities and differences amongst neurofilaments. J. Mol. Biol. **165**, 393–400.
WALKER, J. H., BOUSTEAD, C. M., WITZEMANN, V., 1984: The 43-K protein, v1, associated with acetylcholine receptor containing membrane fragments is an actin-binding protein. EMBO J. **3**, 2287–2290.
WALKER, P. R., WHITFIELD, J. F., 1985: Cytoplasmic microtubules are essential for the formation of membrane-bound polyribosomes. J. Biol. Chem. **260**, 765–770.
WALLACH, D. P., DAVIES, J. A., PASTAN, I., 1978: Purification of mammalian filamin. Similarity to high molecular weight actin-binding protein in macrophages, platelets, fibroblasts, and other tissues. J. Biol. Chem. **253**, 3328–3335.
WALSH, T. P., CLARKE, F. M., MASTERS, C. J., 1977: Modification of the kinetic parameters of aldolase on binding to the actin-containing filaments of muscle. Biochem. J. **165**, 165–167.
— WEBER, A., DAVIS, K., BONDER, E., MOOSEKER, M., 1984: Calcium dependence of villin-induced actin depolymerization. Biochemistry **23**, 6099–6102.
— WINZOR, D. J., CLARKE, F. M., MASTERS, C. J. MORTON, D. J., 1980: Binding of aldolase to actin containing filaments. Biochem. J. **186**, 89–94.
WALTER, M. F., BIESSMANN, H., 1984: Intermediate-sized filaments in *Drosophila* tissue culture cells. J. Cell Biol. **99**, 1468–1477.
WANG, E., CAIRNCROSS, J. G., YUNG, W. K. A., GARBER, E. A., LIEM, R. K. H., 1983: An intermediate filament-associated protein, p 50, recognized by monoclonal antibodies. J. Cell Biol. **97**, 1507–1514.
— GOLDMAN, R. D., 1978: Functions of cytoplasmic fibers in intracellular movements in BHK 21 cells. J. Cell Biol. **79**, 708–726.
WANG, K., 1977: Filamin, a new high-molecular-weight protein found in smooth muscle and non-muscle cells. Purification and properties of chicken gizzard filamin. Biochemistry **16**, 1857–1865.
— ASH, J. F., SINGER, S. J., 1975: Filamin, a new high-molecular-weight protein found in smooth muscle and non-muscle cells. Proc. Nat. Acad. Sci U.S.A. **72**, 4483–4486.
— MCCLURE, J., TU, A., 1979: Titin: major myofibrillar component of striated muscle. Proc. Nat. Acad. Sci. U.S.A. **76**, 3698–3702.

WANG, K., RAMIREZ-MITCHELL, R., PALTER, D., 1984: Titin is an extraordinarily long, flexible, and slender myofibrillar protein. Proc. Nat. Acad. Sci. U.S.A. **81**, 3685–3689.

WANG, L.-L., SPUDICH, J. A., 1984: A 45,000-mol-wt protein from unfertilized sea urchin eggs severs actin filaments in a calcium-dependent manner and increases the steady-state concentration of nonfilamentous actin. J. Cell Biol. **99**, 844–851.

WANG, Y.-L., 1984: Reorganization of actin filament bundles in living fibroblasts. J. Cell Biol. **99**, 1478–1485.

— LANNI, F., MCNEIL, P. L., WARE, B. R., TAYLOR, D. L., 1982: Mobility of cytoplasmic and membrane-associated actin in living cells. Proc. Nat. Acad. Sci. U.S.A. **79**, 4660–4664.

WARNER, F. D., MITCHELL, D. R., 1980: Dynein: the mechanochemical coupling adenosine triphosphatase of microtubule-based sliding filament mechanisms. Int. Rev. Cytol. **66**, 1–43.

WARREN, R. H., 1974: Microtubular organization in elongating myogenic cells. J. Cell Biol. **63**, 550–566.

WEATHERBEE, J. A., MORRIS, N. R., 1983: *Aspergillus nidulans* contains multiple tubulin genes. J. Cell Biol. **97**, 217 a.

— SHERLINE, P., MASCARDO, R. N., IZANT, J. G., LUFTIG, R. B., WEIHING, R. R., 1982: Microtubule-associated proteins of HeLa cells: heat stability of the 200,000 mol wt HeLa MAPs and detection of the presence of MAP 2 in HeLa cell extracts and cycled microtubules. J. Cell Biol. **92**, 155–163.

WEAVER, D. C., MARCHESI, V. T., 1984: The structural basis of ankyrin function. I. Identification of two structural domains. J. Biol. Chem. **259**, 6165–6169.

— PASTERNACK, G. R., MARCHESI, V. T., 1984: The structural basis of ankyrin function. II. Identification of two functional domains. J. Biol. Chem. **259**, 6170–6175.

WEBER, K., BIBRING, T., OSBORN, M., 1975 b: Specific visualization of tubulin-containing structures in tissue culture cells by immunofluorescence. Cytoplasmic microtubules, vinblastine-induced paracrystals, and mitotic figures. Exp. Cell Res. **95**, 111–120.

— GROESCHEL-STEWART, U., 1974: Antibody to myosin: the specific visualization of myosin-containing filaments in nonmuscle cells. Proc. Nat. Acad. Sci. U.S.A. **71**, 4561–4564.

— OSBORN, M., 1982: Microtubule and intermediate filament networks in cells viewed by immunofluorescence microscopy. Cell Surface Rev. **7**, 1–53.

— POLLACK, R., BIBRING, T., 1975 a: Antibody against tubulin: the specific visualization of cytoplasmic microtubules in tissue culture cells. Proc. Nat. Acad. Sci. U.S.A. **72**, 459–463.

WEBSTER, R. E., HENDERSON, D., OSBORN, M., WEBER, K., 1978: Three-dimensional electron microscopical visualization of the cytoskeleton of animal cells: immunoferritin identification of actin- and tubulin-containing structures. Proc. Nat. Acad. Sci. U.S.A. **75**, 5511–5515.

WEEDS, A., 1982: Actin binding proteins—regulators of cell architecture and motility. Nature **296**, 811–816.

WEGNER, A., 1976: Head to tail polymerization of actin. J. Mol. Biol. **108**, 139–150.

— 1982: Kinetic analysis of actin assembly suggests that tropomyosin inhibits spontaneous fragmentation of actin filaments. J. Mol. Biol. **161**, 217–227.

WEHLAND, J., HENKART, M., KLAUSNER, R., SANDOVAL, I. V., 1983: Role of microtubules in the distribution of the Golgi apparatus: effect of taxol and microinjected anti-α-tubulin antibodies. Proc. Nat. Acad. Sci. U.S.A. **80**, 4286–4290.

— OSBORN, M., WEBER, K., 1977: Phalloidin-induced actin depolymerization in the cytoplasm of cultured cells interferes with cell locomotion and growth. Proc. Nat. Acad. Sci. U.S.A. **74**, 5613–5617.

— — —1979: Cell to substratum contacts in living cells, a direct correlation between interference reflexion and indirect immunofluorescence microscopy using antibodies against actin and alpha-actinin. J. Cell Sci. **37**, 257–273.

— — —1980: Phalloidin associates with microfilaments after microinjection into tissue culture cells. Eur. J. Cell Biol. **21**, 188–194.

— SANDOVAL, I. V., 1983: Cells injected with guanosine 5'-(α,β-methylene)triphosphate, an α,β-nonhydrolyzable analog of GTP, show anomalous patterns of tubulin polymerization affecting cell translocation, intracellular movement, and the organization of Golgi elements. Proc. Nat. Acad. Sci. U.S.A. **80**, 1938–1941.

— WEBER, K., 1979: Effects of the actin-binding protein DNase I on cytoplasmic streaming and ultrastructure of *amoeba proteus*. Cell Tissue Res. **199**, 353–372.

— — 1980: Distribution of fluorescently labelled actin and tropomyosin after microinjection in living tissue culture cells observed with TV image intensification. Exp. Cell Res. **127**, 397–408.

— — OSBORN, M., 1980: Translocation of actin from the cytoplasm into the nucleus in mammalian cells exposed to dimethyl sulfoxide. Biol. Cell. **39**, 109–113.

— WILLINGHAM, M. C., 1983: A rat monoclonal antibody reacting specifically with the tyrosylated form of α-tubulin. II. Effects on cell movement, organization of microtubules, and intermediate filaments, and arrangement of Golgi elements. J. Cell Biol. **97**, 1467–1490.

WEIHING, R. R., 1979: The cytoskeleton and plasma membrane. In: Methods and Achievements in Experimental Pathology (JASMIN, G., CANTIN, M., eds.), Vol. 8, pp. 42–109. Basel: Karger.

— 1983: Purification of a HeLa cell high molecular weight actin binding protein and its identification in HeLa cell plasma membrane ghosts and intact HeLa cells. Biochemistry **22**, 1839–1847.

WEINGARTEN, M. D., LOCKWOOD, A. H., HWO, S. Y., KIRSCHNER, M. W., 1975: A protein factor essential for microtubule assembly. Proc. Nat. Acad. Sci. U.S.A. **72**, 1858–1862.

WEISENBERG, R. C., 1972: Microtubule formation *in vitro* in solutions containing low calcium concentration. Science **177**, 1104–1105.

— BORISY, G. G., TAYLOR, E. W., 1968: The colchicine binding protein of mammalian brain and its relationship to microtubules. Biochemistry **7**, 4466–4479.

WEISENBERG, R. C., CIANCI, C., 1984: ATP-induced gelation-contraction of microtubules assembled *in vitro*. J. Cell Biol. **99**, 1527–1533.

— DEERY, W. O., DICKINSON, P. J., 1976: Tubulin-nucleotide interactions during the polymerization and depolymerization of microtubules. Biochemistry **15**, 4248–4259.

— ROSENFELD, A. C., 1975: *In vitro* polymerization of microtubules into asters and spindles in homogenates of surf clam eggs. J. Cell Biol. **64**, 146–158.

WEISS, P., FERRIS, W., 1954: Electron micrographs of larval amphibian epidermis. Exp. Cell Res. **6**, 546–549.

— WANG, H., 1936: Neurofibrils in living ganglion cells of the chick, cultivated *in vitro*. Anat. Rec. **67**, 105–117.

WELSH, M. J., DEDMAN, J. R., BRINKLEY, B. R., MEANS, A. R., 1978: Calcium dependent regulator protein localization in the mitotic apparatus of eukaryotic cells. Proc. Nat. Acad. Sci. U.S.A. **75**, 1867–1871.

WERTH, D. K., NIEDEL, J. E., PASTAN, I., 1983: Vinculin, a cytoskeletal substrate for protein kinase C. J. Biol. Chem. **258**, 11423–11426.

WESTRIN, H., BACKMAN, L., 1983: Association of rabbit muscle glycolytic enzymes with filamentous actin. A counter-current distribution study at high ionic strength. Eur. J. Biochem. **136**, 407–411.

WHEATLEY, D. N., 1982: The Centriole: A Central Enigma in Cell Biology. Amsterdam: Elsevier Biomedical.

WHITE, G. E., GIMBRONE, M. A., FUJIWARA, K., 1983: Factors influencing the expression of stress fibers in vascular endothelial cells *in situ*. J. Cell Biol. **97**, 416–424.

WICHE, G., BAKER, M. A., 1982: Cytoplasmic network arrays demonstrated by immunolocalization using antibodies to a high molecular weight protein present in cytoskeletal preparations from cultured cells. Exp. Cell Res. **138**, 15–29.

— — KINDAS-MUGGE, I., LEICHTFRIED, F., PYTELA, R., 1980: High molecular weight polypeptides from cultured cells and their possible role as mediators of microtubule-intermediate filament interaction. In: Microtubules and Micro-tubule Inhibitors (DE BRABANDER, M., DE MEY, J., eds.), pp. 189–200. Amsterdam: Elsevier.

— KREPLER, R., ARTLIEB, U., PYTELA, R., ABERER, W., 1984: Identification of plectin in different human cell types and immunolocalization at epithelial basal cell surface membranes. Exp. Cell Res. **155**, 43–49.

— — ARTLIEB, U., PYTELA, R., DENK, H., 1983: Occurrence and immunolocalization of plectin in tissues. J. Cell Biol. **97**, 887–901.

WIELAND, T., 1977: Modification of actin by phallotoxins. Naturwiss. **64**, 303–309.

WILDE, C. D., CROWTHER, C. E., CRIPE, T. P., LEE, M. G.-S., COWAN, N. J., 1982: Evidence that a human β-tubulin gene is derived from its corresponding mRNA. Nature **292**, 83–84.

WILKINS, J. A., CHEN, K. Y., LIN, S., 1983: Detection of high molecular weight vinculin binding proteins in muscle and nonmuscle tissues with an electroblot-overlay technique. Biochem. Biophys. Res. Comm. **116**, 1026–1032.

— LIN, S., 1981: Association of actin with chromaffin granule membranes and the effect of cytochalasin B on the polarity of actin filament elongation. Biochim. Biophys. Acta **642**, 55–66.

— — 1982: High-affinity interaction of vinculin with actin filaments *in vitro*. Cell **28**, 83–90.

— SCHWARZ, J. H., HARRIS, D. A., 1983: Brevin, a serum protein that acts on the barbed end of actin filaments. Cell Biol. Int. Rep. **7**, 1097–1104.

WILLARD, M., SIMON, C., 1981: Antibody decoration of neurofilaments. J. Cell Biol. **89**, 198–205.

WILLIAMS, R. C., RUNGE, M. S., 1983: Biochemistry and structure of mammalian neurofilaments. Cell and Muscle Motility **3**, 41–56.

WILLINGHAM, M. C., YAMADA, K. M., YAMADA, S. S., POUYSSEGUR, J., PASTAN, I., 1977: Microfilament bundles and cell shape are related to adhesiveness to substratum and are dissociable from growth control in cultured fibroblasts. Cell **10**, 375–380.

WILSON, E. B., 1928: The cell in development and heredity. New York: Macmillan.

WILSON, J. E., 1978: Ambiquitous enzymes: variation in intracellular distribution as a regulatory mechanism. Trends Biochem. Sci. **3**, 124–126.

WILSON, L., BAMBURG, J. R., MIZEL, S. B., GRISHAM, L. M., CRESWELL, K. M., 1974: Interaction of drugs with microtubule proteins. Fed. Proc. **33**, 158–166.

— MILLER, H. P., PFEFFER, T. A., SULLIVAN, K. F., DETRICH, H. W., 1984: Colchicine-binding activity distinguishes sea urchin egg and outer doublet tubulins. J. Cell Biol. **99**, 37–41.

WISNIEWSKI, H., SHELANSKI, M. L., TERRY, R. D., 1968: Effects of mitotic spindle inhibitors on neurotubules and neurofilaments in anterior cells. J. Cell Biol. **38**, 224–229.

WITMAN, G. B., 1975: The site of *in vivo* assembly of flagellar microtubules. Ann. N.Y. Acad. Sci. **253**, 178–191.

WOJCIESZYN, J., SCHLEGL, R., WU, E. S., JACOBSON, K., 1981: Diffusion of injected macromolecules within the cytoplasm of living cells. Proc. Nat. Acad. Sci. U.S.A. **78**, 4407–4410.

WOLOSEWICK, J. J., PORTER, K. R., 1976: Stereo high voltage electron microscopy of whole cells of the human diploid cell line WI-38. Am. J. Anat. **147**, 303–324.

— — 1979: Microtrabecular lattice of the cytoplasmic ground substance. Artifact or reality? J. Cell Biol. **82**, 114–139.

WONG, A. J., KIEHART, D. P., POLLARD, T. D., 1985: Myosin from human erythrocytes. J. Biol. Chem. **260**, 46–49.

— POLLARD, T. D., HERMAN, I. M., 1983: Actin filament stress fibers in vascular endothelial cells *in vivo*. Science **219**, 867–869.

WONG, J., HUTCHISON, S. B., LIEM, R. K. H., 1984: An isolectric variant of the 150,000-dalton neurofilament polypeptide. J. Biol. Chem. **259**, 10867–10874.

WOODCOCK, C. L. F., 1980: Nucleus-associated intermediate filaments from chicken erythrocytes. J. Cell Biol. **85**, 881–889.

WOODRUM, D. T., LINCK, R. W., 1980: Structural basis of motility in the microtubular axostyle: implications for cytoplasmic microtubule structure and function. J. Cell Biol. **87**, 404–414.

WOODRUM, D. T., RICH, S. A., POLLARD, T. D., 1975: Evidence for a biased bidirectional polymerization of actin filaments using heavy meromyosin prepared by an improved method. J. Cell Biol. **67**, 231–237.

WOODY, R. W., CLARK, D. C., ROBERTS, C. K., MARTIN, S. R., BAYLEY, P. M., 1983: Molecular flexibility in microtubule proteins: proton nuclear magnetic resonance characterization. Biochemistry **22**, 2186–2192.

WU, Y.-J., PARKER, L. M., BINDER, N. E., BECKETT, M. A., SINARD, J. H., GRIFFITHS, C. T., RHEINWALD, J. G., 1982: The mesothelial keratins: a new family of cytoskeletal proteins identified in cultured mesothelial cells and nonkeratinizing epithelia. Cell **31**, 693–703.

WUERKER, R. B., KIRKPATRICK, J. B., 1972: Neuronal microtubules, neurofilaments, and microfilaments. Int. Rev. Cytol. **33**, 45–75.

WULF, E., DEBOBEN, A., BAUTZ, F. A., FAULSTICH, H., WIELAND, T., 1979: Fluorescent phallotoxin, a tool for the visualization of cellular actin. Proc. Nat. Acad. Sci. U.S.A. **76**, 4498–4502.

YAHARA, I., EDELMAN, G. M., 1973 a: Modulation of lymphocyte receptor redistribution by concanavalin A, anti-mitotic agents and alterations of pH. Nature **246**, 152–155.

— — 1973 b: The effects of concanavalin A on the mobility of lymphocyte surface receptors. Exp. Cell Res. **81**, 143–155.

— — 1975 a: Modulation of lymphocyte receptor mobility by concanavalin A and colchicine. Ann. N.Y. Acad. Sci. **253**, 455–469.

— — 1975 b: Modulation of lymphocyte receptor mobility by locally bound concanavalin A. Proc. Nat. Acad. Sci. U.S.A. **72**, 1579–1583.

YAMADA, K. M., OHANIAN, S. H., PASTAN, I., 1976 b: Cell surface protein decreases microvilli and ruffles on transformed mouse and chick cells. Cell **9**, 241–245.

— OLDEN, K., 1978: Fibronectins: adhesive glycoproteins of cell surface and blood. Nature **275**, 179–184.

— YAMADA, S. S., PASTAN, I., 1976 a: Cell surface protein partially restores morphology, adhesiveness, and contact inhibition of movement to transformed fibroblasts. Proc. Nat. Acad. Sci. U.S.A. **73**, 1217–1221.

YAMAMOTO, K., PARDEE, J. D., REIDLER, J., STRYER, L., SPUDICH, J. A., 1982: Mechanism of action of *Dictyostelium* severin with actin filaments. J. Cell Biol. **95**, 711–719.

YANG, H. Y., LIESKA, N., GOLDMAN, A., GOLDMAN, R. D., 1985: A 300,000-mol-wt intermediate filament-associated protein in baby hamster kidney (BHK-21) cells. J. Cell Biol. **100**, 620–631.

YEN, S.-H., DAHL, D., SCHACHNER, M., SHELANSKI, M. L., 1976: Biochemistry of the filaments of brain. Proc. Nat. Acad. Sci. U.S.A. **73**, 529–533.

— FIELDS, K. L., 1981: Antibodies to neurofilaments, glial filaments and fibroblast intermediate filament proteins bind to different cell types of the nervous system. J. Cell Biol. **88**, 115–126.

YERNA, M. J., DABROWSKA, R., HARTSHORNE, D. J., GOLDMAN, R. D., 1979: Calcium-sensitive regulation of actin-myosin interactions in baby hamster kidney (BHK-21) cells. Proc. Nat. Acad. Sci. **76**, 184–188.

YIN, H. L., HARTWIG, J. H., MARUYAMA, K., STOSSEL, T. P., 1981: Ca^{++} control of actin filament length. J. Biol. Chem. **256**, 9693–9697.

— KWIATKOWSKI, D. J., MOLE, J. E., COLE, F. S., 1984: Structure and biosynthesis of cytoplasmic and secreted variants of gelsolin. J. Biol. Chem. **259**, 5271–5276.

— STOSSEL, T. P., 1979: Control of cytoplasmic actin gel-sol transformation by gelsolin, a calcium-dependent regulatory protein. Nature **281**, 583–586.

— — 1980: Purification and structural properties of gelsolin, a Ca-activated regulatory protein of macrophages. J. Biol. Chem. **255**, 9490–9493.

— ZANER, K. S., STOSSEL, T. P., 1980: Ca^{++} control of actin gelation. Interaction of gelsolin with actin filaments and regulation of gelation. J. Biol. Chem. **255**, 9494–9500.

YOKOTA, E., MARUYAMA, K., 1983: Muscle β-actinin inhibits elongation of the pointed end of the actin filaments of brush border microvilli. J. Biochem. **94**, 1897–1900.

YOSHINO, H., MARCHESI, V. T., 1984: Isolation of spectrin subunits and reassociation *in vitro*. Analysis by fluorescence polarization. J. Biol. Chem. **259**, 4496–4500.

YU, J., FISCHMAN, D. A., STECK, T. L., 1973: Selective solublization of proteins and phospholipids from red blood cell membrane by nonionic detergents. J. Supramol. Struct. **1**, 233–248.

ZABRECKY, J. R., COLE, R. D., 1982: Effect of ATP on the kinetics of microtubule assembly. J. Biol. Chem. **257**, 4633–4638.

ZACKROFF, R. V., GOLDMAN, R. D., 1979: *In vitro* assembly of intermediate filaments from baby hamster kidney (BHK-21) cells. Proc. Nat. Acad. Sci. U.S.A. **76**, 6226–2630.

— STEINERT, P. M., AYNARDI-WHITMAN, M., GOLDMAN, R. D., 1981: Intermediate filaments. In: Cytoskeletal Elements and Plasma Membrane Organization (POSTE, G., NICHOLSON, G. L., eds.), pp. 55–97. Amsterdam: Elsevier.

ZAKUT, R., SHANI, M., GIVOL, D., NEUMAN, S., YAFFE, D., NUDEL, U., 1982: Nucleotide sequence of the rat skeletal muscle actin gene. Nature **298**, 857–859.

ZALOKAR, M., 1960: Cytochemistry of centrifuged hyphae of Neurospora. Exp. Cell Res. **19**, 114–132.

ZANER, K. S., FOTLAND, R., STOSSEL, T. P., 1981: Low-shear, small volume, viscoelastometer. Rev. Sci. Instrum. **52**, 85–87.

— STOSSEL, T. P., 1982: Some perspectives on the viscosity of actin filaments. J. Cell Biol. **93**, 987–991.

ZEHNER, Z. E., PATERSON, B. M., 1983: Characterization of the chicken vimentin gene: single copy gene producing multiple mRNAs. Proc. Nat. Acad. Sci. U.S.A. **80**, 911–915.

ZHNANG, Q., ROSENBERG, S., LAWRENCE, J., STRACHER, A., 1984: Role of ABP phosphorylation in platelet cytoskeleton assembly. Biochem. Biophys. Res. Comm. **118**, 508–513.

ZIEVE, G., SOLOMON, F., 1982: Proteins specifically associated with the microtubules of the mammalian mitotic spindle. Cell **28**, 233–242.

ZIEVE, G. E., HEIDEMANN, S. R., McINTOSH, J. R., 1980: Isolation and partial characterization of a cage of filaments that surrounds the mammalian mitotic spindle. J. Cell Biol. **87**, 160–169.

ZUMBE, A., STAEHLI, C., TRACHSEL, H., 1982: Association of a M 50,000 cap-binding protein with the cytoskeleton in baby hamster kidney cells. Proc. Nat. Acad. Sci. U.S.A. **79**, 2927–2931.

Subject Index

Cell Biology Monographs

Cell Biology Monographs

Springer-Verlag Wien New York

Plant Gene Research

Basic Knowledge and Application

Edited by E. S. Dennis, B. Hohn, Th. Hohn (Managing Editor), P. J. King, J. Schell, D. P. S. Verma

The first volume

Genes Involved in Microbe-Plant Interactions

Edited by **D. P. S. Verma** and **Th. Hohn**

1984. 54 figures. XIV, 393 pages. ISBN 3-211-81789-1

For the practical use of new techniques of gene manipulation in plant breeding it is very important to understand the molecular and physiological aspects of the close interactions which occur between higher plants and microorganisms. The nature of these interactions ranges from harmless through symbiotic to parasitic; they also include direct gene transfer from bacteria to plants, which has become a subject of research thanks to the new techniques applied in molecular biology. Knowledge of this gene transfer occurring in nature opens new perspectives for its future utilization in plant breeding. Possibly the strongest influence in making plant gene research a growing field of scientific interest and activity is the expectation that this research will open new and very effective ways for the breeding of agriculturally and biotechnologically important plants. The first volume in the series "Plant Gene Research" provides an overview of the important aspects of plant-microbe interactions and of various research methods.

The second volume

Genetic Flux in Plants

Edited by **B. Hohn** and **E. S. Dennis**

1985. 40 figures. XII, 253 pages. ISBN 3-211-81809-X

This volume gathers together for the first time the most recent information on plant genome instability. The plant genome can no longer be looked upon as a stable entity. Many examples of change and disorder in the genetic material have been reported recently. Chloroplast DNA sequences have been found in nuclei and mitochondria. Mitochondrial DNA molecules can switch between various forms by recombination processes. Stress on plants or on cells in culture can cause changes in chromosome organization. DNA can be inserted into the plant genome by transformation with the Ti plasmid of Agrobacterium tumefaciens, and transposable elements produce insertions and deletions.

Springer-Verlag Wien New York